Recent Trends in Antifungal Agents and Antifungal Therapy

Amit Basak • Ranadhir Chakraborty •
Santi M. Mandal
Editors

Recent Trends in Antifungal Agents and Antifungal Therapy

Springer

Editors
Amit Basak
Department of Chemistry
Indian Institute of Technology Kharagpur
Kharagpur
West Bengal, India

Ranadhir Chakraborty
Department of Biotechnolgy
University of North Bengal
Darjeeling
West Bengal, India

Santi M. Mandal
Department of Microbiology
Vidyasagar University
Midnapore
West Bengal, India

ISBN 978-81-322-2780-9 ISBN 978-81-322-2782-3 (eBook)
DOI 10.1007/978-81-322-2782-3

Library of Congress Control Number: 2016946342

Printed on acid-free paper

This Springer imprint is published by Springer Nature
The registered company is Springer (India) Pvt. Ltd.

Preface

In the history of discoveries of fungal pathogens, the nineteenth century has witnessed two important events. The causal organism of a silkworm disease, *muscardine*, a fungus named later as *Beauveria bassiana*, was revealed by Agostino Bassi in 1835. Six years later, in 1841, the causal agent of the human scalp disease, favus, being a fungus was discovered by David Gruby. The Gruby's unique and innovative method for the isolation of fungus from the infected scalp and on potato slices, repeated infection of the healthy tissues by the isolated fungus (parallel to the Koch's postulate) was left ignored in the pages of science history due to reasons not related to science. The fact remains that even after the seminal researches by Bassi and Gruby, the knowledge of the fungal diseases remained much less than that of bacterial diseases. Compared to bacterial diseases (among which some of them were epidemic) of human beings, diseases caused by fungi were not epidemic in nature and often are occasional but consequences of some mycoses that can be severe to lethal.

Nevertheless, fungal infections are difficult to treat because fungi are eukaryotes with similarity in biochemical composition and phylogenetic nearness to animals. Hence, treatment of an internal infection caused by a fungus is often very complicated as finding a drug that would specifically kill the fungus and not the animal is very difficult. Most fungi are killed by the immune system, and if the host immune system is overpowered by the fungus, the result is most likely death. Abnormalities in the function of neutrophils and neutropenia help the spread of infections caused by *Candida*, *Aspergillus*, and *Mucoraceae* strains, while altered T-lymphocyte mononuclear phagocyte function will allow dissemination of *C. neoformans*, *Histoplasma*, and *Coccidioides*. Treatment and diagnosis of fungal infections in the immunocompromised host are very tricky and difficult, and in obtaining enough tissue for histology and culture, it is most often required to perform invasive procedures. Moreover, fungal infections have taken a new spectrum due to the increased incidence of multidrug-resistant fungal pathogens. The freedom of choice for drugs to treat fungal infections is also narrow because of lesser probability of discovering drugs that would bypass affecting human cells and target fungal cells producing fewer side effects in patients.

The book is edited in such a way that it will serve as an important resource material for not only the students and researchers but also the physicians and infectious disease scientists. It consists of a series of chapters that dealt in details with the development of antifungal compounds; the prospect of finding newer antifungal drugs including natural, synthetic, and designed; the panorama of combinational therapy including immunotherapy, and the susceptibility testing of dermatophytes. Medical relevance is emphasized throughout the text. On a more immediate level, the editors are grateful to all contributing authors for their intelligence, enthusiasm, and cooperation and for their expert and exhaustive scientific review.

Kharagpur, West Bengal, India — Amit Basak
Darjeeling, West Bengal, India — Ranadhir Chakraborty
Midnapore, West Bengal, India — Santi M. Mandal

Contents

About the Editors

Amit Basak, currently Professor of Chemistry and Chairman, School of Bioscience, IIT Kharagpur, obtained his Ph.D. (natural product chemistry) from Calcutta University and D. Phil. (penicillin biosynthesis) from University of Oxford. He then worked on clavulanic acid biosynthesis as a postdoctoral fellow at the Johns Hopkins University. His research interests involve understanding the mechanism of diradical generating reactions and their applications, development of enzyme inhibitors as antimicrobial agents and molecular capture chemistry. He has received several prestigious awards and fellowships for his research contribution.

Ranadhir Chakraborty was born in Darjeeling. He obtained his Ph.D. from Calcutta University. He worked on "Repetitive DNA sequences in Acidithiobacillus ferrooxidans and their role in regulation of sulfur metabolism" under the supervision of Dr. Pradosh Roy, in the Department of Microbiology, Bose Institute. He is at present serving the Department of Biotechnology, University of North Bengal, in the capacity of Professor and Head. He maintains a perfect blend of classical and modern microbiology in his ongoing journey of Science. He probes some basic scientific problems including antimicrobial resistance with cutting edge technology of every passing time period.

Santi M. Mandal obtained his Ph.D. in the field of Molecular Microbiology and continuing research with major focus in Antimicrobial Chemotherapy. He visited UTMB-USA and NUS-Singapore for his postdoctoral training. At present, he is working as an Assistant Professor of Microbiology at Vidyasagar University, India. He has published more than 90 research papers in reputed journals and conferred upon several prestigious awards for his research contribution.

1 Fungi Fights Fungi: Tip-off in Antifungal Chemotherapy

Santi M. Mandal, Anupam Roy, Debarati Paul, Suresh Korpole, Shanker Lal Shrivastava, Ranadhir Chakraborty, and Amit Basak

Abstract

Fungal infections have taken a new spectrum due to the increased incidence of multi-drug resistant fungal pathogens. Freedom of choice for drugs to treat fungal infections is also narrow because of lesser probability of discovering drugs that would bypass affecting human cells and target fungal cells producing fewer side effects in patients. An approach has gained prominence in research is to look for bioactive antifungal compounds from natural sources and discover new classes of antifungals to control the recent emergence of fungal infections. Most of antifungal drugs are originated from fungi. A conservative estimate of total number of fungal species on this planet would exceed 10^6 if taken into account the ones yet to be discovered from diverse habitats ranging from forest land to marine ecosystem. While attempting to summarize the status of reported fungi-derived antifungal compounds discovered since ancient times, the subset of such compounds were found to be anticancer too. Antifungal compounds with the promise of inducing challenge to rediscover the new effective molecules from drug prototype are also discussed.

Anupam Roy and Santi M. Mandal are equally contributed in literature survey.

S.M. Mandal (✉)
Department of Microbiology, Vidyasagar University, Midnapore 721102, West Bengal, India
e-mail: mandalsm@gmail.com

A. Roy • S.L. Shrivastava
Department of Chemistry, Agriculturtal and Food Engineering Department, Indian Institute of Technology Kharagpur, Kharagpur 721302, India

D. Paul
Amity Institute of Biotechnology, Amity University, Sec 125, Noida 201303, India

S. Korpole
CSIR-Institute of Microbial Technology, Sector 39A, Chandigarh 160036, India

R. Chakraborty
Department of Biotechnology, North Bengal University, Siliguri, Darjeeling 734013, India

A. Basak (✉)
Department of Chemistry, Indian Institute of Technology Kharagpur, Kharagpur, West Bengal, India
e-mail: absk@chem.iitkgp.ernet.in

A. Basak et al. (eds.), *Recent Trends in Antifungal Agents and Antifungal Therapy*,
DOI 10.1007/978-81-322-2782-3_1

1.1 Introduction

Diversity in species characterized by unique and unusual biochemical pathways facilitates fungi to offer several bioactive molecules (Keller and Turner 2005). These compounds generally come from fungal cellular components in the form of secondary metabolites (Magdalena et al. 2013). Fungal bioactive compounds suitably combats several diseases in plants and animals. Biological activities such as antibacterial, antifungal, antitumor, anti-cholesterol, cytotoxic, mutagenic, carcinogenic, teratogenic, immunosuppressive, enzyme inhibitory effect, etc. make 'fungal origin' as a potential area of research in natural product discovery. Rapid increase in fungal infections contributing to higher mortality rates has become a major concern. Resistance to currently available antifungal drugs necessitates the discovery of new classes of antifungals from both natural and synthetic approach. In agriculture, infection or contamination from fungi in pre- or post-harvest is a major problem leading to economic loss. Fungal pathogen e.g. *Aspergillus* and *Fusarium* spp. not only relates to economical loss but also creates health problem producing mycotoxins. A details top to bottom outline of fungal derived antifungal compounds with their modifications or synthetic analogues may be helpful to understand the structure-activity relationship, which leads to new compound development in antifungal chemotherapy.

1.2 Fungi-Derived Antifungal Agents

1.2.1 Griseofulvin

Griseofulvin, a metabolic product of *Penicillium griseofulvum*, was first isolated by Oxford et al. in 1939 (1939). In 1946, Brain et al. reported that *Penicillium janczewski* was found to produce a substance capable of shrinking and stunting of fungal hyphae (Brian and Curtis 1946). The physical, chemical and biological identity of the isolated compound was established by several researchers (Brian 1949; Grove and McGowan 1947; Brian and Curtis 1949). Subsequent studied had established that *P. platulum* and *P. raistrickii* also produces Griseofulvin (Brian and Curtis 1949, 1955). Thereafter, the production of griseofulvin from various fungi has been thoroughly studied (Brian 1949; Araujo et al. 1990; Petit et al. 2004; Oxford et al. 1939; Wright 1955; Brian and Curtis 1955; Clarke and Mckenzie 1967). The molecule offers in vitro fungistatic action against dermatophytes, such as Microsporum, Epidermophyton and Trichophyton, whereas activity was restricted to yeast, actinomyces and Nocardia. The minimum inhibitory concentrations (MIC) were observed as low as 5–20 μg/ml and bind to microtubules comprising the spindles and inhibit mitotic cell division (Huber and Gottlieb 1968). Initially the structure of griseofulvin was reported as 7-chloro-4,6 dimethoxy-3-cumarone-2-spiro-1′,2′-methoxy- 6′-methyl-2′-cyclohexen-4′-one (Fig. 1.1). The structures proposed by several groups are inconsistent (Oxford and Raistrick 1939; Grove et al. 1951, 1952). Recently approaches for strain improvement by mutation are studied to enhance griseofulvin production (Aytoun and Mcwilliam 1957; Songgang and Yunshen 1983; Kommmunarskaya 1969, 1970).

1.2.2 Strobilurins

Strobilurins (methyl (E)-3-methoxy-2-(5-phenylpenta-2,4-dienyl) acrylate) are another class of fungal metabolites reported by Anke et al. (1977). *Strobilurus tenacellus*, a basidiomycetes fungus produce strobilurins A and B, showed high activity against yeasts and filamentous fungi but inactive against bacteria (Anke et al. 1977). Strobilurins inhibits the mitochondrial respiration in fungi and binds at the Qo-centre on cytochrome b which blocks the electron transfer between cytochrome b and cytochrome c1 (Balba 2007; Bartlett et al. 2002). Therefore, it is called as Qo inhibitors (QoI), or Quinone outside inhibitors. Anke and his

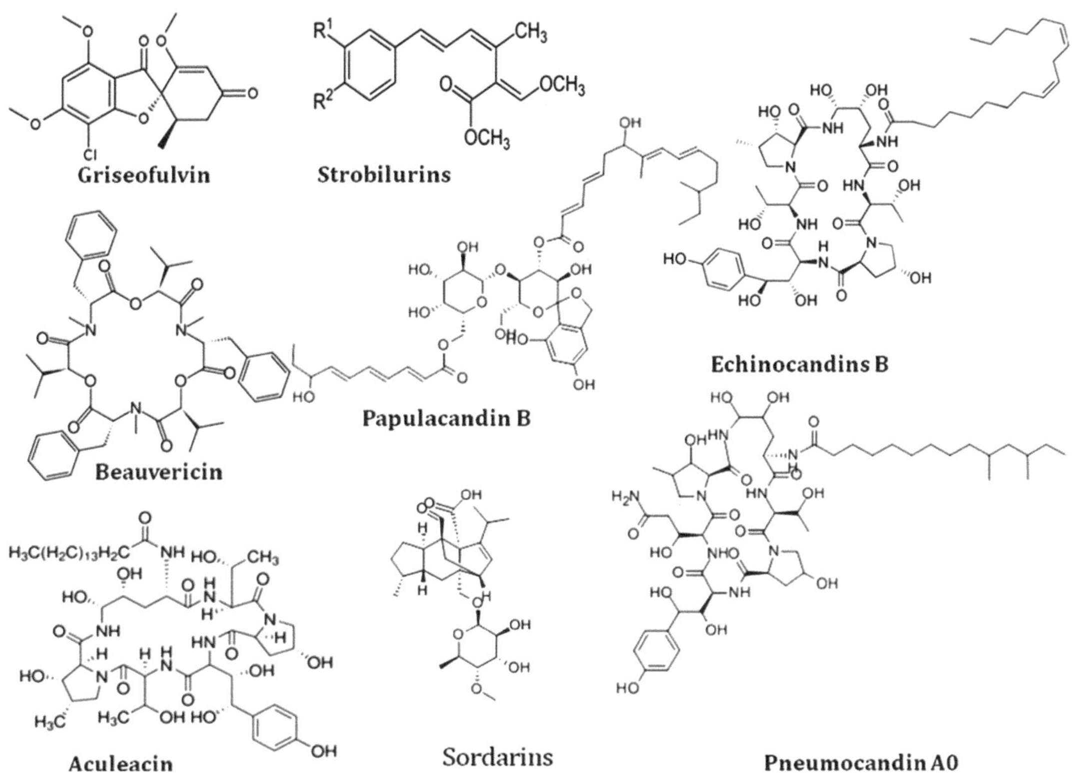

Fig. 1.1 The representative chemical structure of some fungi-derived antifungal agent

coworker first attempt to resolve the structure and variable structure of strobilurins are listed in Table 1.1. The structure of strobilurin may vary in only in the aromatic ring substitutions at 3 and 4 positions. Strobilurin in natural form break down easily under light (Dolores et al. 2010).

1.3 Echinocandins, Pneumocandin and Papulacandin

In 1970s, two structurally important antifungals were screened. The first one belonged to the lipopeptide class termed as echinocandins and second one was glycopeptides class as papulacandin, affecting cell wall components are the prime target of fungal inhibition. Fungi-derived echinocandins and pneumocandin are the antifungal compounds having inhibitory effect on the synthesis of glucan by noncompetitive inhibition of the enzyme 1,3-β glucan synthase (Morris and Villmann 2006a, b). Besides that, these molecules are recognized as potent backbones or as a basic molecular structure for synthesis and developing analogues.

Echinocandins are cyclic antifungal hexapeptides core N-acylated with different aliphatic carboxylic acids. The first report of echinocandin discovery was in early 1974. Researchers of Ciba-Geigy, Sandoz and Eli Lilli isolated echinocandins B from the fermentation broth of *Aspergillus nidulans* var. *echinolatus*, *Aspergillus nidulansvar roseus* and *Aspergillus rugulosus* in random screening of the available strain collections (Benz et al. 1974; Keller-Juslén et al. 1976; Nyfeler and Keller 1974; Geiser et al. 2007). Afterword, a series of fungi now has come into existence having ability of synthesizing natural echinocandin (Nyfeler and Keller 1974; Geiser et al. 2007; Traber et al. 1979) (Table 1.2). The presence of different substituents in the hexapeptide ring or a distinct fatty acid chain makes echinocandins different

Table 1.1 Natural strobilurins with their respective structure

Natural isolated Strobilurin	Structure	Substitution in R1 and R2	Name of the fungus	Reference
General structure	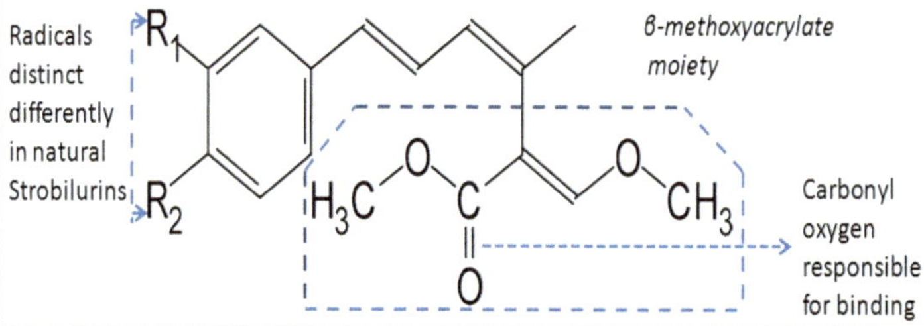			
Strobilurin A	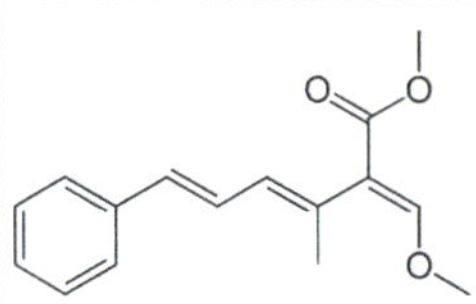	H– in both R1 and R2	*Strobilurus tenacellus*	Anke et al. (1977), Balba (2007), and Schramm et al. (1978)
Strobilurin B	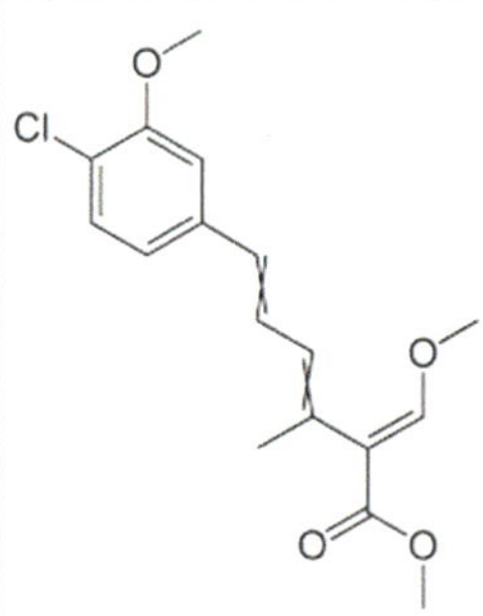	CH_3—O in R1 and Cl– in R2	*Strobilurus tenacellus*	Anke et al. (1977), Balba (2007), and Schramm et al. (1978)
Strobilurin C	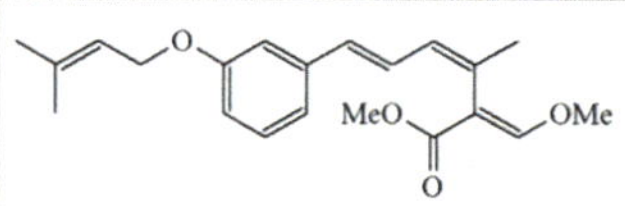	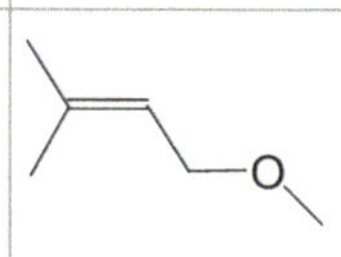 in R1 and Cl– in R2	*Xerula* sp. (agaricales)	Balba (2007) and Anke et al. (1983)
Strobilurin D	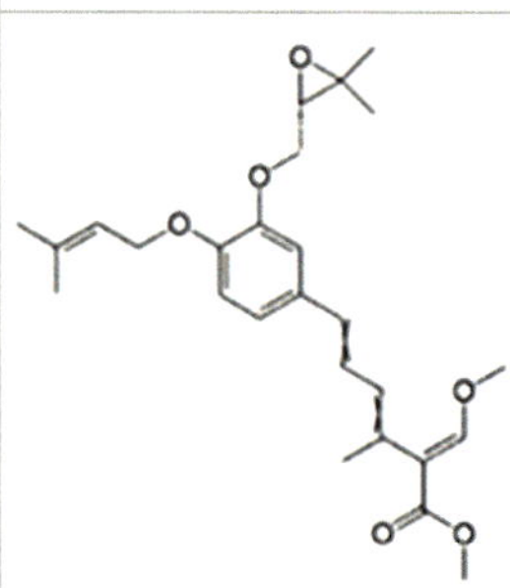	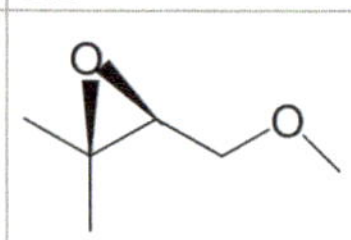 in R1 and 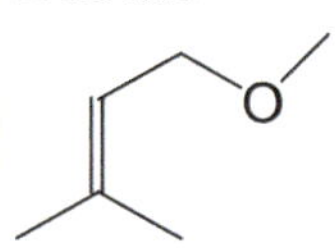 in R2	*Cyphellopsis anomala*	Balba (2007) and Weber et al. (1990b)
Strobilurin E	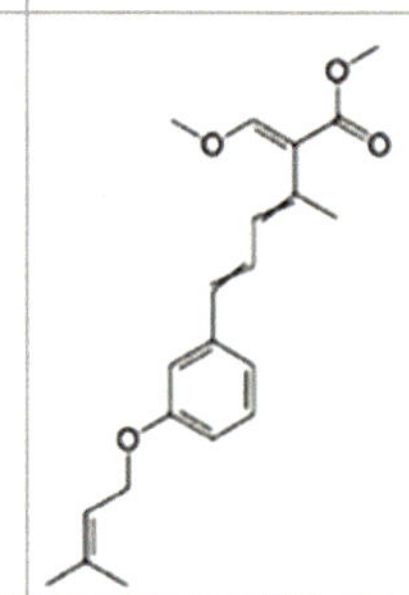		*Crepidotus fulvotomentosus*.	Weber et al. (1990a)

(continued)

Table 1.1 (continued)

Natural isolated Strobilurin	Structure	Substitution in R1 and R2	Name of the fungus	Reference
Strobilurin F	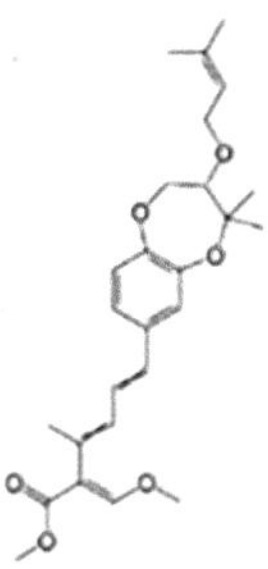	−OH in R1 and 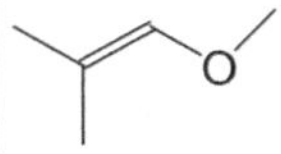 in R2	*Cyphellopsis anomala* and *Bolinea lutea*. **I.**	Balba (2007), Weber et al. (1990b), Fredenhagen et al. (1990a, b)
Strobilurin G	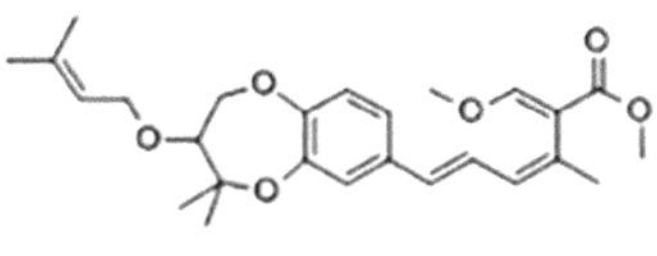	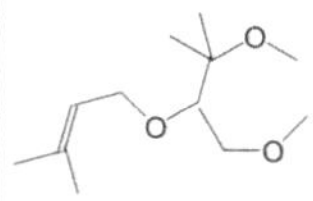 in R2	*Bolinea lutea*. **I.**	Balba (2007), Weber et al. (1990a), and Fredenhagen et al. (1990a, b)
Strobilurin H	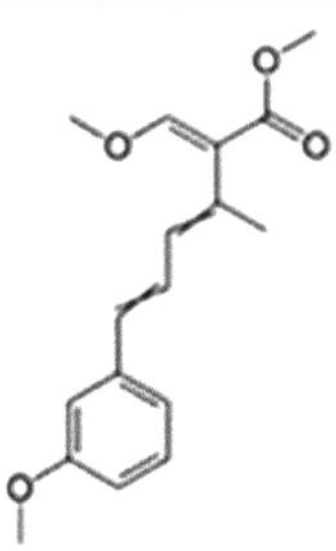	−OH in R1 and −H in R2	*Bolinea lutea*. **I.**	Balba (2007), and Fredenhagen et al. (1990a b)

from each other (Fig. 1.2). Several unusual amino acids like dihydroxyornithine, 4-hydroxyproline, dihydrooxy homotyrosine and 3-hydroxy-4-methylproline, as well as two threonine component of hexapeptide nucleus are reported (Kurtz and Rex 2001).

In 1977, a new antifungal antibiotic, *Aculeacin* is isolated from the mycelial cake of *Aspergillus aculeatus* M-4214 (Mizuno et al. 1977b). Subsequently, another six new antibiotics were isolated as the minor components related to aculeacin A from the same culture named as aculeacins B, C, D, E, F and G. The structure of Aculeacin is similar to echinocandin B but differs in the acyl moiety. Their acyl moiety is either the myristoyl (aculacin Aα–Dα) or palmytoyl (aculacin Aγ–Dγ) group. Physicochemical properties aculeacins B, C, D, E, F and G were analogous to those of aculeacin A and they all showed significant activity against fungi (Satoi et al. 1977).

Pneumocandin is another fungi-mediated antifungal compound. Pneumocandin has a sulfate moiety in the molecule and is differentiated from echinocandins by their structural difference (Fig. 1.2). The first member of pneumocandin class was pneumocandin B0 and was isolated from *Glarea lozoyensis* in 1985 at CIBE, a subsidiary of Merck located in Madrid, Spain. Subsequently, pneumocandin Ao was also reported from same culture by the same research group (Schwartz et al. 1989, 1992). Pneumocandin Ao is less haemolytic than other member of naturally occurring echinocandins (Boeck et al. 1989), whereas pneumocandin Bo appears to be the most potent glucan synthase inhibitor compared to other pneumocandin and in vitro and in vivo. Pneumocandin Bo differs from pneumocandin Ao only by the absence of a methyl on one of

Table 1.2 Natural Echinocandin with fungal in origin

Echinocandins	Fungus origin	Reference
Echinocandin A	*Aspergillus nidulans* *A. rugulosus*	Geiser et al. (2007)
Echinocandin B	*Aspergillus nidulans, A. rugulosus* *A. nidulans var. roseus A. rugulosus*	Nyfeler and Keller (1974), Geiser et al. (2007), and Traber et al. (1979)
Echinocandin c	*A. rugulosus*	Traber et al. (1979)
Echinocandin d	*A. rugulosus*	Traber et al. (1979)
Aculeacin A–G	*Aspergillus aculeatus*, *A. japonicus var. aculeatus*	Mizoguchi et al. (1977a), Mizuno et al. (1977a), Satoi et al. (1977), and Hino et al. (2001)
Mulundocandin deoxymulundocandin	*Aspergillus sydowi*	Roy et al. (1987), Mukhopadhyay et al. (1992), and Hawser et al. (1999)
Cryptocandin	*Cryptosporiopsis quercina*	Strobel et al. (1999)
"Catechol-sulfate" echinocandins (have thesame peptide nucleus as echinocandin B but an N-palmitoyl side chain (FR901379, FR901381-82, FR190293, FR209602-4, FR220897, FR220899, FR227673)	*Coleophoma empetri*, *Coleophoma crateriformis*	Iwamoto et al. (1994), and Kanasaki et al. (2006a, b, c)
Pneumocandin A–E	*Glarea lozoyensis*, *Pezicula carpinea*, *Cryptosporiopsis G. lozoyensis*, *Pezicula carpinea*, *Cryptosporiopsis* sp., *Aspergillus aculeatus*	Satoi et al. (1977), Schwartz et al. (1989, 1992), Nobel et al. (1991), Morris et al. (1994), Bills et al. (1999), Mizoguchi et al. (1977)
Sporiofungin A–C	*Cryptosporiopsis* sp.	Tscherter and Dreyfuss (1982)
WF11899s	*Coleophoma empetri*	Hino et al. (2001)
WF738s	*Coleophoma crateriformis*	Hino et al. (2001)
WF14573s	*Coleophoma empetri*	Hino et al. (2001)
WF16616	*Tolypocladium parasiticum*	Hino et al. (2001)
WF22210	*Chalara* sp.	Hino et al. (2001)
Aculeacin A–G	*A aculeatus*, *Aspergillus japonicus var. aculeatus*	Mizuno et al. (1977a, b), Satoi et al. (1977), Hino et al. (2001)

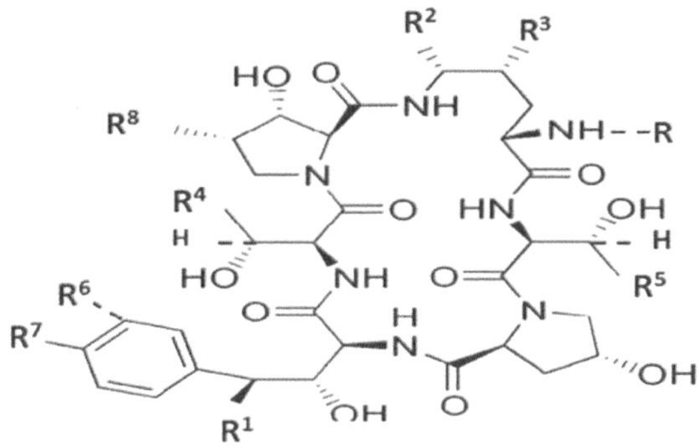

Echinocandin Lipopeptide family	R	R1	R2	R3	R4	R5	R6	R7	R8
Echinocandin B	linoleoyl	OH	OH	OH	CH_3	CH_3	H	OH	-
Echinocandin C	linoleoyl	H	OH	OH	CH_3	CH_3	H	OH	
Echinocandin D	linoleoyl	H	H	H	CH_3	CH_3	H	OH	
Aculeacin Ay	palmitoyl	OH	OH	OH	CH_3	CH_3	H	OH	
Mulundocandin	12-methylmyristoyl	OH	OH	OH	H	H	H	OH	
sporiofungin A	10,12-dimethylmyristoyl	OH	OH	OH	CH_2CONH_2	H	H	OH	
Pneumocandin Ao	10,12-dimethylmyristoyl	OH	OH	OH	$CONH_2$	CH3	H	OH	CH3
Pneumocandin Bo	10,12-dimethylmyristoyl	OH	OH	OH	$CONH_2$	CH3	OH	OH	H
Pneumocandin Co		OH	OH	OH	$CONH_2$	-	OH	OH	H
WF11899A	palmitoyl	OH	OH	OH	$CONH_2$	CH3	OSO3H	OH	H
L693, 989	10,12-dimethylmyristoyl	OH	OH	OH	$CONH_2$	CH3	-	OPO(OH)2	H
L 705, 589	10,12-dimethylmyristoyl	OH	$OCH_2CH_2NH_2$	OH	$CONH_2$	CH3	-	OH	H
L 731, 373	10,12-dimethylmyristoyl	OH	OH	OH	CH_2NH_2	CH3	-	OH	H
L733, 560	10,12-dimethylmyristoyl	OH	$OCH_2CH_2NH_2$	OH	CH_2NH_2	CH3	-	OH	H

Fig. 1.2 Structure of echinocandin compounds. Structural skeleton (*above picture*) with variable functional groups (listed at *lower side*)

the proline ring (Lóránd and Kocsis 2007) (Lorand). Both the drug shows low water solubility that makes it difficult to formulate. Pneumocandin Co is another structural isomer of pneumocandin Bo with a hydroxyl group at the c-4 of proline. Pneumocandin Do possess hydroxyl group at both c-3 and c-4 of proline. Pneumocandin Eo has no hydroxyl groups on the proline at position 1. Sporiofungin is another echinocandins type of antifungal antibiotic isolated from *Coleophoma empetri*. Three sporiofungin antibiotics exist in the form of Sporiofungin A, B and C. This compound poses potent antifungal action against *Candida*. Sporiofungins do not contain Thr but they have 3R-hydroxyl-L-Gln, and L-Ser residues at position 5 and 2, respectively. Besides the 3R-hydroxyl-LGln moiety, they also have a 10,12-dimethylmyristoyl acyl group, similar to pneumocandins (Emri et al. 2013). Eli Lilly Company has it under clinical trials by the name LY-303366.

Mulundocandins a lipopeptide antibiotic belonging to the echinocandin class. The first report of isolation was done in 1987 from *Aspergillus sydowi* strain no. Y-30462 (Roy et al. 1987). This compound has 10,12-dimethylmyristoyl acyl moiety and they contain L-Ser (instead of LThr) at position 5 from the N-terminus Orn (Mukhopadhyay et al. 1992). Pneumocandins differ from mulundocandins by the 3R-hydroxyl-L-Gln at position 5 (Mukhopadhyay et al. 1992). This group of antifungal agent shows broad spectrum activity against *Candida albicans* (MIC range 0.5–4.0 μg/ml), *C. glabrata* (2.0–4.0 μg/ml) and

C. tropicalis (1.0–8.0 μg/ml) (Mizuno et al. 1977b).

Cryptocandin is also a similar class of echinocandin. It was extracted from the strain of *Cryptosporiopsis* cf.*quercina*, isolated from an endophyte of *Tripterigeum wilfordii* stems (Strobel et al. 1999). More related compound like WF11899A came to discovery from fungal endophytes. This compound possess sulphate group at the para or meta position of the homotyrosine in the hexapeptide ring. This compound offers minimal inhibitory concentration values of 0.03–0.07 μg/ml against *Candida albicans*, *Trichophyton mentagrophytes* and *Trichophyton rubrum*. It contains palmitoyl moiety and also Gln residue at position 5 which makes the difference from echinocandin B.

Catechol-sulfate echinocandins are the fungal metabolites having antifungal activity and have drawn a remarkable attention nowadays. Their acyl-moiety is palmitoyl. They have a catechol-sulfate core in the homoTyr residue and contain 3R-hydroxyl-L-Gln at position 5. In some cases, the second amino acid was L-Thr and the others are L-Ser (Kanasaki et al. 2006b, c). They also show some heterogeneity (both meta and para positions may occur) in the position of sulfate group (Emri et al. 2013). Several synthetic analogues are reported as FR901379, FR901381-82, FR190293, FR209602-4, FR220897, FR220899, FR227673, etc.

1.4 Fungi-Derived Antifungal Compounds Having Anticancer Activities

1.4.1 Polyketides

Polyketides represent one of the major classes of natural products known to have members possessing both anti-cancer and antifungal activities were shown in Table 1.3. Some well-known fungal polyketides e.g compactin and lovastatin having anticancer activity belong to the statin family (Qiao et al. 2007; Chamilos et al. 2006; Macreadie et al. 2006). A dicyclohexene ring system linked to a side chain with a closed lactone ring or an open acid form constitutes the basic structure of statin. Small antifungal polyketides, brefeldin A and terrain were isolated from *P. brefeldianum* and *A. terreus*, respectively (Betina et al. 1962; Singleton et al. 1958; Ghisalberti and Narbey 1990; Raistrick and Smith 1935). Griseofulvin, another famous polyketide isolated from *P. griseofulvum*, described earlier in details, exhibits both anticancer and antifungal activities. One of the antifungal γ-pyrones, 3-*O*-methylfunicone, produced by *Talaromyces pinophilus* was also seen to have anticancer activity (Hector 2005). Nowadays, several allied azaphilones belonging to antifungal chaetomugilin family were isolated from *Chaetomium globosum* (Buommino et al. 2007; Baroni et al. 2009). The hypocrellins of the perylenequinone family was isolated from *Hypocrella bambusae*, and one of the potent analog hypocrellin D showed both antifungal and anticancer properties (Cai et al. 2011).

1.4.2 Terpenes

Wortmannin, produced by *T. wortmannii*, is an antifungal compound capable of inhibiting the activity of leukemia HL-60 and K-562 (OH et al. 2001). The fungal genera *Bipolaris*, *Aspergillus*, *Sarocladium* and *Drechslera* are the known sources of sesterterpenes including ophiobolins (Table 1.4). In addition to anticancer activity, the ophiobolin family also shows antifungal activity against a wide range of fungi (Li et al. 1995; Krizsán et al. 2010). The isolation of taxol from the bark of *Taxus brevifolia* and identification of taxol as one of the most efficacious drug against cancer followed by the high demand of raw material led to the search for an alternative producer strain. The fungus *Taxomyces andreanea* and *P. raistrickii* produces taxol (Visalakchi and Muthumary 2010; Niedens et al. 2013). Terrecyclic acid A is an example of another sesquiterpene, isolated from *A. terreus*, which exhibit both antifungal and anticancer activities (Nakagawa and Hirota 1982). Among the several trichothecenes, AETD

Table 1.3 Polyketides having both anticancer and antifungal actvities

Producer organism	Compound	Structure
Talaromyces pinophilus (*Penicillium pinophilum*)	3-O-methylfunicone	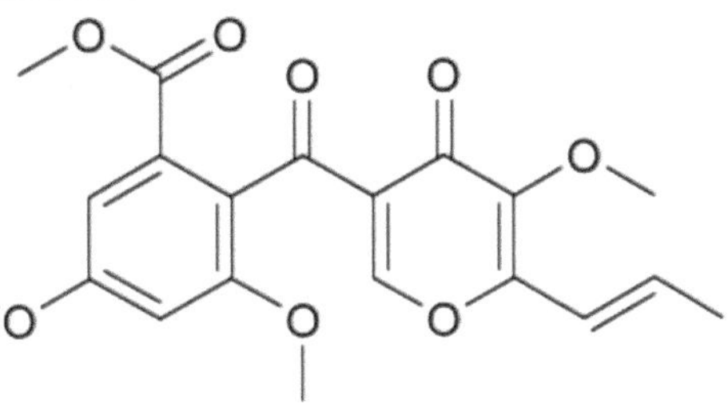
P. brefeldianum	Brefeldin A	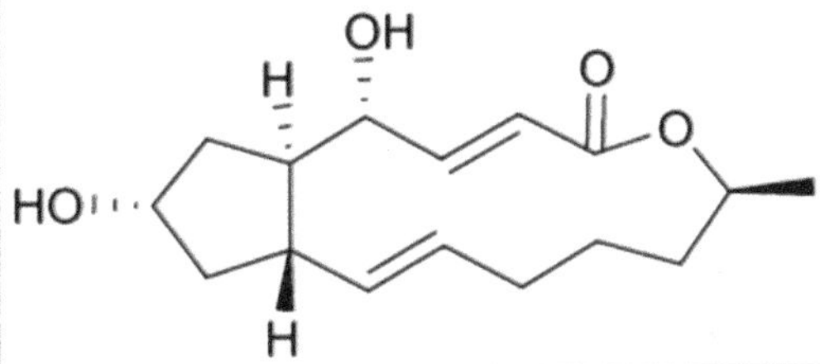
Chaetomium globosum	11-epichaetomugilin I	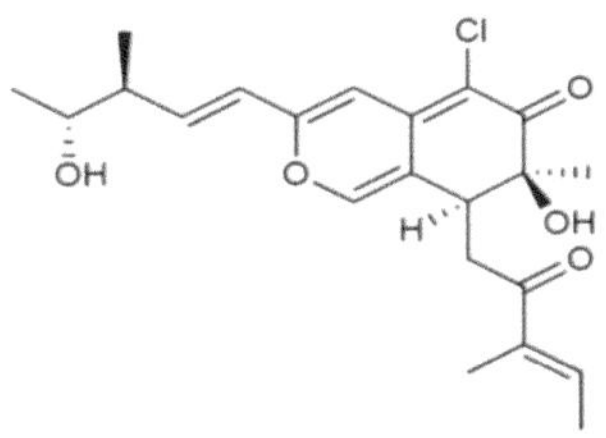
P. griseofulvum	Griseofulvin	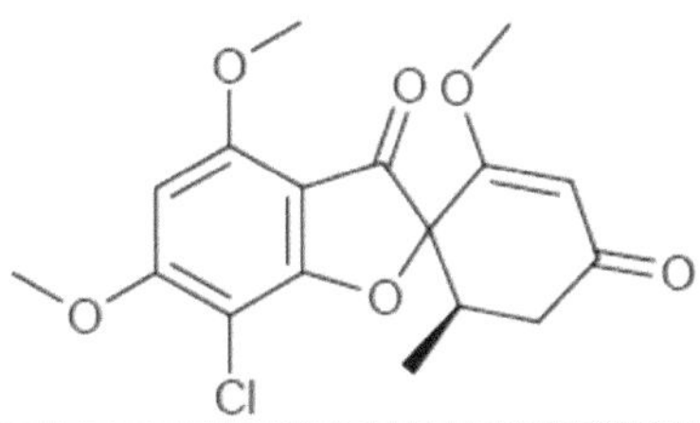
Hypocrella bambusae	Hypocrellin D	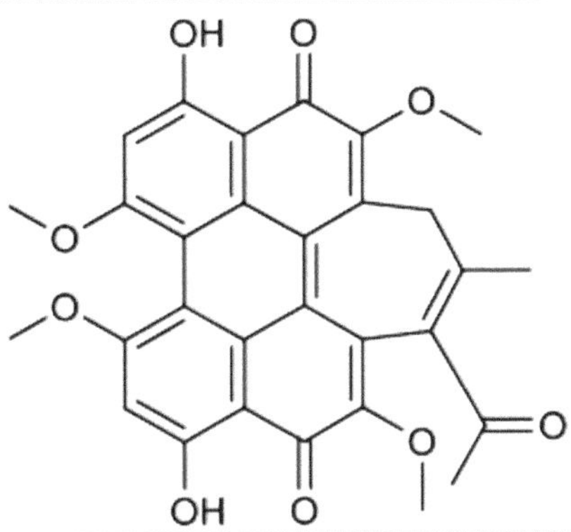

(continued)

Table 1.3 (continued)

Producer organism	Compound	Structure
P. solitum; P. citrinum	Compactin	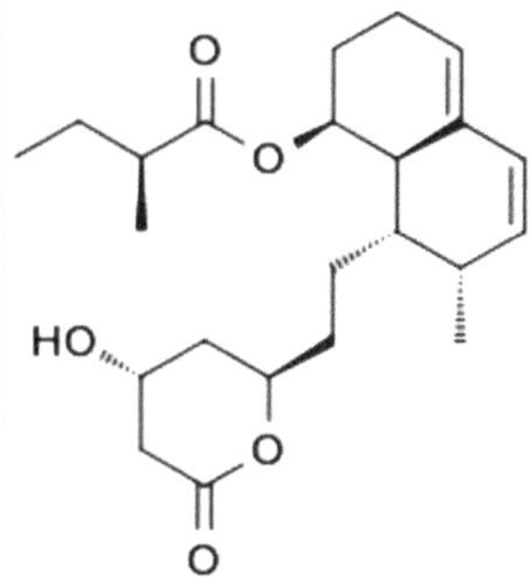
Aspergillus terreus Monascus sp.	Lovastatin	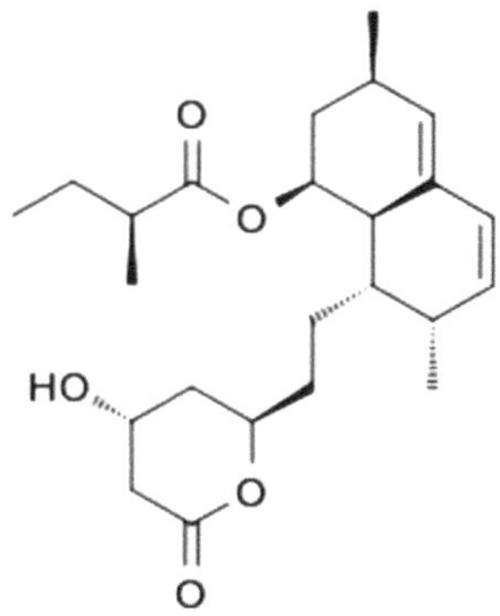
A. terreus	Terrein	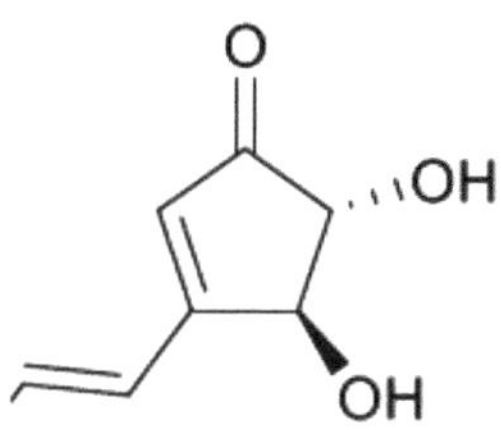

has also characteristic cytotoxic activity against both fungi and cancer cell lines (Woloshuk and Shim 2013). Another terpene, wentilactones, isolated from the marine alga-derived endophytic fungus, *Aspergillus wentii*, has shown potential antifungal activity beside its anticancer property (Sun et al. 2012).

1.5 Fungal Nitrogenous Compounds Including Non-ribosomal Peptides (NRPs)

A large group of natural products constituted nitrogen containing compounds of fungal origin that integrate amino acid building blocks into often complex heteroaromatic compounds such as diketopiperazines, quinazolines and benzodiazepines known to have biological activities (Table 1.5). These compounds contain primary, secondary or tertiary amine functionalities and because of their basic nature are often called as alkaloids. However, compounds only containing amide bonds, that are essentially neutral, are also referred to as alkaloids. Several nitrogen containing compounds are produced biochemically with the aid of non-ribosomal peptide synthases called non-ribosomal peptides (NRPs) even though they are also called alkaloids. Xanthocillin X, an

Table 1.4 Terpenes having both anticancer and antifungal activities

Producer organism	Compound	Structure
T. wortmannii	Wortamannin	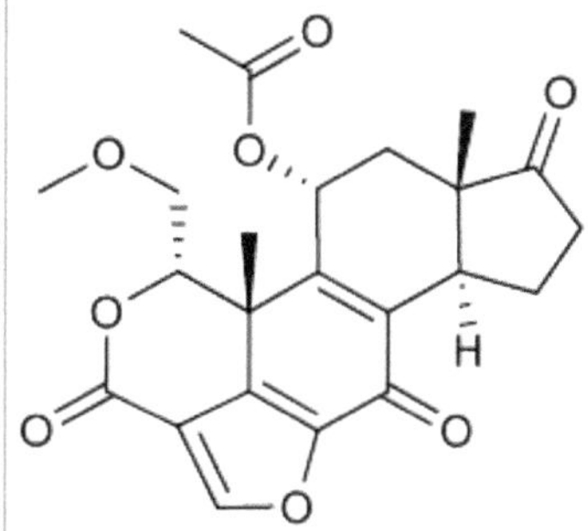
Bipolaris sp. *A. calidoustas* *A. ustus*	Ophiobolin A	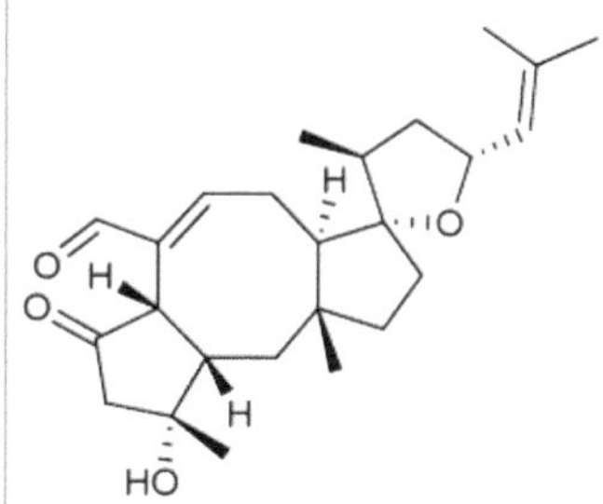
Taxomyces andreanea	Taxol	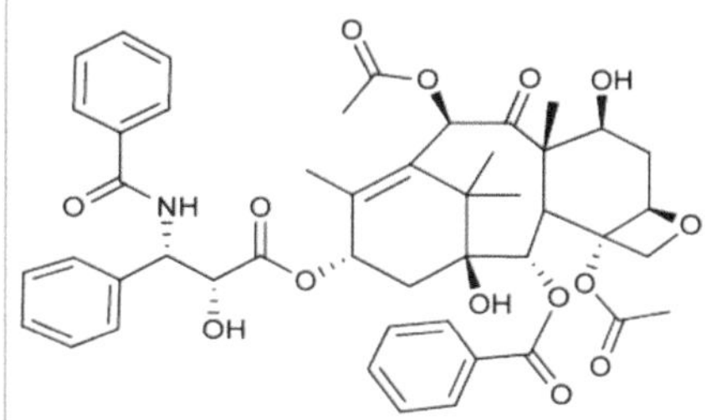
A. terreus	Terrecyclic acid A	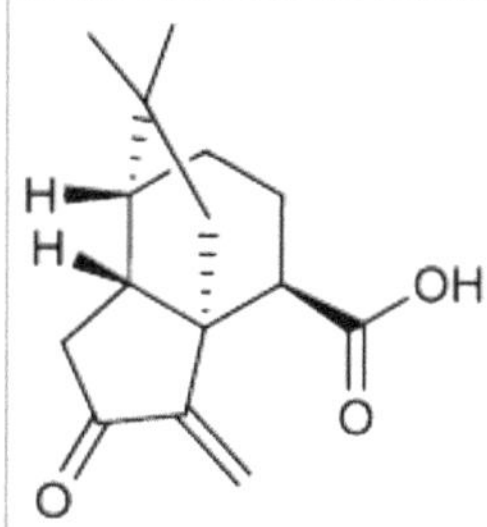
Fusarium spp.	4-Acetyl-12,13-epoxy-9-trichothecene-3,15-diol [AETD]	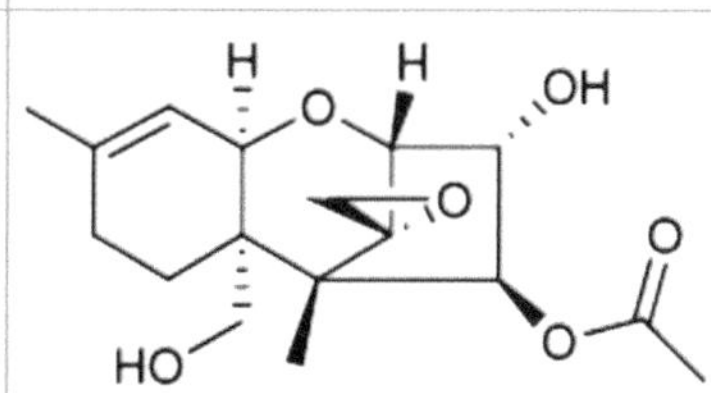
A. wentii	Wentilactones	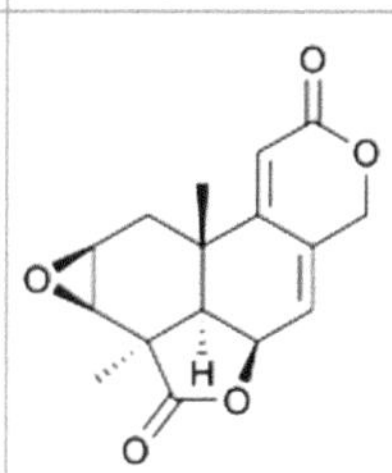

Table 1.5 Fungal nitrogenous compounds including non-ribosomal peptides (NRPs) having both anticancer and antifungal activities

Producer organism	Compound	Structure
A. fumigatus	(Diketopiperazines with a disulfide bridge) Gliotoxin	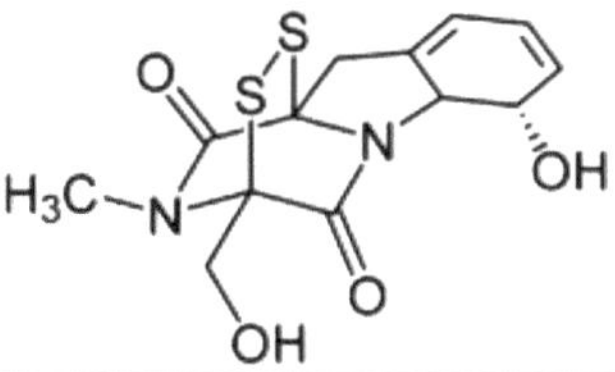
Emericella striata	(Diketopiperazines with a disulfide bridge) Emestrin A	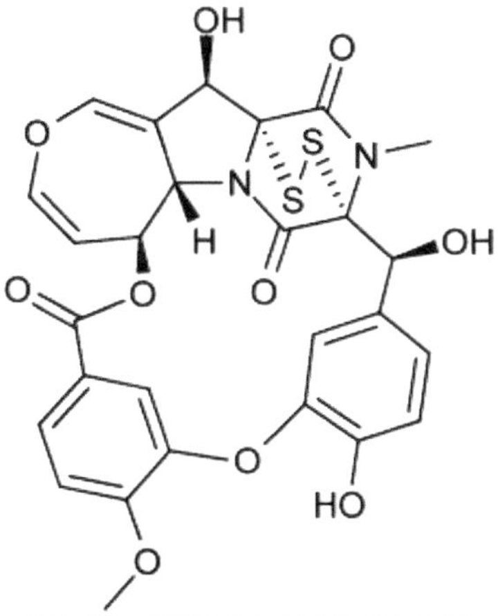
A. fumigates *A. fischeri*	(Tryptophan/proline diketopiperazines) Fumitremorgin C	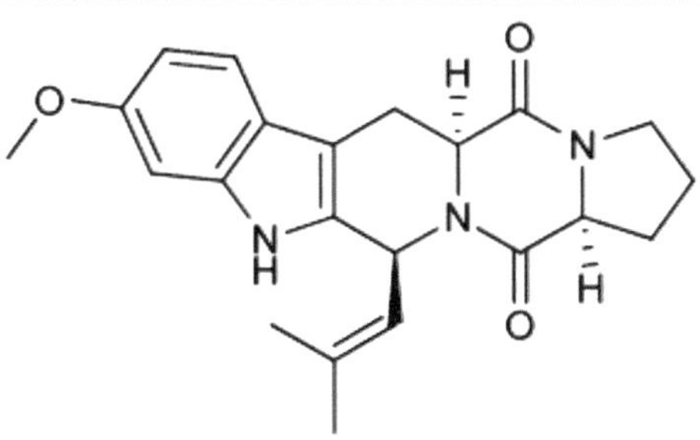
P. chrysogenum	Xanthocillin X	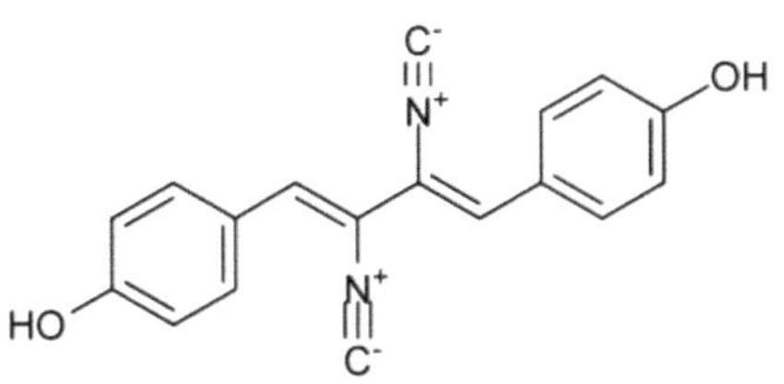
Purpureocillium lilacinum	(Peptide) Leucinostatin A	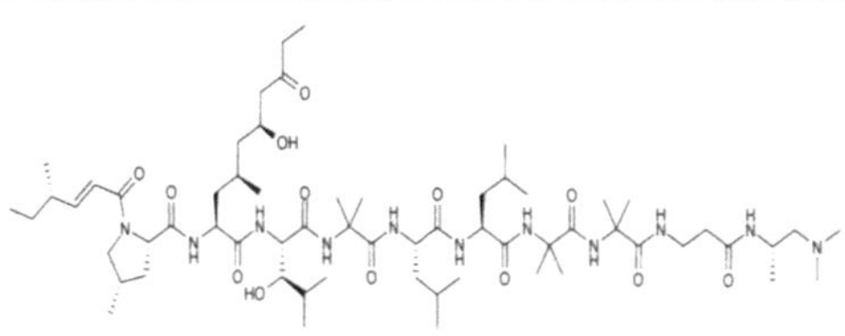

antifungal compound, produced by *P. chrysogenum* was also found to inhibit several cancer cell lines (Frisvad et al. 2004). Diketopiperazines are basically cyclic dipeptides that are reported to inhibit cell cycle at G2/M phase (Borthwick 2012). The fumitremorgin group of compounds among tryptophan/proline diketopiperazines is produced by *A. fumigatus* and *A. fischeri* (Abraham and Arfmann 1990). Fumitremorgin C, a potent antifungal compound, was also found active against human carcinoma cell lines (Zhao et al. 2002). Among another group of diketopiperazines containing a di-sulfide bridge in the diketopiperazine ring, antifungal compounds, gliotoxin and emestrin A, were shown to be potent inhibitor of human cancer cell lines (Seya et al. 1986; Finefield et al. 2012). The peptide leucinostatin A, isolated from *Purpureocillium lilacinum*, was found to be active against a number of fungi and Gram-positive bacteria as well as several cancer cell lines (Arai et al. 1973; Kawada et al. 2010).

1.6 Antifungal Metabolites from Coprophilous Fungi

Coprophilus fungi are the type of saprobic fungi prefer to grow on animal dung shares a lot of fruitful metabolites having antifungal activity and may lead some new in near future. They have been generally isolated from animal dung particularly from herbivorous mammals. The interference competition among coprophilous fungi in dung environment offers the production of secondary metabolites by one species that deter the growth of competitors (Bills and Gloer 2013). Here are some of the coprophilous fungi with their antifungal metabolites (Table 1.6).

1.7 Mechanism of Antifungal Action

Antifungal drug from natural or synthetic in origin share some common strategies which facilitate to inhibit the fungal cells. Very often the strategies are cell wall biosynthesis, sphingolipid synthesis, protein synthesis, electron transport, membrane integrity, etc. Fungal cell wall composition varies among species to species but three major polymeric components are glucan, chitin and mannoproteins. Finding inhibitors of such polymeric components are the prime objective of cell wall biosynthesis inhibition (Oxford and Raistrick 1939). Sphingolipids and their metabolites are present in relatively small proportion modulate various cellular events including proliferation, differentiation and apoptosis (Brian and Curtis 1946). Inhibition of sphingolipid synthesis results in disturbance in fungal growth and subsequent cell death (Brian 1949; Grove and McGowan 1947; Brian and Curtis 1949).

Apparently, the similarity in human and fungal sphingolipid biosynthetic pathway may seem that it is difficult to develop a controlling point. But major enzymatic deviations like serine palmitoyltranferase, ceramide synthase and IPC synthase from mammalian system make this as active target of sphingolipid synthesis. Although protein synthesis is the primary and most important choice of bacterial inactivation, fungal and mammalian protein synthesis machinery is almost same which makes it difficult to target (Brian and Curtis 1949). Several approaches are made to overcome such problems. Membrane integrity and electron transport are the targets of several antifungal agents. Compounds that target membrane integrity bind with the common fungal sterols, causing membrane permeability and leakage of cytoplasmic content yielding cell

Table 1.6 Other antifungal compounds from fungi special focus to coprophilous fungi

Name of the compound	Biosynthetic family	Name of the fungus	Year	References
Appenolides A–C	Polyketide	*Podospora appendiculata*	1993	Wang et al. (1993)
Apiosporamide	Polyketide-amino acid	*Apiospora montagnei*	*1994*	Alfatafta et al. (1994)
Coniochaetones A and B:	Polyketide antifungal	*Coniochaeta saccardoi.*	1995	Wang et al. (1995a)
Terezines A–D	Peptide	*Sporormiella teretispora*	1995	Wang et al. (1995b)
Petriellin A	Peptide	*Petriella sordida*	1995	Lee et al. (1995)
Polytolypin	Terpenoid	*Polytolypa hystricis*	1995	Gamble et al. (1995)
Cercophorins A–C:	Polyketide	*Cercophora areolata*	*1996*	Whyte et al. (1996)
Anserinones A and B	Polyketide	*Podospora anserina*	*1997*	Wang et al. (1997)
Coniochaetones and Cerdarin	Both polyketide	*Cercophora sordarioides*	1997	Whyte et al. (1997)
Arugosin F:	Polyketide	*Ascodesmis sphaerospora*	*1998*	Hein et al. (1998)
Sporovexins A–C	Mixed preussomerin analog	*Sporormiella vexans*	1999	Soman et al. (1999)
Bombardolides	Modified polyketide	*Bombardioidea anartia*	2001	Hein et al. (2001)
Sordarins	Terpenoid glycoside	*Lasiosphaeriaceae pleiospora*	2001	Odds (2001)
Pseudodestruxins A, B and Ascochlorin	Peptide and Terpenoid/polyketide	*Nigrosabulum globosum*	2001	Che et al. (2001)
Decipinin A and decipienolides A and B	Benzopyrans.	*Podospora decipiens.*	*2002*	Che and Gloer (2002)
Tulasnein and podospirone	Terpenoid	*Podosordaria tulasnei*	2004	Ridderbusch et al. (2004)
Antiamoebins, myrocin B	Peptide and Terpenoid	*Stilbella erythrocephala*	2006	Lehr et al. (2006)
Others				
Gymnoascolides A–C	Shikimate	*Gymnoascus reessii*	*2005*	Clark et al. (2005)
Pestalachlorides A–C		*Pestalotiopsis adusta*	*2008*	Li et al. (2008)
Solanapyrone analogues	Solanapyrone	Hawaiian fungicolous fungus	*2007*	Schmidt and Gloer (2007)
Naphthoquinone spiroketal	Naphthoquinone	*Endophytic fungus, Edenia gomezpompae.*	*2008*	Macías-Rubalcava et al. (2008)
zaragozic acids	Polyketide-TCA-amino acid	*S. intermedia,* and *L. elatius*	1995	Bergstrom et al. (1995)
australifungin	Polyketide	*Sporormiella australis*	1995	Mandala et al. (1995)
Sporminarins A and B	Polyketide	*Sporormiella minimoides*	2006	Mudur and Gloer (2006)
Similins A and B	Polyketide	*Sporormiella similis*	1992	Weber and Swenson (1992)
Fusidic acid		Marine isolate of the fungus, *Stilbella aciculosa*	2001	Kuznetsova et al. (2001)

death (Fig. 1.3). Mitochondrial electron transport is now used as a target to fungal control. Compounds (UK2A, UK3A) are structurally related to actinomycin A, with nine member dilactone ring offers broad spectrum antifungal activities (Brian and Curtis 1955; Brian 1949; Araujo et al. 1990).

1.8 Advancement of Biosynthesized Antifungal Agents

Fungal origin requires critical care in producing compound in a large scale that includes from designing bioreactor to finding optimum

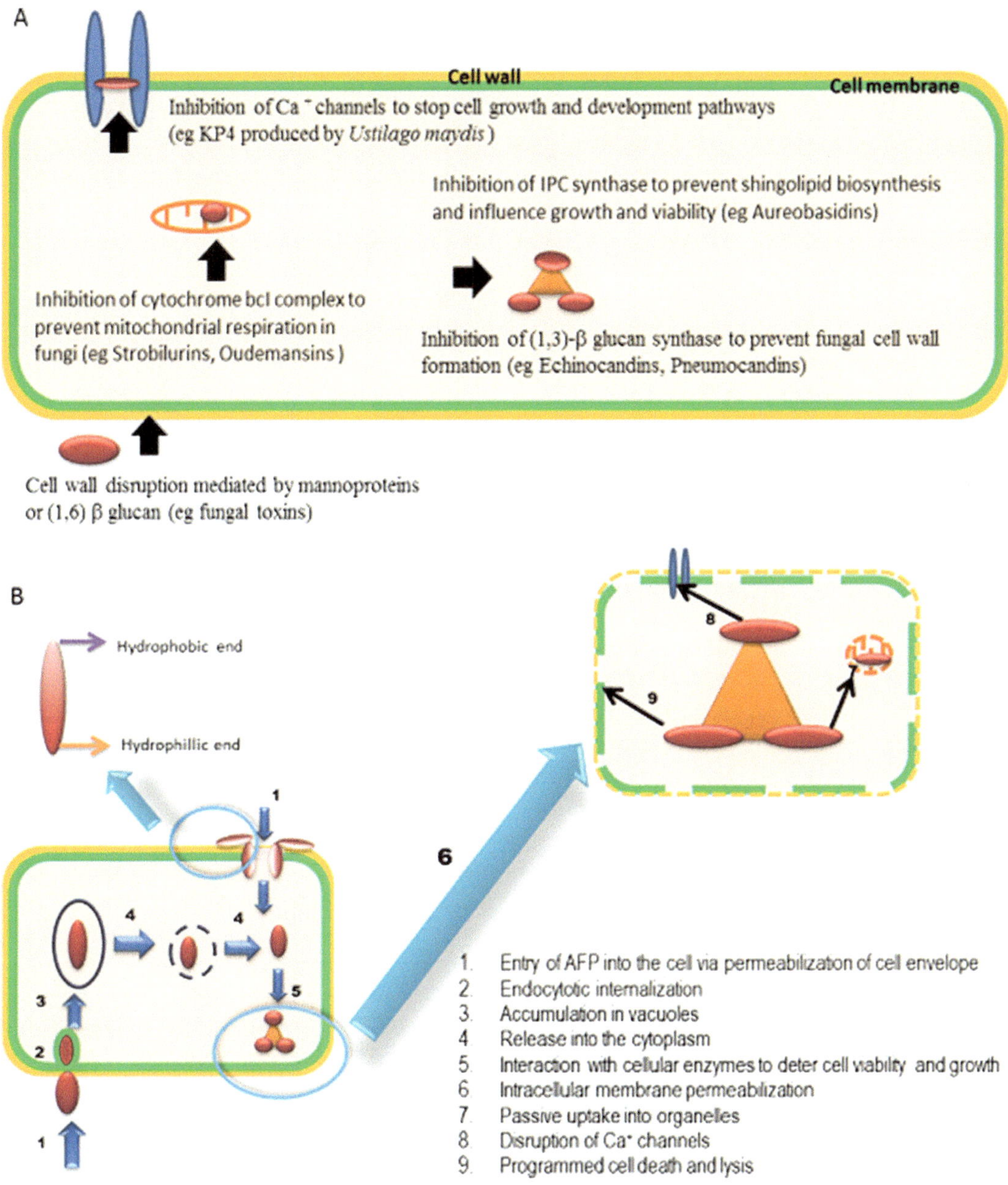

Fig. 1.3 Schematic diagram on the mechanism of action of antifungal agents

condition of substrate, nitrogen source, temperature, inoculum dose, pH, etc. Several studies have been carried out including the development of high-yielding strains through genetic modification (Keller and Turner 2005; Magdalena et al. 2013; Oxford and Raistrick 1939; Brian and Curtis 1946). Besides, the improvement of biological activity, absorption, distribution, metabolism and excretion are the prime object of antifungal medicinal chemistry research. Synthesis of derivatives and analogues of natural products are the most important sources for new drug candidates and tools for medicinal chemistry. Therefore, need for the development of efficient chemical synthesis methods for accessing the natural products and their modified derivatives are inspiring to control the resistance mechanism in great importance. Therefore, need for the development of efficient chemical synthesis methods for accessing the natural products and their modified derivatives are inspiring to control the resistance mechanism in great importance.

1.9 Echinocandin Past and Present: A Possible Clue

FR901379, a cyclic lipopeptide, showed in vivo antifungal activity against *Candida albicans* by inhibiting the synthesis of 1,3-β-glucan, a key component of fungal cell wall (Louise and Neil 2010). The producer strain of FR901379 was a soil isolate collected at Iwaki City, Fukushima Prefecture, Japan, and identified as *Coleophoma empetri* F-11899 (Ascomycota) (Iwamoto et al. 1994; Iwamoto et al. 1993; Yamashita et al. 2005). However, this compound had haemolytic activity and was also less active against *Aspergillus fumigatus*. To overcome these shortcomings, a series of strategies are followed that opens up new directions of antifungal drug discovery. In 2001, Fujie et al. have reported to synthesized FR131535 by doing some structural modifications as octyloxybenzoyl acyl side chain was added instead of a fatty acid. The new compound FR131535 retained the original activity displayed by FR901379, acquired potent anti-*Aspergillus* activity, and its hemolytic activity was significantly reduced (Fujie et al. 2001). Further extensive chemical modification of FR901379 has led to the discovery of micafungin (FK463), which is effective against both *Candida* and *Aspergillus* spp. Micafungin has been marketed in Japan and in United States as a candin-class parenteral antifungal agent for life-threatening mycoses (Akihiko 2007). Similarly, Tomishima et al. reported a series of novel acylated analogs of the echinocandin, FR901379. They demonstrated relationship between antifungal activity and lipophilicity of acyl side chain, expressed as *C*log *P*. Their analog was reported to have higher efficiency than previously disclosed 4-(*n*-octyloxy)benzoyl side chain analog, FR131535 (Tomishima et al. 2008a). They further optimized side chain analogs of the natural product FR901379 and led to the discovery of a new modified compound. This compound is reported to have reduced hemolytic potential with a well-balanced profile and was selected as the clinical candidate (Tomishima et al. 2008b).

Kanda et al. (2009) has tried to improve the production of FR901379 by mutant selection and medium optimization. They have performed an interesting task and evaluated seven generation's strain-breeding, beginning with a wild type. Medium for selection of screening and large-scale production were designed (Kanda et al. 2010). In the beginning of year 2011, Ueda et al. purified FR901379 acylase, an enzyme that catalyzes the hydrolysis of the palmitoyl moiety of the antifungal lipopeptide FR901379 from the culture broth of *Streptomyces* sp. no. 6907 (FERM BP-5809). They transferred it and found 250-fold increase in FR901379 acylase activity in recombinant *Streptomyces lividans* 1326 carrying the cloned gene (Ueda et al. 2011a). In the same year, they again reported that the same species after mutagenesis using UV-irradiation capable of hyperproducing the FR901379-acylase enzyme (Ueda et al. 2011b).

Echinocandin B, a lipopeptide with antifungal activity, was first discovered in 1974 in *Aspergillus nidulans* var. echinulatus A 32204 by a group of German scientist (Nyfeler and Keller-Schierlein 1974). In the fermentative production of echinocandin B, a potent carcinogen sterigmatocystin is produced in significant amount. To overcome the production of such carcinogen, Hodges et al. (1994) worked out the genetic modification of an echinocandin B-producing strain of *Aspergillus nidulans* capable to produce mutants blocked in sterigmatocystin biosynthesis. In 1999, Butchko et al. (1999) reported the regulation strategies by the pathway-specific transcription factor, aflR, clustered on chromosome IV of *A. nidulans* and involved in the biosynthesis of sterigmatocystin (Butchko and Adams 1999). Hodges et al. (2000) reported characterization of a mutant designated A42355-OC-1 (OC-1), which is blocked in ST biosynthesis, was the result of a chromosomal translocation (Niedens et al. 2013).

The synthetic approaches are also carried out in parallel. Debono et al. (1988) has modified and prepared new analogs of echinocandin B by enzymatic deacylation and chemical reacylation of echinocandin B. The strategy carried out here was the incorporation of phenyl group into the fatty acid chain. Synthesized antifungal agent cilofungin (LY121019) offered excellent anticandidal activity and low toxicity and was superior to other available antifungal antibiotics (Debono et al. 1988, 1989). Hence incorporation of phenyl group becomes a pioneer breakthrough in synthetic modification. Therefore, research was mainly focused in incorporating more phenyl group in fatty acid chain. One more phenyl to the cilofungin side chain offered biphenyl compound LY 298095, which showed tenfold increased in vitro activity. Addition to another phenyl compound yields rigid side chain analog of LY280949, which offered a similar antifungal activity and efficacy in animal model similar to that of LY 298095. This was the first echinocandin compound having oral efficacy in an animal model of fungal infection. In this group of compound, the side chain is crucial for their desired property. Besides, the incorporation of flexible alkyl section maximizes the antifungal potency, possibly by a crucial adjustment of their lipophilicity (Turner and Rodriguez 1996).

Subsequently in 1995, chemical modification lead to an important compound of this group LY303366, termed as anidulafungin. Side chain of LY303366 has a rigid terphenyl head and a flexible C5 tail. It proved to be effective by oral route of administration (Debono et al. 1995) and was chosen as new generation clinical candidate for antifungal therapy. The semisynthetic analog, LY303366 of echinocandin B showed potential activity against growing yeast by inhibiting the beta-D glucan synthase (Lisa et al. 1999; Petraitiene et al. 1999), which also revealed their significant efficacy in animal model experiments (Keller and Turner 2005; Oxford and Raistrick 1939). β-D-glucan synthase, exhibits efficacy in animal models of human fungal infections and works best against actively growing yeasts (Lisa et al. 1999; Petraitiene et al. 1999). In 2007, Jamison et al. (1997) have reported the synthesis strategies and biological activity of a series of N-alkylated derivatives of echinocandin B. In reductive alkylation conditions, echinocandin nucleus was treated with a slight excess of aldehyde in the presence of sodium cyanoborohydride and refluxed with methanol/dimethylformamide for 24 ~ 48 h had yield N-alkylated derivatives of echinocandin B. Mulder et al. (2011) studied details on the synthetic strategies to make novel echinocandins analogues by on-resin ring closing to metathesis or disulfide formation. This was to explore the influence of cyclic peptide backbone on the antifungal activity. They concluded that the ring size was an important factor for antifungal activity (Mulder et al. 2011).

1.10 Problem, Strategies and Prospects of Griseofulvin

Griseofulvin is an orally acting antifungal antibiotic with low water solubility. After the discovery of Griseofulvin, several efforts were made to synthesize its derivative with more efficient activity and increased solubility than its original one. But research carried out till are seems to be ineffective to produce the appropriate derivative. Crosse et al. tested more than 300 analogues compound of Griseofulvin. But unfortunately none of them exceed the potency of original molecule (Crosse and McWillium 1964). Further, Fischer and Riegelman (1967) prepared some derivatives of the 4′-keto group of griseofulvin but again they were failed to sufficiently improve the biological action. But they succeeded in another way by increasing the water solubility and absorption potential of the drugs. In 1970, Fields and Newman (1970) prepared analogs 5′-diazogriseofulvin of antifungal drug griseofulvin. But in that case, again the activity was reduced. But in that case, again the activity was reduced (Fields and Newman 1970). In the meantime, some other synthesis strategies of griseofulvin were followed (Danishefsky et al. 1978, 1979; Danishefsky and Etheredge 1978), but didn't work properly. In 1990, Ko and Oritani relate possible structure-activity relationship of griseofulvin (Ko and Oritani 1990). They concluded that the presence of 6′-methyl group and 4′-oxo group on the C ring is very important factor in the biological activities of griseofulvin. Yamato and co-workers with their three consecutive publications explored details for the possible structure activity relationship of griseofulvin (Tomozane et al. 1990; Takeuchi et al. 1997a, b). According to them, the relationship between the antifungal activity and the position or kind of substituents on the benzene ring of griseofulvin is important. This might serve as a clue for modern synthetic modification of griseofulvin. In 1992, Gaoxiong et al. also tried to modify the griseofulvin but they didn't succeed to improve the biological activity. Among their synthesized derivatives, only oxime of griseofulvin was shown to have the most potent ability than griseofulvin. But they were able to increase the water solubility of some compounds (Delgado et al. 1992). In a recent study in 2012, 53 analogues compound of griseofulvin also followed the same trend with the majority being less active than griseofulvin and none had more than twice the potency of parent compound (Pan et al. 2012). In the same year, 4′-thiosemicarbazonegriseofulvin, a new thiosemicarbazide derivative of griseofulvin, was synthesized and evaluated for its potential in the control of enzymatic browning and postharvest disease of fruits. This derivative showed strong inhibition of the mycelial growth of the target fungi (Rønnest et al. 2012). Recently, an enantioselective Michael-aldol tandem reaction was done with respect to prochiral 2-substituted benzofuran-3-ones and enones by a facile primary amine catalyst, rapid access to the desired pharmaceutically active griseofulvin analogues is achieved (Dong et al. 2013). Thus, earlier development may serve the key issues and trends which must be effectively handled in near future in the way to discover efficient antifungal.

1.11 Strobilurins Synthesis: Success and Future Threat

Strobilurins in natural form are volatile and unstable in nature. Using Quantitative Structure Activity Relationship (QSAR) over natural strobilurins, many pesticide companies were able to discover many synthetic analogues which are more potential and stable fungicides (Balba 2007). Simpler structure in chemical synthesis leads strobilurins A as the preferred choice to the researchers. The first patent on strobilurins derivative came in 1999 and after that a number of patents were issued with structural modifications only. The major objective of chemical modification was to improve photostability and to reduce the volatility and chemical systematic properties (David and Geoffrey 1980). Thus this objective is quite effectively being solved. Anke et al. were pioneers in strobilurins synthesis and modifications. The initial modification starts with the work of Wolfgang (Hubert and Wolfgang 1999). Further, a lot of modifications are carried out by themselves which led a new way in strobilurins research (Bartlett et al. 2002). Several chemical companies like Zeneca Research Team, BASF, Novartis, Sinochem Shanghai, Shenyang Research Institute of Chemistry Industry, Kumiai Chemical Industry, etc. had done a lot of independent research programmes. Modification of the α-substitution in the (E)-β-methoxyacrylate group was the primary focus of the structural modifications. BASF research programme leads to the replacement of the basic toxiphoric group of (E)-β-methoxyacrylate with a methoxyiminoacetate group. Photostability and solubility was improved by adding diphenyl ether and benzene ring, respectively. In 1992, Zeneca (now Syngenta) and BASF announced QoI fungicides azoxystrobin and kresoxim-methyl, respectively. These fungicides become commercially available in 1996. Whereas further modifications by replacing (E)-β-methoxyacrylate group with 2-methoxyiminoacetamide leads to the development of metominostrobin (Dolores et al. 2010). Recently, several products are available in the market and some of them are presented in Table 1.7. The drugs were acquired 10–15 % by 2005 of world's fungicide market and now are the second largest fungicidal chemicals used in agricultural sector (David and Geoffrey 1980). Several reviews and works on synthesis strategies, assigned problem have been published (Balba 2007; Hubert and Wolfgang 1999; Vladimir and Leonid 1998; Sauter and Ammermann 1996; Dave et al. 2001). The optimistic and successful discovery leads strobilurins to be considered as one of the most valuable classes of single-site fungicides ever discovered by the agrochemical industries. But the recent problem of strobilurins fungicide resistance is a recent threat to researchers. This group of drug inhibits respiration by blocking the ATP synthesis and is prone to potential resistance. Just one mutation at the target site can develop resistant strain. In this prospect, fungicides with multiple mode of action are necessary to being developed. This requires a detailed knowledge on the existing structures that might open new avenue in antifungal chemotherapy.

1.12 Pneumocandin B0 as the Starting Material of Synthetic Antifungal Caspofungin Acetate

Pneumocandin B0 served as the starting material for effective semisynthetic antifungal drug caspofungin acetate (CANCIDAS®). This drug inhibits the β -D glucan that interferes with fungal cell wall synthesis (Keller and Turner 2005; Oxford and Raistrick 1939). The development of such effective drug with improved potency, water solubility, half-life and stability faces several challenges. The fermentive material, pneumocandin B0 remained with other 20 closely related compounds made difficulties in initial approach of purification. The prime objective in caspofungin acetate synthesis was surface modification of the peptide core of the pneumocandin B0. The strategy followed was reduction of the primary amide of 3-hydroxyglutamine to an amine and condensation of the hemiaminal moiety with ethylenediamine. Although several difficulties like instability of caspofungin and its intermediates in varied pH and freezed condition have been attempted to address. Problem in isolation, purification and crystalization of final product is yet to be solved (Balkovec et al. 2014).

1.13 Future Perspectives

Epidemiologic studies have showed increased mortality rate due to infections with different fungal species belonging to the genera like *Aspergillus*, *Fusarium*, *Candida* and *Zygomycetes*. Most of the antifungal agents are currently being used to act by altering the functions of cellular membrane or cell wall. The development of new antifungal compounds with broad spectrum of activity with diverse mechanism of action is essentially required. Simultaneous usage of two or more antifungal compounds to treat the diseases caused by resistant varieties is another way to achieve maximum antifungal activity. In the present era of novel and improved medical practices, the treatment of fungal diseases has also attained newer heights and has better perspectives. New diagnostic measures are now replacing the use of blind antifungal therapy that was normally used for patients suffering from mycoses. The diagnosis needs to be involved with fungal antibodies, antigens or metabolites for fungal detection.

However, both fungus and humans are eukaryotic and effective treatment depends upon locating and exploiting other cellular targets. Current antifungals are limited to a range of cellular targets and so are either toxic or become less effective or ineffective due to natural or induced resistance. To achieve the development of highly specific antifungal agent,

Table 1.7 Synthetic Strobilurins analog group with example

Natural structure of strobilurins	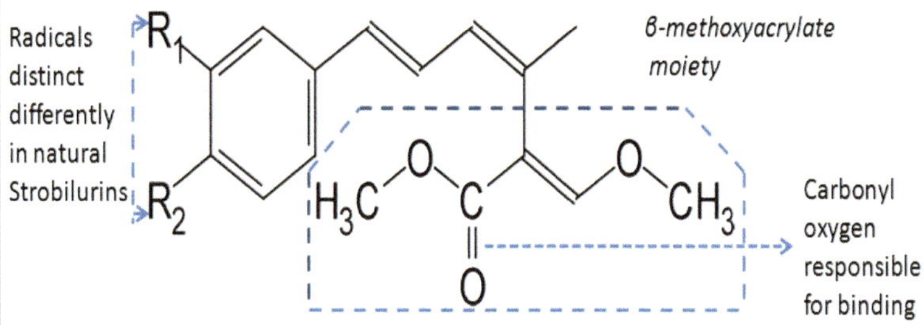		
Class	Methoxyacrylates		
Example	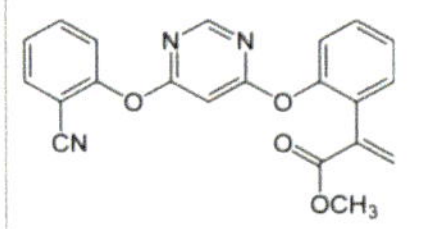	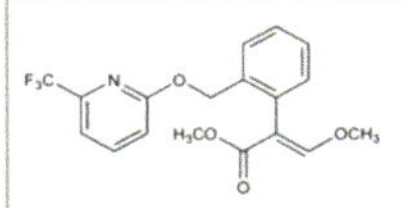	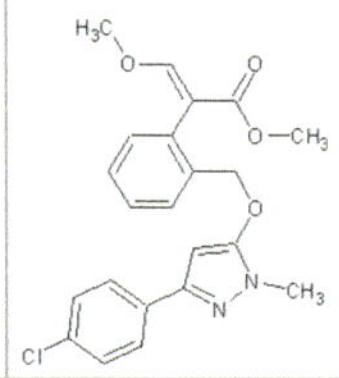
	Azoxystrobin	Picoxystrobin	Pyraoxystrobin
Class	Oximinoacetates		
Example	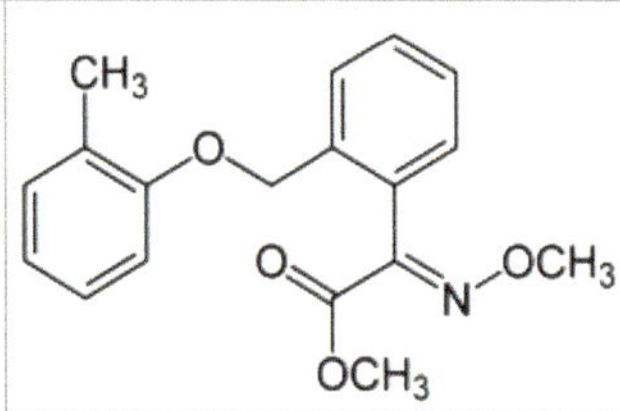		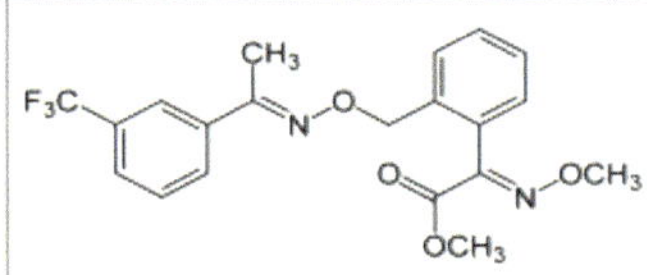
	Kresoxim-methyl		Trifloxystrobin
Class	Oximinoacetamides		
Example	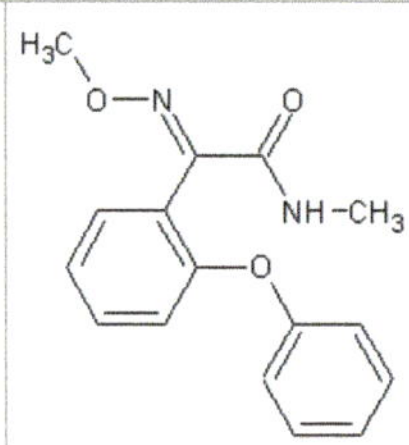	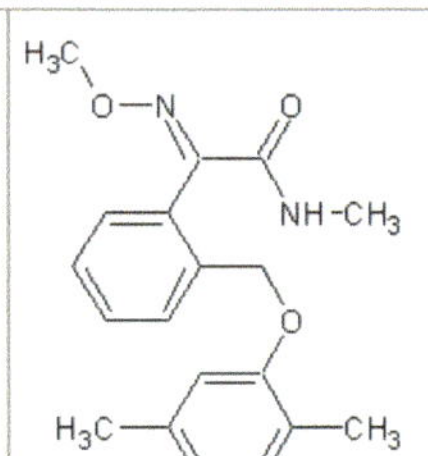	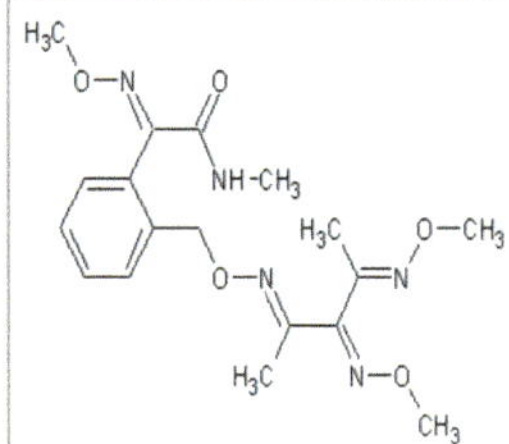
	Metominostrobin	Dimoxystrobin	Orysastrobin
Class	Methoxycarbamates		
Example	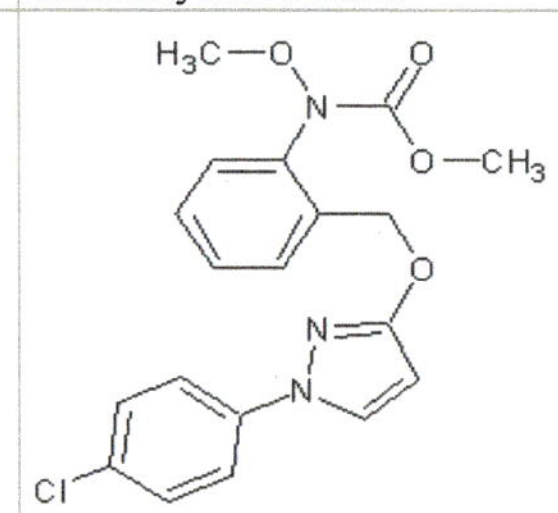		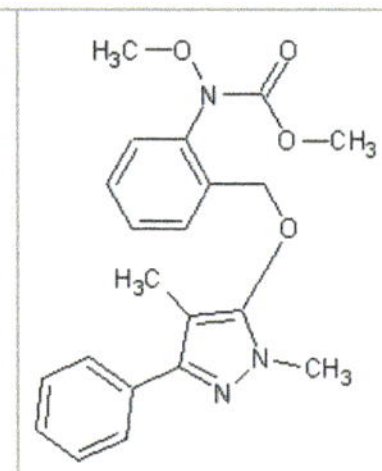
	Pyraclostrobin		Pyrametostrobin

(continued)

Table 1.7 (continued)

Natural structure of strobilurins	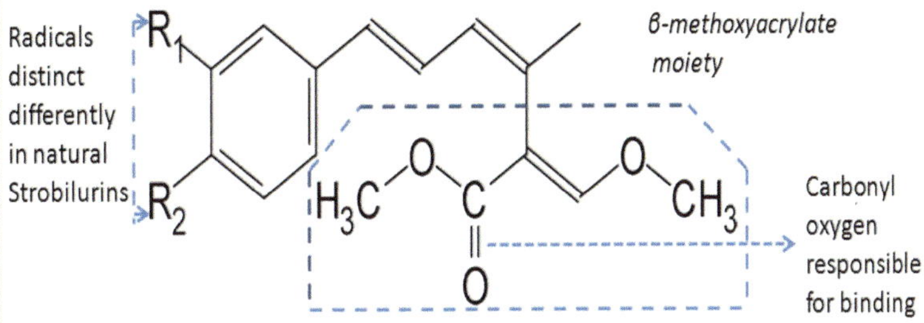
Class	Dihydrodioxazines
Example	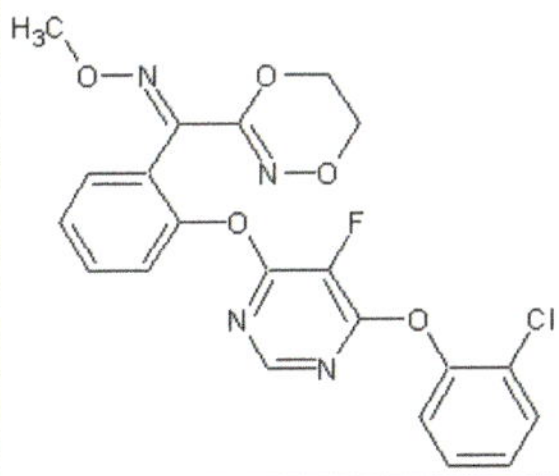
	Fluoxastrobin
Class	Benzylcarbamates
Example	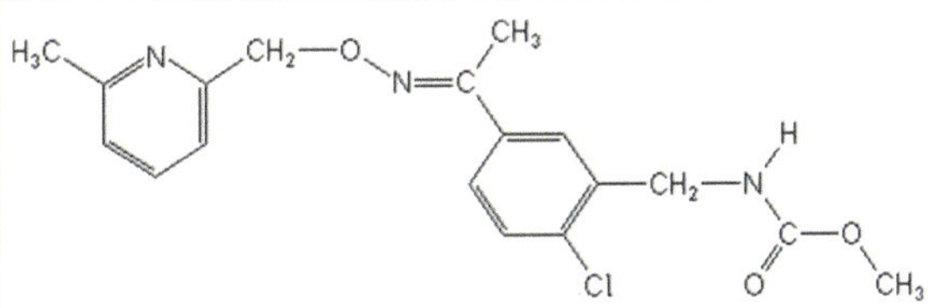
	Pyribencarb

the combined application of antifungal screening procedures and rational antifungal design is necessary. The systemic fungal infections are often caused by one opportunistic fungus at a time and not a 'mix' or consortium of cultures; therefore, the uses of 'broad spectrum' antifungals are harmful and unnecessary for treatment. In this light, fungal ecology is of particular interest and exploring and exploiting endophytic fungi to develop antifungal agents might be considered as a unique option.

Endophytic fungi might serve as an important source of antifungal agents because fungal antagonism has been reported in many fungal ecosystems. Endophytic fungi inhabiting spaces within plant cells are capable of producing certain antifungal agents that eventually ward off plant pathogens. The endophytic basidiomycete fungus KG146A found in rosemary shrub demonstrated antifungal activity against several strains of *Candida* (human pathogen). This pathogen affects several immune compromised patients and causes systemic mycoses that tend to prevail even with antifungal therapies.

Several fungal-derived antifungal compounds have anticancerous activity which may remind us their promiscuousity, in which multiple targets are associated with a common structure. The promiscuous activity of an antifungal compound should be used in the development of novel biopharmaceuticals. Emphasis will be given to their structural skeleton responsible for selectivity against additional targets and molecular mechanisms. The development of novel synthetic analogs of molecule is important for enhancing their multipurpose activities. Secondary infections with fungal species are very much common in immune compromised cancer patients. In such cases, the molecules having promiscuous (both anti-cancerous and antifungal) activities should be an excellent options as chemotherapeutics agents.

Many fungal-derived compounds with immense biodiversity have been isolated and used as antifungal and antitumor therapeutic agents. With the onset of this golden era of mycology and antifungal chemotherapy, it is more imperative that our future provides novel, enhanced anti-fungal agents for safer treatment of fungal diseases. Nowadays, it might be said that 'no fungi is non-pathogenic fungi' and so we have to enter into a battle against fungi using our ammunition of 'improved or novel antifungals'.

Transparency Declarations None to declear.

Funding Anupam Roy is thankful to CSIR, India, for providing CSIR Individual Fellowship to Anupam Roy (CSIR sanction No- 9/1103 (0001)2 k13-EMR-I).

References

Abraham WR, Arfmann H (1990) 12,13-Dihydroxy-fumitremorgin C from *Aspergillus fumigatus*. Phytochemistry 29:1025–1026

Akihiko F (2007) Discovery of micafungin (FK463): a novel antifungal drug derived from a natural product lead. Pure Appl Chem 79(4):603–614. doi:10.1351/pac200779040603 © IUPAC

Alfatafta AA, Gloer JB, Scott JA et al (1994) Apiosporamide, a new antifungal agent from the coprophilous fungus Apiospora montagnei. Nat Prod 57:1696–1702

Anke T, Oberwinkler F, Steglich W et al (1977) The strobilurins–new antibiotics from the basidiomycete *Strobilurus tenacellus* (Pers. ex Fr.) Sing. J Antibiot 30:806–810

Anke T, Besl H, Mocek U et al (1983) Antibiotics from basidiomycetes. XVIII. Strobilurin C and oudemansin B, two new antifungal metabolites from Xerula species (agaricales). J Antibiot 36:661–666

Arai T, Mikami Y, Fukushima K et al (1973) A new antibiotic, leucinostatin, derived from Penicillium lilacinum. J Antibiot 26:157–161

Araujo OE, Flowers FP, King MK (1990) Griseofulvin: a new look at an old drug. DICP. Ann Pharmacother 24:851–854

Aytoun RSC, Mcwilliam RW (1957) Mutants of the genus Penicillium. British Patent 788,118

Balba H (2007) Review of strobilurin fungicide chemicals. J Environ Sci Health B 42:441–451

Balkovec JM, Hughes DL, Masurekar PS et al (2014) Discovery and development of first in class antifungal caspofungin (CANCIDAS®) – a case study. Nat Prod Rep 31:15–34

Baroni A, de Luca A, de Filippis A et al (2009) 3-*O*-methylfunicone, a metabolite of *Penicillium pinophilum*, inhibits proliferation of human melanoma cells by causing G(2) + M arrest and inducing apoptosis. Cell Prolif 42:541–553

Bartlett DW, Clough JM, Godwin JR (2002) The strobilurin fungicides. Pest Manag Sci 58:649–662

Benz F, Knüsel F, Nüesch J et al (1974) Stoffwechselprodukte von Mikroorganismen 143. Mitteilung. Echinocandin B, ein neuartiges Polypeptid-Antibioticum aus Aspergillus nidulans var. echinulatus: Isolierung und Bausteine. Helv Chim Acta 57:2459–2477

Bergstrom JD, Dufresne C, Bills GF et al (1995) Discovery, biosynthesis, and mechanism of action of the zaragozic acids: potent inhibitors of squalene synthase. Annu Rev Microbiol 49:607–639

Betina V, Nemec P, Dobias J et al (1962) Cyanein, a new antibiotic from *Penicillium cyaneum*. Folia Biol 7:353–357

Bills GF, Gloer JB (2013) Coprophilous fungi: antibiotic discovery and functions in an underexplored arena of microbial defensive mutualism. Curr Opin Microbiol 16:549–565. doi:10.1016/j.mib.2013.08.001. Epub 2013 Aug 23

Bills GF, Platas G, Peláez F et al (1999) Reclassification of a pneumocandin-producing anamorph, Glarea lozoyensis gen. et sp. nov., previously identified as Zalerion arboricola. Mycol Res 103:179–192

Boeck LD, Fukuda DS, Abott BJ et al (1989) Deacylation of echinocandin B by Actinoplanes utahensis. J Antibiot 42:382

Borthwick AD (2012) 2,5-Diketopiperazines: synthesis, reactions, medicinal chemistry, and bioactive natural products. Chem Rev 112:3641–3716

Brian PW (1949) Studies on the biological activity of griseofulvin. Ann Bot – London, N.S. XIII 49: 59–77

Brian PW, Curtis PJ (1946) A substance causing abnormal development offungal hyphae produced by *Penicillium janczewskii* Zal. I. Biological assay, production and isolation of 'curling factor'. Trans Br Mycol Soc 29:173–187

Brian PW, Curtis PJ (1949) A substance causing abnormal development offungal hyphae produced by *Penicillium janczewskii* Zal. III. Identity of 'curling factor' with griseofulvin. Trans Br Mycol Soc 32:30–33

Brian PW, Curtis PJ (1955) Production of griseofulvin by *penicillium raistrickii*. Trans Br Mycol Soc 38:305–308

Buommino E, Boccellino M, de Filippis A et al (2007) 3-*O*-methylfunicone produced by *Penicillium pinophilum* affects cell motility of breast cancer cells, downregulating avb5 integrin and inhibiting metalloproteinase-9 secretion. Mol Carcinog 46:930–940

Butchko RA, Adams TH (1999) Aspergillus nidulans mutants defective in stc gene cluster regulation. Genetics 153:715–720

Cai S, Sun S, Zhou H et al (2011) Prenylated polyhydroxy-pterphenyls from *Aspergillus taichungensis* ZHN-7-07. J Nat Prod 75:1106–1110

Chamilos G, Lewis RE, Dimitrios P et al (2006) Lovastatin has significant activity against zygomycetes and interacts synergistically with voriconazole. Antimicrob Agents Chemother 50:96–103

Che Y, Gloer JB (2002) Decipinin A and decipienolides A and B: new bioactive metabolites from the coprophilous fungus Podospora decipiens. J Nat Prod 65:916–919

Che Y, Swenson DC, Gloer JB et al (2001) Pseudodestruxins A and B: new cyclic depsipeptides from the coprophilous fungus Nigrosabulum globosum. J Nat Prod 64:555–558

Clark B, Capon RJ, Lacey E et al (2005) Gymnoascolides A-C: aromatic butenolides from an Australian isolate of the soil ascomycete Gymnoascus reessii. J Nat Prod 6:1226–1230

Clarke SM, Mckenzie M (1967) *Penicillium sclerotigenum:* a new source of Griseofulvin. Nature 213:504–505. doi:10.1038/213504b0

Crosse R, McWillium R (1964) Some relations between chemical structure and antifungal effects of griseofulvin analogues. J Gen Microbiol 34:51

Danishefsky S, Etheredge SJ (1978) Synthesis of dl-epigriseofulvin. J Org Chem 43:4604

Danishefsky S, Singh RK, Gammill RE (1978) Diels-Alder reactions of 1,1-dimethoxy-3-trimethylsilyloxy-1,3-butadiene. J Org Chem 43:379

Danishefsky S, Yan CF, Singh RK et al (1979) Derivatives of 1-methoxy-3-trimethylsilyloxy-1,3-butadiene for Diels-Alder reactions. J Am Chem Soc 101:7001–7008

Dave WB, John MC, Chris RAG (2001) Understanding the strobilurin fungicides. Pestic Outlook 12:143–148

David M, Geoffrey DR (1980) 21st century guidebook to fungi 2011 G. Schramm, Dissertation, Universität Bonn

Debono M, Abbott BJ, Turner JR et al (1988) Synthesis and evaluation of LY121019, a member of a series of semisynthetic analogues of the antifungal lipopeptide echinocandin B. Acad Sci 544:152–167

Debono M, Abbott BJ, Fukuda DS et al (1989) Synthesis of new analogs of echinocandin B by enzymatic deacylation and chemical reacylation of the echinocandin B peptide: synthesis of the antifungal agent cilofungin (LY121019). J Antibiot (Tokyo) 42:389–397

Debono M, Turner WW, LaGrandeur L et al (1995) Semisynthetic chemical modification of the antifungal lipopeptide echinocandin B (ECB): structure-activity studies of the lipophilic and geometric parameters of polyarylated acyl analogs of ECB. J Med Chem 38:3271–3281

Delgado L, De Croos PZ, Lu MC et al (1992) Structure modification and biological activity of some griseofulvin derivatives. Gaoxiong Yi Xue Ke Xue Za Zhi 8:632–639

Dolores F, Juan AT, Antonio DV (2010) Chapter 10: The QoI fungicides, the rise and fall of a successful class of agricultural fungicides. In: Agricultural and biological sciences "Fungicides", book edited by Odile Carisse, ISBN 978-953-307-266-1, Published: December 14, 2010 under CC BY-NC-SA 3.0 license

Dong N, Li X, Wang F (2013) Asymmetric Michael-aldol tandem reaction of 2-substituted benzofuran-3-ones and enones: a facile synthesis ofgriseofulvin analogues. Org Lett 15:4896–4899. doi:10.1021/ol402346c. Epub 2013 Sep 3

Emri T, Majoros L, Tóth V et al (2013) Echinocandins: production and applications. Appl Microbiol Biotechnol 97:3267–3284. doi:10.1007/s00253-013-4761-9. Epub 2013 Mar 6. Review

Fields TL, Newman H (1970) 5′-diazogriseofulvin. J Med Chem 13:1242–1243

Finefield JM, Frisvad JC, Sherman DH et al (2012) Fungal origins of the bicyclo[2.2.2]diazaoctane ring system of prenylated indole alkaloids. J Nat Prod 75:812–833

Fischer LJ, Riegelman S (1967) Absorption and activity of some derivatives of Griseofulvin. J Pharm Sci 56 (4):469–476

Fredenhagen A, Kuhn A, Peter HH et al (1990a) Strobilurins F, G and H, three new antifungal metabolites from Bolinea lutea. I. Fermentation, isolation and biological activity. J Antibiot 43:655–660

Fredenhagen A, Hug P, Peter HH et al (1990b) Strobilurins F.G. Three new antifungal metabolites from *Bolinea lutea.* II. Structure determination. J Antibiot 43:661–667

Frisvad JC, Smedsgaard J, Larsen TO et al (2004) Mycotoxins, drugs and other extrolites produced by species in *Penicillium* subgenus *Penicillium.* Stud Mycol 49:201–241

Fujie A, Iwamoto T, Sato B et al (2001) FR131535, a novel water-soluble echinocandin-like lipopeptide: synthesis and biological properties. Bioorg Med Chem Lett 11:399–402

Gamble WR, Gloer JB, Scott JA et al (1995) Polytolypin, a new antifungal triterpenoid from the coprophilous fungus Polytolypa hystricis. J Nat Prod 58:1983–1986

Geiser DM, Klich MA, Frisvad JC et al (2007) The current status of species recognition and identification in Aspergillus. Stud Mycol 59:1–10

Ghisalberti EL, Narbey MJ (1990) Metabolites of *Aspergillus terreus* antagonistic towards the take-all fungus. J Nat Prod 53:520–522

Grove JF, McGowan JC (1947) Identity of griseofulvin and 'curling-factor'. Nature 160:574–574. doi:10.1038/160574a0

Grove JF, Ismay D, Macmillan J, Mulholland TPC et al (1951) The structure of griseofulvin. Chem Ind 11:219–220

Grove JF, Macmillan J, Mulholland TPC et al (1952) Griseofulvin I. J Chem Soc 27:3949–3958

Hawser S, Borgonovi M, Markus A et al (1999) Mulundocandin, an echinocandin-like lipopeptide antifungal agent: biological activities in vitro. J Antibiot (Tokyo) 52:305–310

Hector RF (2005) An overview of antifungal drugs and their use for treatment of deep and superficial mycoses in animals. Clin Tech Small Anim Pract 20:240–249

Hein SM, Gloer JB, Koster B et al (1998) Arugosin F: a new antifungal metabolite from the coprophilous fungus Ascodesmis sphaerospora. J Nat Prod 61 (12):1566–1567

Hein SM, Gloer JB, Koster B et al (2001) Bombardolides: new antifungal and antibacterial gamma-lactones from the coprophilous fungus Bombardioidea anartia. J Nat Prod 64:809–812

Hino M, Fujie A, Iwamoto T et al (2001) Chemical diversity in lipopeptide antifungal antibiotics. J Ind Microbiol Biotechnol 27:157–162

Hodges RL, Hodges DW, Goggans K et al (1994) Genetic modification of an echinocandin B-producing strain of Aspergillus nidulans to produce mutants blocked in sterigmatocystin biosynthesis. J Ind Microbiol 13:372–381

Hodges RL, Kelkar HS, Xuei X et al (2000) Characterization of an echinocandin B producing strain blocked for sterigmatocystin biosynthesis reveals atranslocation in the stcW gene of the aflatoxin biosynthetic pathway. J Ind Microbiol Biotechnol 25:333–341

Huber FM, Gottlieb D (1968) The mechanism of action of griseofulvin. Can J Microbiol 14:111–118. PMID: 5689440 [PubMed - indexed for MEDLINE]

Hubert S, Wolfgang S (1999) Strobilurins: evolution of a new class of active substances. Angew Chem Int Ed 38:1328–1349

Iwamoto T, Sakamoto K, Yamashita M et al (1993) Abstract 33rd ICAAC, No.F37, New Orleans, USA

Iwamoto T, Fujie A, Sakamoto K et al (1994) WF11899A, B and C, novel antifungal lipopeptides, I: taxonomy, fermentation, isolation and physico-chemical properties. J Antibiot 47:1084–1091

Jamison JA, Zeckner DJ (1997) The synthesis and antifungal activity of N alkylated analogs of echinocandin B. J Antibiot (Tokyo) 50:562–566

Kanasaki R, Abe F, Kobayashi M et al (2006a) FR220897 and FR220899, novel antifungal lipopeptides from Coleophoma empetri no. 14573. J Antibiot 59:149–157

Kanasaki R, Kobayashi M, Fujine K et al (2006b) FR227673 and FR190293, novel antifungal lipopeptides from Chalara sp. No. 22210 and Tolypocladium parasiticum No. 16616. J Antibiot 59:158–167

Kanasaki R, Sakamoto K, Hashimoto M et al (2006c) FR209602 and related compounds, novel antifungal lipopeptides from Coleophoma crateriformis no.738. I. Taxonomy, fermentation, isolation and physico-chemical properties. J Antibiot 59:137–144

Kanda M, Tsuboi M, Sakamoto K et al (2009) Improvement of FR901379 production by mutant selection and medium optimization. J Biosci Bioeng 107:530–534. doi:10.1016/j.jbiosc.2009.01.002

Kanda M, Yamamoto E, Hayashi A et al (2010) Scale-up fermentation of echinocandin type antibiotic FR901379. J Biosci Bioeng 109:138–144. doi:10.1016/j.jbiosc.2009.07.019. Epub 2009 Aug 27

Kawada M, Inoue H, Ohba SI et al (2010) Leucinostatin A inhibits prostate cancer growth through reduction of insulin-like growth factor-I expression in prostate stromal cells. Int J Cancer 126:810–818

Keller NP, Turner G (2005) Fungal secondary metabolism – from biochemistry to genomics. Nat Rev Microbiol 3:937–947

Keller-Juslén C, Kuhn M, Loosli HR et al (1976) Struktur des cyclopeptid-antibiotikums sl 7810 (= echinocandinb). Tetrahedron Lett 17:4147–4150.2

Ko BS, Oritani T (1990) Synthesis and biological activities of griseofulvin analogs. Agric Biol Chem 54:2199–2204

Kommmunarskaya AD (1969) Abramson VS. Manufacture of griseofulvin free of toxic by products. Fr. Pat. 1 1969; 565–661

Kommmunarskaya AD (1970) Sulfur-35 induced variation in Penicillium nigricans producing griseofulvin. Antibiotiki 15:216–220

Krizsán K, Bencsik O, Nyilasi I et al (2010) Effect of the sesterterpene-type metabolites, ophiobolins A and B, on zygomycetes fungi. FEMS Microbiol Lett 313:135–140

Kurtz MB, Rex JH (2001) Glucan synthase inhibitors as antifungal agents. Adv Protein Chem 56:423–475

Kuznetsova TA, Smetanina OF, Afiyatullov SS et al (2001) The identification of fusidic acid, a steroidal antibiotic from marine isolate of the fungus Stilbella aciculosa. Biochem Syst Ecol 29:873–874

Lee KK, Gloer JB, Scott JA et al (1995) Petriellin A: a novel antifungal depsipeptide from the coprophilous fungus Petriella sordida. J Org Chem 60:5384–5385

Lehr NA, Meffert A, Antelo L et al (2006) Antiamoebins, myrocin B and the basis of antifungal antibiosis in the coprophilous fungus Stilbella erythrocephala (syn. S. fimetaria). FEMS Microbiol Ecol 55:105–112

Li E, Clark AM, Rotella DP et al (1995) Microbial metabolites of ophiobolin A and antimicrobial evaluation of ophiobolins. J Nat Prod 58:74–81

Li E, Jiang L, Guo L et al (2008) Pestalachlorides A-C, antifungal metabolites from the plant endophytic fungus Pestalotiopsis adusta. Med Chem 16:7894–7899. Epub 2008 Jul 30

Lisa JG, Philip M, Larry L et al (1999) LY303366 exhibits rapid and potent fungicidal activity in flow cytometric assays of yeast viability. Antimicrob Agents Chemother 43:830–835

Lóránd T, Kocsis B (2007) Recent advances in antifungal agents. Mini Rev Med Chem 7:900–911

Louise AW, Neil ARG (2010) Fungal echinocandin resistance. Fungal Genet Biol 47: 117–126. doi:10.1016/j.fgb.2009.09.003. PMCID: PMC2812698

Macías-Rubalcava ML, Hernández-Bautista BE, González JM, Anaya AL (2008) Naphthoquinone spiroketal with allelochemical activity from the newly discovered endophytic fungus Edenia gomezpompae. Phytochemistry 69:1185–1196. Epub 2008 Jan 29

Macreadie IG, Johnson G, Schlosser T et al (2006) Growth inhibition of *Candida*species and *Aspergillus fumigatus* by statins. FEMS Microbiol Lett 262:9–13

Magdalena J, Monika OJ, Grzegorz J et al (2013) New bioactive fungal molecules with high antioxidant and antimicrobial capacity isolated from Cerrena unicolor idiophasic cultures. BioMed Res Int 2013:497492

Mandala SM, Thornton RA, Frommer BR et al (1995) The discovery of australifungin, a novel inhibitor of sphinganine N-acyltransferase from Sporormiella australis. Producing organism, fermentation, isolation, and biological activity. J Antibiot (Tokyo) 48:349–356

Mizoguchi J, Saito T, Mizuno K et al (1977) On the mode of action of a new antifungal antibiotic, Aculeacin A: inhibition of cell wall synthesis in Saccharomyces cerevisiae. J Antibiot 30:308–313

Mizuno K, Yagi A, Satoi S et al (1977a) Studies on aculeacin. I. Isolation and characterization of aculeacin A. J Antibiot 30:297–302

Mizuno K, Yagi A, Satoi S et al (1977b) Studies on aculeacin. I. Isolation and characterization of aculeacin A. J Antibiot (Tokyo) 30:297–302

Morris MI, Villmann M (2006a) Echinocandins in the management of invasive fungal infections, Part 1. Am J Health Syst Pharm 63:1693–1703. doi:10.2146/ajhp050464.p1

Morris MI, Villmann M (2006b) Echinocandins in the management of invasive fungal infections, Part 2. Am J Health Syst Pharm 63:1813–1820

Morris SA, Schwartz RE, Sesin DF et al (1994) Pneumocandin D0, a new antifungal agent and potent inhibitor of Pneumocystis carinii. J Antibiot 47:755–764

Mudur SV, Gloer JB (2006) Sporminarins A and B: antifungal metabolites from a fungicolous isolate of Sporormiella minimoides. J Antibiot 59:500–506

Mukhopadhyay T, Roy K, Bhat RG et al (1992) Deoxymulundocandin-a new echinocandin type antifungal antibiotic. J Antibiot 45:618–623

Mulder MP, Kruijtzer JA, Breukink EJ et al (2011) Synthesis and evaluation of novel macrocyclic antifungal peptides. Bioorg Med Chem 19:6505–6517. doi:10.1016/j.bmc.2011.08.034. Epub 2011 Aug 22

Nakagawa M, Hirota A (1982) Terrecyclic acid A, a new antibiotic from *Aspergillus terreus*. I. Taxonomy, production, and chemical and biological properties. J Antibiot 35:778–782

Niedens BR, Parker SR, Stierle DB et al (2013) First fungal aromatic L-amino acid decarboxylase from a paclitaxel-producing *Penicillium raistrickii*. Mycologia 91:619–626. 140, 141

Nobel HM, Langley D, Sidebottom PJ et al (1991) An echinocandin from an endophytic Cryptosporiopsis sp. and Pezicula sp. in Pinus sylvestris and Fagus sylvatica. Mycol Res 95:1439–1440

Nyfeler R, Keller SW (1974) Metabolites of microorganisms, 143: echinocandin B, a novel polypeptide-antibiotic from Aspergillus nidulans var echinulatus—isolation and structural components. Helv Chim Acta 57:2459–2477

Odds FC (2001) Sordarin antifungal agents. Expert Opin Ther Pat 11:283–294

OH GS, Hong KH, OH H et al (2001) 4-Acetyl-12,13-epoxyl-9-trichothecene-3,15-diol isolated from the fruiting bodies of *Iaria japonica* YASUDA induces apoptosis of human leukemia cells (HL-60). Biol Pharm Bull 24:785–789

Oxford AE, Raistrick H (1939) Griseofulvin, C17H17O6Cl, a metabolic product of *Penicillium griseofulvum* Dierckx. Biochem J 33:240–248

Pan ZZ, Zhu YJ, Yu XJ et al (2012) Synthesis of 4-′-thiosemicarbazonegriseofulvin and its effects on the control of enzymatic browning and postharvest disease of fruits. J Agric Food Chem 60:10784–10788. doi:10.1021/jf302356x. Epub 2012 Oct 18

Petraitiene R, Petraitis V et al (1999) Antifungal activity of LY303366, a novel Echinocandin B, in experimental disseminated Candidiasis in rabbits. Antimicrob Agents Chemother 43:2148–2155

Petit KE, Mondeguer F, Roquebert MF, Biard JF, Pouchus YF (2004) Detection of griseofulvin in a marine strain of Penicillium waksmanii by ion trap mass spectrometry. J Microbiol Methods 58:59–65

Qiao J, Kontoyiannis DP, Wan Z, Li R et al (2007) Antifungal activity of statins against *Aspergillus* species. Med Mycol 45:589–593

Raistrick H, Smith G (1935) LXXI. Studies in the biochemistry of micro-organisms. XLII. The metabolic products of *Aspergillus terreus* Thom. A new mould metabolic product-terrein. Biochem J 29:606–611

Ridderbusch DC, Weber RW, Anke T et al (2004) Tulasnein and podospirone from the coprophilous xylariaceous fungus Podosordaria tulasnei. Naturforsch C 59:379–383

Rønnest MH, Raab MS, Anderhub S et al (2012) Disparate SAR data of griseofulvin analogues for the dermatophytes Trichophyton mentagrophytes, T. rubrum, and MDA-MB-231 cancer cells. J Med Chem 55:652–660. Epub 2012 Jan 17

Roy K, Mukhopadhyay T, Reddy GC et al (1987) Mulundocandin, a new lipopeptide antibiotic. I. Taxonomy, fermentation, isolation and characterization. J Antibiot 40:275–280

Satoi S, Yagi A, Asano K et al (1977) Studies on aculeacin. II. Isolation and characterization of aculeacins B, C, D, E, F and G. J Antibiot 30:303–307

Sauter H, Ammermann E (1996) Strobilurins- from natural products to a new class of fungicides. In: Copping LG (ed) Crop protection agents from nature: natural products and analogues. Royal Society of Chemistry, Cambridge, pp 50–81

Schmidt LE, Gloer JB (2007) Solanapyrone analogues from a Hawaiian fungicolous fungus. J Nat Prod 70:1317–1320. Epub 2007 Jul 31

Schramm G, Steglich W, Anke T, Oberwinkler F (1978) Antibiotics from basidiomycetes, III. Strobilurin A and B, antifungal metabolites from *Strobilurus tenacellus*. Chem Ber 111:2779–2784

Schwartz RE, Giacobbe RA, Bland JA et al (1989) L-671,329, a new antifungal agent. I. Fermentation and isolation. J Antibiot 42:163–167

Schwartz RE, Sesin DF, Joshua H et al (1992) Pneumocandins from Zalerion arboricola. Discovery and isolation I. J Antibiot 45:1853–1866

Seya H, Nozawa K, Nakajima S et al (1986) Studies on fungal products. Part 8. Isolation and structure of emestrin, a novel antifungal macrocyclic epidithiodioxopiperazine from *Emericella striata*. X-Ray molecular structure of emestrin. J Chem Soc Perk T 1 1986:109–116

Singleton V, Bohonos N et al (1958) Decumbin, a new compound from a species of *Penicillium*. Nature 181:1072–1073

Soman AG, Gloer JB, Koster B et al (1999) Sporovexins A-C and a new preussomerin analog: antibacterial and antifungal metabolites from the coprophilous fungus Sporormiella vexans. J Nat Prod 62:659–661

Songgang W, Yunshen B (1983) Study on the characterstics of Penicillium patulum strains with high ability of producing griseofulvin. Weishengwuxue Tongbao 10:204–206

Strobel GA, Miller RV, Martinez-Miller C et al (1999) Cryptocandin, a potent antimycotic from the endophytic fungus Cryptosporiopsis cf. quercina. Microbiology 145:1919–1926

Sun HF, Li XM, Meng L et al (2012) Asperolides A-C, tetranorlabdane diterpenoids from the et al. marine alga-derived endophytic fungus *Aspergillus wentii* EN-48. J Nat Prod 75:148–152

Takeuchi Y, Tomozane H, Misumi K et al (1997a) Syntheses and antifungal activity of dl-griseofulvin and its congeners. II. Chem Pharm Bull (Tokyo) 45:327–332

Takeuchi Y, Watanabe I, Misumi K et al (1997b) Syntheses and antifungal activity of dl-griseofulvin and it congeners. III. Chem Pharm Bull (Tokyo) 45:2011–2015

Tomishima M, Ohki H, Yamada A et al (2008a) Novel echinocandin antifungals. Part 1: Novel side-chain analogs of the natural product FR901379. Bioorg Med Chem Lett 18:1474–1477. doi:10.1016/j.bmcl.2007.12.062. Epub 2007 Dec 28

Tomishima M, Ohki H, Yamada A et al (2008b) Novel echinocandin antifungals. Part 2: Optimization of the side chain of the natural productFR901379. Discovery of micafungin. Bioorg Med Chem Lett 18:2886–2890. doi:10.1016/j.bmcl.2008.03.093. Epub 2008 Apr 8. PMID:18424132[PubMed - indexed for MEDLINE]

Tomozane H, Takeuchi Y, Choshi T et al (1990) Syntheses and antifungal activities of dl-griseofulvin and its congeners. I. Chem Pharm Bull (Tokyo) 38:925–929

Traber R, Keller-Juslén C, Loosli HR et al (1979) Cyclopeptid-Antibiotika aus Aspergillus-Arten. Struktur der Echinocandine C und D. Helv Chim Acta 62:1252–1267

Tscherter H, Dreyfuss MM (1982) New metabolites. Processes for their production and their use. Internat. Patent Appl PCT/EP8/ 00121

Turner WW, Rodriguez MJ (1996) Recent advances in medicinal chemistry of antifungal agent. Curr Pharm Des 2:209–224

Ueda S, Shibata T, Ito K et al (2011a) Cloning and expression of the FR901379 acylase gene from Streptomyces sp. no. 6907. J Antibiot (Tokyo) 64:169–175. doi:10.1038/ja.2010.151. Epub 2010 Dec 1

Ueda S, Kinoshita M, Tanaka F et al (2011b) Strain selection and scale-up fermentation for FR901379 acylase production by Streptomyces sp. no. 6907. J Biosci Bioeng 112:409–414. doi:10.1016/j.jbiosc.2011.06.002. Epub 2011 Jul 12

Visalakchi S, Muthumary J (2010) Taxol (anticancer drug) producing endophytic fungi: an overview. Int J Pharm Bio Sci 1:1–9

Vladimir VZ, Leonid VK (1998) Natural compounds of the strobilurin series and their synthetic analogues as cell respiration inhibitors. Russ Chem Rev 67:535. doi:10.1070/RC1998v067n06ABEH000426

Wang Y, Gloer JB, Scott JA et al (1993) Appenolides A-C: three new antifungal furanones from the coprophilous fungus Podospora appendiculata. J Nat Prod 56:341–344

Wang HJ, Gloer JB, Scott JA et al (1995a) Coniochaetones A and B: new antifungal benzopyranones from the coprophilous fungus Coniochaeta saccardoi. Tetrahedron Lett 36:5847–5850

Wang Y, Gloer JB, Scott JA et al (1995b) Terezines A-D: new amino acid-derived bioactive metabolites from the coprophilous fungus Sporormiella teretispora. J Nat Prod 58:93–99

Wang H, Gloer KB, Gloer JB et al (1997) Anserinones A and B: new antifungal and antibacterial benzoquinones from the coprophilous fungus Podospora anserina. J Nat Prod 60:629–631

Weber HA, Swenson DC (1992) Similins A and B: new antifungal metabolites from the coprophilous fungus Sporormiella similis. Tetrahedron Lett 33:1157–1160

Weber W, Anke T, Steffan B et al (1990a) Antibiotics from basidiomycetes. XXXII. Strobilurin E: a new cytostatic and antifungal (E)-beta-methoxyacrylate antibiotic from Crepidotus fulvotomentosus Peck. J Antibiot (Tokyo) 43:207–212

Weber W, Anke T, Bross M et al (1990b) Strobilurin D and strobilurin F: two new cytostatic and antifungal

(E)-β-methoxyacrylate antibiotics from *Cyphellopsis anomala* (1). Planta Med 56:446–450

Whyte AC, Gloer JB, Scott JA (1996) Cercophorins A-C: novel antifungal and cytotoxic metabolites from the coprophilous fungus Cercophora areolata. J Nat Prod 59:765–769

Whyte AC, Gloer KB, Gloer JB et al (1997) New antifungal metabolites from the coprophilous fungus Cercophora sordarioides. Can J Chem 75:768–772

Woloshuk CP, Shim WB (2013) Aflatoxins, fumonisins, and trichothecenes: a convergence of knowledge. FEMS Microbiol Rev 37:94–109

Wright JM (1955) The production of antibiotics in soil: ii. Production of griseofulvin by penicillium nigricans. Ann Appl Biol 43:288–296

Yamashita M, Matsuda M, Oohata N (2005) Study of industrial manufacturing methods for Micafungin (FK463). Seibutsu-kogaku 83:123–131

Zhao S, Smith KS, Deveau AM et al (2002) Biological activity of the tryprostatins and their diastereomers on human carcinoma cell lines. J Med Chem 45:1559–1562

Essential Oil and Antifungal Therapy 2

Mohammad Moghaddam and Leila Mehdizadeh

Abstract

Essential oils are complex mixtures of odorous principles stored in different parts of special plants that play an important role in plant defense mechanisms against pathogenic fungi. They are composed of different chemical compounds which have various biological activities and therapeutic effects that can be used in different industries as well as control molds and various fungi such as food-borne and phytopathogenic fungi which cause different postharvest diseases and animal and human disorders. The antifungal activity of essential oil is associated with phytochemical components. In addition to this, percentage inhibition of mycelia growth depends on some other factors such as the antifungal activity method, the day of observation, oil concentration, and examined fungal species. In order to reduce the hazardous effects of synthetic fungicidal products, the increasing interest in the possible application of essential oils for pathogen management has induced to investigate new sources of biologically active natural products in antifungal therapy.

2.1 Introduction

Plants have been used medicinally for thousands of years by cultures all over the world. Medicinal plants include herbs, herbal materials, and products that contain different parts of plants or other plant materials as active ingredients and are the most popular forms of traditional plants which are used to maintain health, as well as prevent or treat physical and mental disorders. Herbal medicines are referring to herbal remedies, products, phytomedicines, phytotherapeutic agents, and phytopharmaceuticals. Recently medicinal plants and bioactive phytocompounds are applied worldwide (WHO 2005). There is an increasing interest on the use of them in the industrial and pharmaceutical applications. Nonedible purpose of plant usage is dating back to prehistory. Since ancient time, natural products have been displayed as an

M. Moghaddam (✉) • L. Mehdizadeh
Department of Horticulture, Faculty of Agriculture, Ferdowsi University of Mashhad, P.O. Box 91775-1163, Mashhad, Iran
e-mail: m.moghadam@um.ac.ir; moghaddam75@yahoo.com

A. Basak et al. (eds.), *Recent Trends in Antifungal Agents and Antifungal Therapy*,
DOI 10.1007/978-81-322-2782-3_2

important source of drugs, and today lots of practical drugs are derived from natural sources (Lang and Buchbauer 2012).

Traditionally, plants have been the main source of materials to maintain health and prevent ill health, and it is only comparatively recently that they have been replaced by synthetics. The study of plant structure and function should not be regarded as too simple. The chemical compounds of plants can be classified into two main groups. The primary metabolites are those that are distributed worldwide across the plants as basic building blocks of life. Four major subgroups of these metabolites include proteins, carbohydrates, nucleic acids, and lipids. Besides the primary metabolites, plants produce a great variety of secondary metabolites related to mechanisms of adaptation and are important in mediating interactions between plants and biotic environment. The secondary metabolites existing in some plant species are classified into terpenoids, shikimates, polyketides, and alkaloids. The most important ones which are related to essential oils are the terpenoids and shikimates (Baser and Buchbauer 2010). Secondary metabolites obtained from plants have diverse functions such as defensive and protective process in the plants itself. The phytochemical studies are directed with an aim to find biopesticides of plant origin. Previously other researches demonstrated that diverse secondary metabolites have different sites of action and different molecular targets when they interact with enzymes and metamorphosis processes (Torres et al. 2003; Silva and Fernandes 2010; Baser and Buchbauer 2010).

The increasing interest in the possible application of secondary metabolites for pathogen management has aimed to investigate new sources of biologically active natural products, with new and specific action (Edris 2007; Lang and Buchbauer 2012). Microbial infections including bacterial, fungal, and viral infections are the most commonly encountered illnesses worldwide. There is a widespread effort to find new fungicidals, and currently it is focused on natural compounds. Various experimental works have been performed with natural products, which are potential models for defensive substances against fungal predators (Tabassum and Vidyasagar 2013; Stević et al. 2014). Among the secondary metabolites, essential oils are very important natural products which have different therapeutic and biological activity. Therefore in this chapter, we investigate the effects of essential oils on pathogenic fungi.

2.2 What Is Essential Oil?

Essential oils (volatile oils) are complex mixtures of odorous principles stored in special plant cells, glands, glandular hairs, oil ducts, or resin ducts in any part of a plant which are obtained by living organisms and isolated by pressing and hydro or steam distillation from a whole plant or different parts of plants such as leaves, flowers, fruits, grass, root, wood, bark, gum, and blossom in some plant families. The most important plant families which are used to extract essential oil are Lamiaceae, Myrtaceae, Rutaceae, and Apiaceae. Essential oils are soluble in alcohol and fats but only slightly soluble in water. Most essential oils are colorless, but some of them are colored (e.g., *Matricaria chamomilla* oil has **azulene**, which is blue). Upon exposure to light and air, they readily oxidize and resinify. The total essential oil content of plants is generally very low (<1 %). Plant essential oils have various applications such as fragrances, flavorings, preservative agents, condiments or spices, as well as medicinal uses in food, cosmetic, and pharmaceutical industries. In addition, they are the most concentrated part of a plant's vital force or energy, and they have antimicrobial and insecticidal properties (Ahmad et al. 2006; Rai and Cecilia Carpinella 2006; Baser and Buchbauer 2010; Silva and Fernandes 2010).

2.3 Essential Oil Extraction and Composition

There are several methods for producing essential oils which include steam distillation, solvent extraction, expression, enfleurage, supercritical carbon dioxide extraction, and phytonic process. Essential oils comprise over 100 chemical compositions; however, many compounds are present in small quantities which makes it hard to detect them. Generally the therapeutic effects of essential oils reflect their constituents. Plants have specialized mechanisms that are capable of synthesizing complex carbohydrates from simple primary materials such as hydrogen, carbon, and oxygen. The required energy for this action is obtained from the sun by photosynthesis (Mc Guinness 2007). While there are obvious variations in essential oil composition between oils from different species of the same genus (e.g., peppermint and spearmint) or same plants from different regions, there are many factors that influence the composition of oils. The most influential factors that affect the ratios of essential oil molecules made by the plants include growth conditions, climate, altitude, soil type, agricultural methods, phenological stage of growth, plant part extracted, harvesting time even the time of day when harvesting is done, and method of extraction (Bowles 2003; Mc Guinness 2007).

2.4 Chemistry of Essential Oil

Essential oils are composed of different chemical compounds including terpene, terpenoid, phenolic and phenylpropanoid, non-terpenoid aliphatic, and heterocyclic molecules (Bowles 2003).

2.4.1 Terpene and Terpenoid Molecules

Terpene and terpenoid molecules are the most essential oil molecules and are applied by plants to repel predators and prevent bacterial and fungal infections as well as used in flowers to attract insect pollinators. Terpenoids, or terpenes, are the most important group of natural products comprising one of the most important groups of active compounds in plants with over 20,000 known structures. They are determined as constituents composed of isoprene (five carbons) units containing two unsaturated bonds. The isoprene unit is the starting point for manufacture of terpenoid compounds, the main type of molecule found in essential oils. They are synthesized from acetate via the mevalonic acid biosynthetic pathway. In essential oil, terpenoid has two or three isoprene units which are easily linked together to form long, branched chains of carbon atoms (Baser and Buchbauer 2010). Isoprene unit and aromatic rings are two main building blocks of essential oils chemical structure. Aromatic rings consist of six carbon atoms joined together in a ring form (Mc Guinness 2007) (Fig. 2.1).

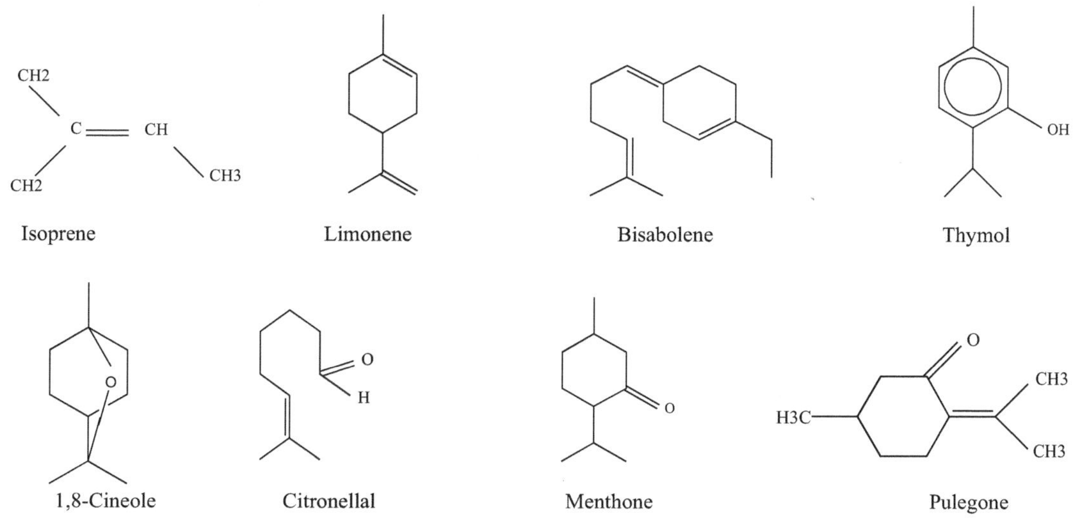

Fig. 2.1 Terpenes and terpenoid molecules

There are two different classifications of mono- and sesquiterpene compositions including hydrocarbons and oxygenated compounds. Hydrocarbons are the first major category of compounds that only contain hydrogen and carbon molecules and classified as terpenes. The second major category of components is oxygenated compounds which contain hydrogen, carbon, and oxygen and classified under various chemical types such as acids, alcohols, aldehydes, esters, ketones, lactones, oxides, and phenols and classified as terpenoids (Mc Guinness 2007).

Most essential oils are highly complex mixtures of mono- (C_{10}) and sesquiterpenes (C_{15}), including biogenetically related phenols (phenylpropenes and cinnamates), along with carbohydrates, alcohols, ethers, aldehydes, and ketones that are responsible for the characteristic properties.

2.4.1.1 Monoterpenes

Monoterpenes are the simplest molecule where two isoprene units are joined together. They have monoterpenoid carbon chain of 10 carbons and may contain a hydroxyl group anywhere along the chain. At this point, these molecules are called *monoterpenes* (from the Latin *mono*, "one"). Limonene is a common monoterpene in citrus and peppermint essential oils (Fig. 2.1).

2.4.1.2 Sesquiterpenes

Terpene molecules with three isoprene units, that is, 15 carbon atoms, are known as sesquiterpenes (from the Latin sesqui, "one and a half"). Bisabolene is an open-chain sesquiterpene found in myrrh oil and German chamomile (Fig. 2.1).

Both monoterpenes and sesquiterpenes can be modified by the addition of functional groups. When a terpene has a functional group added to it, it is known as a terpen**oid**, either monoterpen**oid** or sesquiterpen**oid**. The **oid** ending means "like," or "derived from"; hence, "terpenoid" can refer to all molecules with a terpene-like structure. Some authors refer to all terpenoid molecules by the general term (Bowles 2003; Mc Guinness 2007).

Furthermore, there are sometimes trace amounts of diterpenes in essential oils with four isoprene units (Bowles 2003; Baser and Buchbauer 2010). The direction of joining of isoprene units is in one direction. The branched end of the chain is related to as the head of the molecules and the other as the tail. Therefore it is called head-to-tail joining. This pattern of coupling is explained by the biosynthesis of terpenoids (Bu'Lock 1965; Croteau 1987; Mann et al. 1994).

The predominant group, the monoterpenes, are often cyclic, with a saturated (e.g., pulegone) or unsaturated hexacycle (e.g., limonene) or an aromatic system (e.g., thymol). Bicyclic (e.g., 1,8 cineole) and acyclic examples (e.g., citronellal) are not uncommon. Functional groups include alcohols (e.g., thymol), aldehydes (e.g., citronellal), and ketones (e.g., menthone), but simple hydrocarbons also occur (e.g., limonene). Generally due to their low boiling points (140–180 °C for monoterpenes, >200 °C for sesquiterpenes), all are relatively volatile, accounting for their use as fragrances and as fungistatic and fungicidal properties (Rai and Cecilia Carpinella 2006) (Fig. 2.1).

2.4.2 Functional Groups

2.4.2.1 Acids

Organic acids are rare component of essential oils. They are based on the carboxyl group and have the chemical grouping COOH. Acids have a low volatility rate. Phenylacetic acid in neroli is an example of acids found in essential oils (Fig. 2.2).

2.4.2.2 Alcohols

Terpenic alcohols are found in many essential oils. Their name all end in –ol. They have a hydroxyl group joined to one carbon atom. Monoterpenic alcohols (monoterpenols) are good antiseptics with antifungal properties. Linalool is a monoterpenol found in some essential oil like basil and coriander (Fig. 2.2).

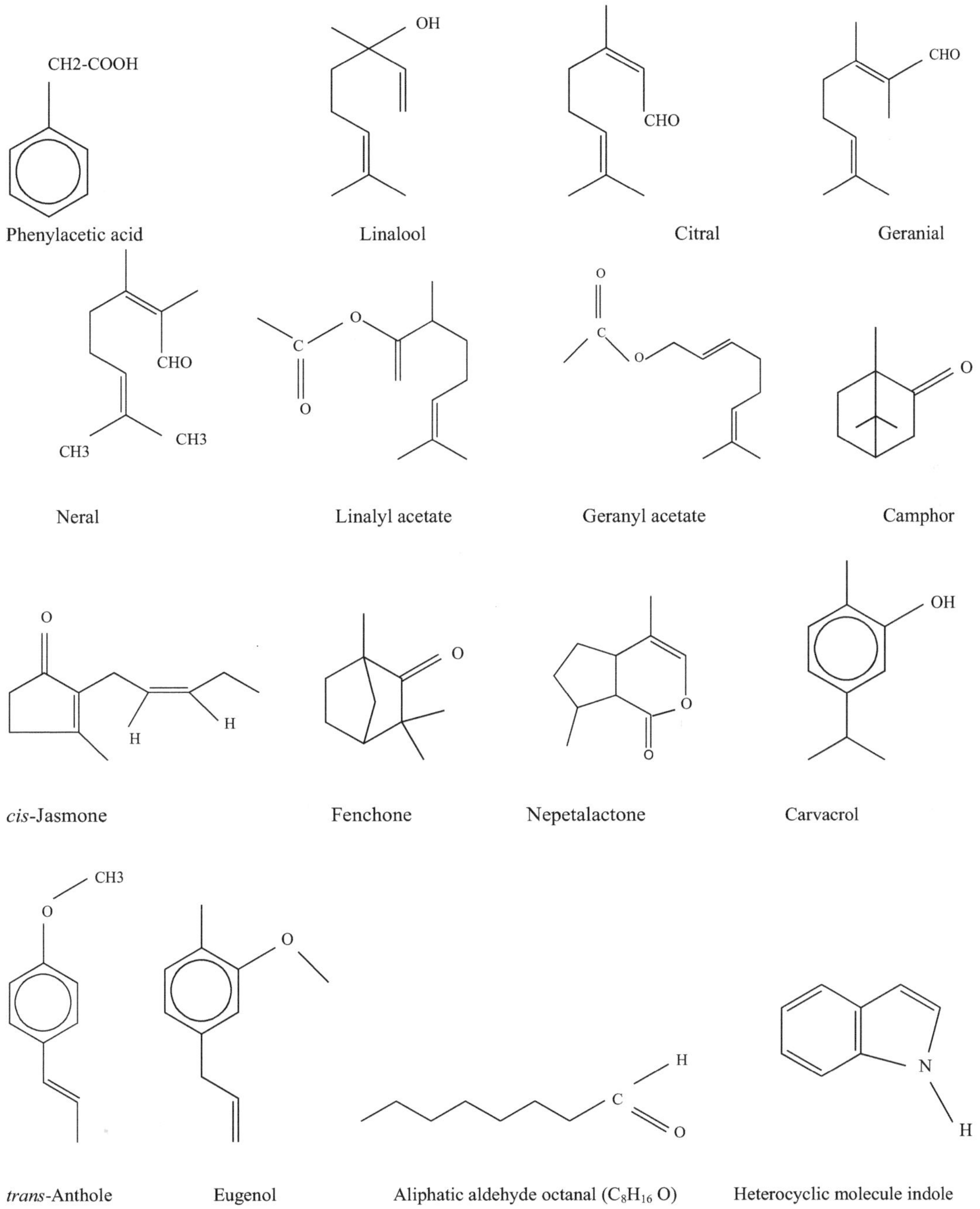

Fig. 2.2 Functional groups in essential oils

2.4.2.3 Aldehydes

Aldehydes have an oxygen atom double bonded to a carbon atom at the end of a carbon chain. The fourth bond is usually a hydrogen atom (Bowles 2003). They all end in –al and are important to the aroma of the plant. Citral in lemon balm (*Melissa officinalis*), citronella in lemongrass (*Cymbopogon citratus*), geranial in lemon eucalyptus (*Eucalyptus citriodora*), and neral in lemon verbena (*Aloysia triphylla*) are some examples of them (Fig. 2.2).

2.4.2.4 Esters

Esters are a combination of an acid and alcohol and possess fruity smell. Their names end in "–ate," and some of them have antifungal property. Some famous esters are linalyl acetate found in lavender (*Lavandula angustifolia*), clary sage (*Salvia sclarea*), bergamot (*Citrus aurantium* subsp. *bergamia*), and geranyl acetate found in sweet marjoram (*Origanum majorana*) (Fig. 2.2).

2.4.2.5 Ketones

A ketone is extracted from an alcohol by oxygenation and has oxygen atom double bonded to a carbon atom, which is also bonded to two other carbon atoms (Bowles 2003). The name of ketones ends in "– one," camphor being the exception. Camphor in rosemary (*Rosmarinus officinalis*), *cis*-jasmone in jasmine (*Jasminum officinale*) (International School of Aromatherapy 1993), fenchone in fennel (*Foeniculum vulgare* var. dulce), and menthone in peppermint (*Mentha piperita*) are some important ketones (Fig. 2.2).

2.4.2.6 Oxides

An oxide has an oxygen atom in a chain of carbons, which form a ring. The most common oxides include 1,8 cineole that is found in blue gum (*Eucalyptus globulus*), rosemary (*Rosmarinus officinalis*), and bay laurel (*Laurus nobilis*) (Fig. 2.2).

2.4.2.7 Lactones

Lactones always have an oxygen atom double bonded to a carbon atom. The carbon atom is attached to another oxygen atom that is part of a closed ring. The name of this group ends in "–lactone" or "–ine." Nepetalactone in catnip (*Nepeta cataria*) is a common lactone (Fig. 2.2).

2.4.2.8 Ethers

Ethers occur when a methyl or ethyl group is attached to a benzene ring via an oxygen molecule; *trans*-anethole fennel (*Foeniculum vulgare*) is an example of ethers (Fig. 2.2).

2.4.3 Phenolic and Phenylpropanoid Molecules

A few phenolic compounds are found in essential oils. They are most likely used by the plant to repel predators. Phenols and phenylpropanoids are made via the shikimic acid pathway in plants which also make compounds like the tannins found in tea. They have a distinctive benzene or aromatic ring. Phenols have a hydroxyl (–OH) group attached to the benzene ring and usually an isopropyl tail (3-carbon chain bonded to the ring at the middle carbon atom). The names end in "–ol" like alcohols. The common phenols in essential oils are thymol and carvacrol in thyme (*Thymus vulgaris*) and oregano (*Origanum vulgare*) and eugenol in clove (*Syzygium aromaticum*). Phenylpropanoids usually have a methyl ether functional group attached to the ring and a propenyl tail (3-carbon chain with one C–C bonded to the ring by one end). Anthole is an example of phenylpropanoid (Fig. 2.2).

2.4.4 Non-terpenoid Aliphatic Molecules

These are likely to repel predators, as they are found in citrus peel oils. The word "aliphatic" describes molecules made up of carbon chains that are in a straight line and do not have a closed or aromatic ring. Examples of aliphatic molecules are the acrid-smelling C_8 (8-carbon), C_9, and C_{10} aldehydes found in small amounts in citrus oils and the green-leafy smelling C_6 compounds found in some floral oils like jasmine and rose. A molecule of the aldehyde octanal ($C_8H_{16}O$) is found in sweet orange oil. Aliphatic molecules are usually found only in trace amounts in essential oils, but if they have oxygenated functional groups attached, their odors are usually noticeable despite this (Fig. 2.2).

2.4.5 Heterocyclic Molecules

These molecules contain atoms other than carbon in closed rings. They are found in only a few oils and include molecules such as the nitrogen-containing indole and methyl anthranilate and the oxygen-containing lactones, coumarins, and furanoid compounds. Heterocyclic compounds are made up of carbon atoms arranged in a ring, with either a nitrogen or oxygen atom included as part of the ring. These molecules are uncommon in essential oils, occurring mainly in heady floral oils like jasmine, neroli, and narcissus. Indole is found in jasmine oil. Alkaloids are heterocyclic compounds that feature a nitrogen atom as part of the closed ring. However, alkaloid molecules are seldom found in steam-distilled essential oils as they are soluble in water (Fig. 2.2).

2.5 The Biosynthesis Pathway of Essential Oils

Essential oils are organic compounds where carbon, hydrogen, and oxygen are their building blocks (Mc Guinness 2007). There are three biosynthetic pathways for deriving the main compounds of essential oil: the mevalonate pathway which leads to sesquiterpenes, the methyl-erythritol-pathway leading to mono- and diterpenes, and the shikimic acid pathway along the way to phenylpropenes (Baser and Buchbauer 2010).

2.5.1 Mevalonic Acid Pathway

Terpenoid molecules are made by mevalonic acid pathway. The steps of pathway include (1) to make mevalonic acid with 6-carbon atoms, (2) to rearrange mevalonic acid by a series of enzymatic transformations to form isopentenyl pyrophosphate consisting of a branched 5-carbon molecule (isoprene unit), and (3) isoprene joined to two phosphate groups. Isoprene is the main constituent of essential oil that starts the manufacture of terpenoid compounds (Bowles 2003) (Fig. 2.3).

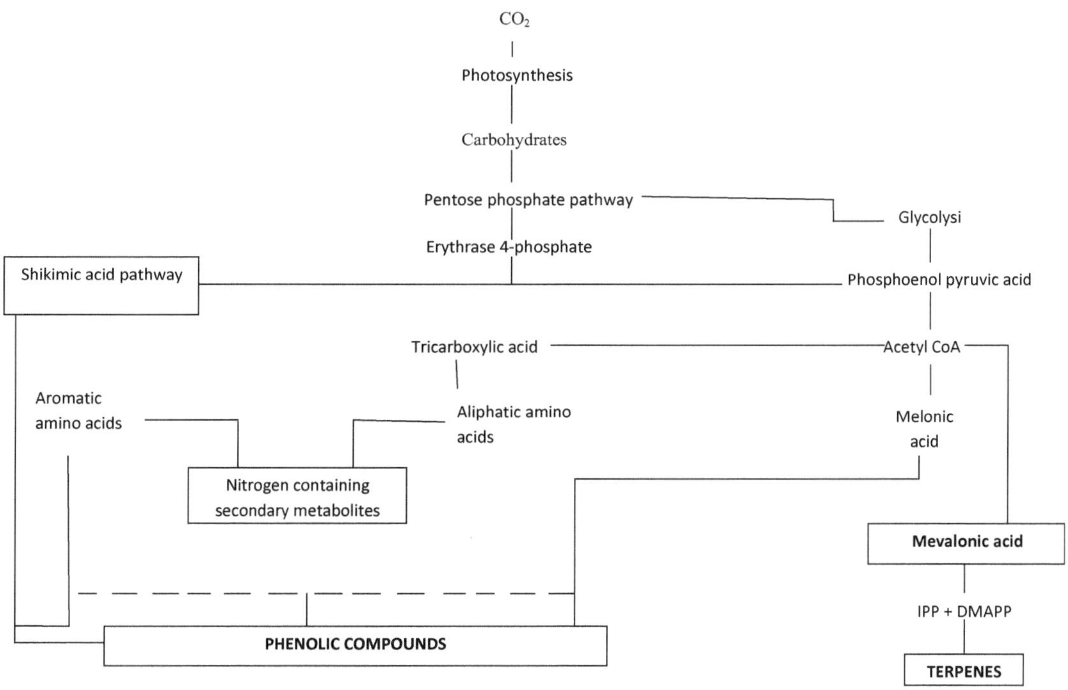

Fig. 2.3 Secondary metabolite synthesis within the plant (Mc Guinness 2007)

2.5.2 Shikimic Acid Pathway

Shikimic acid is a key synthetic intermediate for plants since it is the key precursor for both the flavonoids and lignin (Bu'Lock 1965; Mann et al. 1994). Flavonoids act as antioxidant, colors, and protective agents against ultraviolet light; also, lignin is a key compound of the structural substances of plants (Baser and Buchbauer 2010) (Fig. 2.3).

2.6 The Pharmacology of Essential Oils

The study of the actions and effects of essential oil in living organisms comes under pharmacology. The two main branches of pharmacology include pharmacokinetics and pharmacodynamics. In the present context, pharmacokinetics and pharmacodynamics means the action of the body toward the essential oil and the action of the essential oil on the body, respectively.

Pharmacodynamics is related to the method of essential oil binding to drug targets, biochemical, physiological, and side effects of them. All molecules of essential oil activate by interacting with the target molecules in the body. There are various types of target molecules and sites for essential oil molecule interaction such as cell membranes, neuronal and muscular ion channels, and different receptors.

Individual patients respond in varying ways to the same dose of the same oil. This variation is thought to be directly related to pharmacokinetics. Pharmacokinetics means what the body should do to the essential oil. It is the certain description of the rate and extent of absorption, distribution, metabolism, and excretion of essential oil (Gwilt 1994). These processes describe the movements of essential oil within the body.

2.6.1 Absorption

Inhalation and dermal applications are two major ways of essential oil usage that convey topical dosage of essential oils to respiratory membranes or the skin. In each method, essential oil components are absorbed in different speeds and varying amounts into the circulatory system. Absorption depends on the mode of delivery. Transdermal path is thought to be slower and more controlled. It is thought to decrease the variation between the maximum and minimum oil concentrations achieved during a dosing interval (Blaschke and Bjornsson 1995).

2.6.2 Distribution

Distribution is determined as the movement of essential oil substances between two locations in the body (Bryant et al. 2003). The distribution of any essential oil is controlled by the bloodstream to the tissue or organ, as well as by the oil's ability to bind to plasma proteins. The pathway of essential oil from absorption to excretion is illustrated in Fig. 2.4.

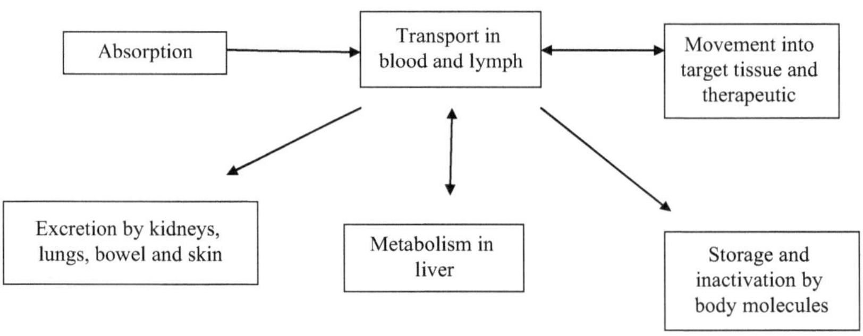

Fig. 2.4 The pathway of essential oils through different parts of the body from absorption to excretion (Bowles 2003)

2.6.3 Metabolism

Essential oils are mainly soluble in alcohol. The main function of essential oil metabolism is to alter the alcohol solubility of their molecules to excrete via urine; however, some essential oil molecules can be directly excreted in the urine without metabolism. The chemical reactions to transform essential oil molecules in to various metabolism are usually carried out by enzymes which take place in the liver or epidermal layers of the skin (Rawlins 1989). Essential oils that are inhaled or applied topically do not go through the first stage of metabolism by the liver (Price and Price 1999). Essential oils are lipid soluble, and their components are easily accessible to the brain. While being carried by the bloodstream, the constituents travel readily to the adrenal glands and kidneys (Tisserand and Balacs 1995). The rate of essential oil removal from the body is related to the concentration of the oil in the bloodstream. It is complicated because the essential oil begins to be eliminated while it is still being absorbed.

2.6.4 Excretion

Excretion refers to the ability of an organ or the whole body to remove essential oil molecules and its metabolites from the bloodstream, which is named as clearance. Essential oils are excreted via the kidneys, lungs, skin, and feces. The time taken for the oil concentration in the blood to decrease by one half is called half-life. It depends on both the distribution degree of the oil and the rate at which that drug is removed from the body (Gwilt 1994).

2.7 Therapeutic Effects of Essential Oils and Their Constituents

Nature is a rich source of products with interesting and useful biological activities. Plants' secondary metabolites evidencing therapeutic, flavoring, odoriferous, stimulant, poisoning, hallucinogenic, or coloring properties still play an important role in human activities.

Essential oils have various biological activities and therapeutic effects which can be used in different industries and treat several disorders in humans, animals, and plants and also used as foods. Some activities of essential oils such as acaricidal, anticarcinogenic, antioxidant, and antimicrobial including antibacterial, antifungal, and antiviral activities were reported earlier. In previous sections, we understood the mechanisms and processes of essential oils in the body. The target sites or receptors of essential oils are related to the diseases or conditions that could benefit from binding their molecules to each type of receptor or site. Every cell is surrounded by a phospholipid membrane, which keeps its substances separated from the fluid between the cells. This membrane can be damaged by lipophilic solvents or free radicals and oxidation. Essential oil constituents have activities in each of these areas, mainly due to their lipophilic nature (Bowles 2003).

2.8 Antifungal Therapy with Essential Oils

Essential oils are volatile compounds produced in many plant species that play an important role in plant defense mechanisms against pathogenic microorganisms (Liu and Chu 2002). As natural substances, they represent a potential source of new antifungal agents. During the past years, a number of studies have been carried out concerning the application of essential oils as antifungal agents. Particularly, essential oils obtained from aromatic herbs and spice plants are known to be inhibitive or lethal to fungi. Essential oils have been reported to control molds and various fungi such as food-borne and phytopathogenic fungi which cause different postharvest diseases and animal and human disorders (Banihashemi and Abivardi 2011). Various essential oils could be useful alternative substances replacing synthetic fungicides in the plant disease management (Gwinn et al. 2010; Nguefack et al. 2013). Aromatic plants such as members of the Asteraceae, Lamiaceae, Rutaceae, and Verbenaceae contain essential oils which have bioactivity against several fungi (Lahlou 2004; Edris 2007).

2.9 Methods for Testing the Antifungal Activity of Essential Oil

As mentioned in the previous section, pharmacological researches of essential oils and their antifungal activities have been done before. So the information from other studies were collected and examined to obtain the variability of test parameters, data variation, and the significance of such factors in the interpretation of results. There are several methods for testing in vitro antifungal activity of essential oils including disk diffusion, serial broth or agar dilution, the vapor phase tests, airtight boxes, spore germination, and poisoned food test.

Recording of growth curves and determination of the amount of a compound being effective to inhibit growth to 50 % (IC_{50}) of the test organisms, poisoned food techniques in which the delay of microbial growth is determined in the presence of growth inhibitors, spore germination, and short contact time studies in fungi are some other tests which comprise different kill-time study are in practice (Friedman et al. 2004; Lang and Buchbauer 2012; Jing et al. 2014).

2.9.1 Broth or Agar Dilution Method

The agar dilution method is consist of adding different amounts of pure essential oils dissolved in absolute ethyl alcohol, 5 % Tween 80 (v/v), to potato dextrose agar (PDA) before pouring it into Petri dishes while the medium is still warm (40–50°°C), to obtain the desired concentrations of the oil. A 5 mm diameter disk of inoculums of the tested fungi with a cork borer, cut from the periphery of an actively growing culture on PDA plates, is placed at the center in each Petri plate. Control group consists of just PDA. Then the colony radius is measured at different incubation times at 25 °C. The percentage of inhibition of mycelia growth is calculated from mean values of colony diameter in treated and control Petri dishes. The size of inhibition zone is regarded as measure for the antifungal potency of an essential oil (Lang and Buchbauer 2012).

There are some differences between the broth and the agar dilution test. In the agar dilution test, a concentration gradient of the antifungal substance is placed onto an agar plate; also the agar with supplements is able to adequately support fungus growth, while in the broth dilution test, a concentration series of the tested substance is obtained by a broth medium, which is seeded with fungal strains. In this method, lack of or poor growth of many anaerobic microorganisms is observed because of insufficient supplements (Varela et al. 2008; Jiang 2011; Lang and Buchbauer 2012).

2.9.2 Disk Diffusion Method

Sterile filter paper disks soaked in desired concentrations of pure essential oil were placed in the inner surface of the Petri dish lid. The dishes are sealed with parafilm and incubated upside down at 25 °C colony radius (in mm) is measured at different time intervals. The control consists of sterile distilled water-soaked filter paper in the lid of Petri dish (Jiang 2011).

2.9.3 Vapor Phase Test (VP)

In the vapor phase, a standardized method does not exist to study antifungal activity of essential oils. In most of the examinations, a reservoir (paper disk, cup, and glass) contains the sample of essential oil and a seeded agar plate inverted over the reservoir. After incubation, an inhibition zone is formed, which is the measure of activity (Cavanagh 2007).

2.9.4 Airtight Boxes

Because these methods allow only the creation of relative values, a few examiners defined the MIC in atmosphere (MIC_{air}) by using airtight boxes, which contained a seeded agar plate and the essential oil on the glass or paper. Otherwise, the results are estimated with + + + = normal growth; −+ = reduced growth; + − = visible growth; and NG = no growth. The test parameters are growth medium, incubation time, incubation temperature, and activity evaluation or MIC in μg mL^{-1} or ppm (Baser and Buchbauer 2010; Pauli and Schilcher 2010).

2.9.5 Spore Germination Assay

It is a method described by Rana et al. 1997 for spore germination assay. Different concentrations of the essential oil and control are tested for the ability to inhibit spore germination of the fungi. Aliquots of 0.1 mL from each sample are mixed with fungal spores obtained from cultural fungi and placed on separate glass slides. Slides containing the spores are incubated in a moist chamber at 28 °C for 24 h. Then each slide is fixed in lactophenol cotton blue and observed under the microscope for spore germination. A spore is considered to germinate when the germ tube length is 1.5 times the spore diameter (Ibrahim et al. 2015).

2.9.6 Poisoned Food Technique

In this method, the fungitoxicity is expressed with respect to the percentage of mycelia growth (Sharma and Tripathi 2008).

2.9.7 Percent Inhibition of Mycelia Growth

When the growth of the colony in the control treatment reaches the edge of the plate, the diameters are measured of both treatments and control. Fungistatic effect is expressed in terms of mycelia growth inhibition (%). The percentage inhibition of fungal growth is determined on the growth in test plates compared to the respective control plates.

The inhibition rate could be estimated according to the formula of previous studies:

$$I = (D_T{}_{-}D_t)/D_{T\times 100}$$

where I is the inhibition rate of mycelia growth of tested fungal isolates (%), D_T the diameter of the mycelia growth area in the control medium (mm), and D_t the diameter of the mycelia growth area in the treated medium (mm) (Schadler and George 2006; Jing et al. 2014; Messgo-Moumene et al. 2015)

2.9.8 Estimation of Minimal Inhibitory Concentration (MIC) and Minimum Fungicidal Concentration (MFC)

A method has been described by Thompson (1989) to determine the nature of toxicity (fungistatic and/or fungicidal effect) of the essential oils against fungi and to identify whether the test essential oil only had fungistatic activity on the pathogens or also could it have fungicidal effect. The inhibited fungal mycelia plugs of the essential oil-treated PDA plates were reinoculated into fresh medium and observed for revival of their growth to determine whether the essential oil used had fungicidal effect on the tested fungi.

The MIC and MFC of the most effective essential oils against tested fungi was described by Plodpai et al. (2013). The PDA plates are amended with various concentrations of essential oils. To enhance the solubility of the essential oil, Tween 20, 0.01 % (v/v), was added. Each plate was to be inoculated with a mycelia plug of the fungus. All plates with respective concentration were to be incubated at 28 °C for 72 h. Plates with only Tween 20, without any essential oil were used as control. The MIC values were determined as the lowest concentration of each essential oil that completely prevented the visible fungal growth.

To determine MFC, the mycelia plugs obtained from each Petri dish was treated with the oil concentrations higher than MIC and cultured on PDA at 28 °C for 72 h. MFC is defined as the lowest concentration at which no colony growth is observed after subculturing into fresh PDA medium (Khaledi et al. 2015).

2.10 Antifungal Effects of Essential Oils

The use of herbs for treating plant disease has been intensified. Plant volatile oils may play an important role in the preservation of foodstuffs against fungi, in the fungicidal application against plant diseases, and in the fight against various human fungal infections. Various studies have shown the biological activities of essential oils and their individual pure constituents and have documented the inhibitory activity of these substances against the growth of various fungi (Dwivedi and Singh 1999; Abraham and Prakasam 2000; Rao et al. 2000; Rawal and Thakore 2000; Tsao and Zhou 2000; Agarwal et al. 2001; Agnese et al. 2001; Alvarez-Castellans et al. 2001; Abou-Jawdah et al. 2002; Lee et al. 2005; Romagnoli et al. 2005; Talas-Ogras et al. 2005).

2.11 Importance of Natural Products

Crop protection from competing weeds, insects, and diseases is the prime requirement for the higher yields in agriculture. Natural chemical materials can be useful for crop and food protection. Medicinal plants are potential sources of antifungal components, which could be used in the management of plant diseases with low hazardous and negative effects on human health (Balbi-Peoa et al. 2006). There are numerous plant diseases created by pathogenic microbes, which are caused to reduce crop production. The application of chemical compounds for controlling diseases brings about several problems such as residues in edible plant parts, resistant strains, and environmental pollution. From the 1970s, multiplied interests on the study of the plants' autoprotective capacity against pathogenic agents or non-conducive environmental conditions were observed (Bailey et al. 1992). For controlling fungi, several methods such as biological control (Sreedevi et al. 2011; Huang et al. 2012), chemical control (Mahmoud et al. 2006), and application of plant essential oils (Seema and Devaki 2010; Javaid and Naqvi 2012) were recommended. In addition, consumers prefer to use more natural products rather than synthetic ones in different industries including food, drink, cosmetic, agricultural, and pharmaceuticals (Lang 1998; Walton and Brown 1999; Bansal and Gupta 2000; Singh et al. 2000; Kamanzi Atindehou et al. 2002). For these reasons, there has been an increasing interest for using eco-friendly production to control pests and diseases of various crops (Thind et al. 2005). Hence, the development and use of safe antifungal agents such as plant-based essential oils to control phytopathogens is rising in the agriculture sector (Adebayo et al. 2013).

2.12 Pathogenic Fungi

Fungal growth has been demonstrated to be responsible for food spoilage and plant disease, which leads to significant economic losses as well as disorders in animal and human system. Microbial infection increases the risk of food-borne diseases which are a growing public health problem worldwide, in addition to its role in reducing the shelf life of foods (Scallan et al. 2011). *Penicillium*, *Rhizopus*, *Aspergillus*, and *Fusarium* species are the most important fungi that cause spoilage of foodstuffs and rapid deterioration of fresh commodities. Fungi are also responsible for off-flavor formation and production of allergenic compounds and mycotoxins that affect their quality and shorten the shelf life (Lichter et al. 2002; Tejeswini et al. 2014). Molds – in most cases *Aspergillus* species – can lead to invasive infections especially in patients with a weakened immune system. *Aspergillus* species cause a number of severe diseases, both in normal and immunocompromised hosts, clubbed under the name aspergillosis which includes allergic disease, saprophytic disease, superficial infections, and invasive infections (Leyngold et al. 2014; Zhang et al. 2014). Among the *Aspergillus* species, *A. fumigatus* is the most common human infectious agent, followed by *A. flavus*, *A. niger*, and *A. terreus* (Latgé and Steinbach 2009; Gustot et al. 2014). *A. versicolor* produces many toxic compounds which can cause severe symptoms in humans and animals infected through inhalation or other forms of contact with debris or spores (Dagenais and Keller 2009; Veraldi et al. 2010).

The designation "dermatophytic fungi" comprises different kinds of *Epidermophyton*, *Microsporum*, and *Trichophyton* species. These pathogens are responsible for the generation of fungal infections concerning the human skin, nails, and hair (Woodfolk 2005).

Trichophyton species are pathogenic fungi causing superficial mycoses, commonly known as tinea infections, in various tissues of humans and other animals (Czaika 2013). *T. rubrum* is a predominate cause of dermatophytosis, followed by *T. mentagrophytes* and *T. tonsurans* (Hryncewicz-Gwóźdź 2010; Molenberg et al. 2010; Mapelli et al. 2012; Wu et al. 2013).

Yeasts are unicellular, eukaryotic organisms that belong to the kingdom of fungi. The two most well-known yeasts are *Saccharomyces cerevisiae* and *Candida albicans* which have been studied frequently in order to gain knowledge about the eukaryotic cell (Kurtzman and Piškur 2006). The yeasts of the genus *Candida* are the most frequently recovered from human fungal infections. The *Candida* genus comprises over 150 heterogeneous species (Calderone 2002), but only a minority (about 20 species) are known to be etiological agents of human and animals' infections (Pfaller and Diekema 2004; Silva et al. 2012). *Candida* species are a natural part of the human flora in the gastrointestinal tract, in the genitourinary system, and on the skin. *Cryptococcus neoformans* enters the human body through inhalation.

Field crops and stored grain are susceptible to attack by different pathogenic fungal species, one of the most serious agents of disease in plants worldwide (Woodfolk 2005). The genera *Aspergillus* and *Fusarium* are examples of known phytopathogenic fungi reported to be widespread and harmful toxin producers (Harris and Mantle 2001; Harvey et al. 2001; Hussein and Brasel 2001).

Macrophomina phaseolina (Tassi) Goid causes dry root rot, charcoal rot, seedling blight, and ashy stem blight in several hosts, mainly bean plants (Su et al. 2001; Gachitua et al. 2009). It is one of the most destructive necrotrophic soilborne pathogens, infecting about 500 plant species (Mihail and Taylor 1995). Outbreaks of charcoal rot disease associated with *M. phaseolina* caused reduction of seed production and yield losses ranging in many bean-growing regions worldwide (Amusa et al. 2007; Mayek-Perez et al. 2003). *R. solani* is an important necrotrophic pathogen, with a broad host range and little effective resistance in crop plants (Nikraftar et al. 2013). Stem and root rot caused by *R. solani* is a major constraint leading to severe yield losses in bean production.

2.13 Antifungal Mode of Action of Essential Oil

The mechanism of the inhibitory action of essential oil on the fungi is not well understood. As yet, there are a limited number of studies describing the mode of action of many essential oils. The antifungal property of plant essential oil has been supposed to be because of their lipophilic nature, responsible for disruption of plasma membrane, loss of membrane integrity, and leakage of cellular material and thereby negatively affecting the main cellular components particularly mitochondria as has been supported by the transmission electron microscopy studies on essential oil-treated *Aspergillus flavus* culture (Tian et al. 2011, 2012; Da Silva Bomfim et al. 2015). The data obtained by microscopic observations are in accordance with previous studies in which essential oils of aromatic plants caused morphological alterations on the fungal hyphae (Romagnoli et al. 2005; Soylu et al. 2010).

In general, the antifungal activity of essential oils against pathogenic fungi can be attributed to morphological changes in the cell wall and interference in enzymatic reactions of wall synthesis, which affect fungal growth and morphogenesis (Sharma and Tripathi 2008). This causes either an increase in ion permeability and leakage of vital intracellular constituents or impairment of the fungal enzyme systems (De Billerbeck et al. 2001; Romagnoli et al. 2005).

2.14 Interaction Effects Between Essential Oil Composition and Fungi

Essential oils contain different kinds of volatile molecules such as terpenes and terpenoids and phenol-derived aromatic and aliphatic components. Different constituents of essential oils are responsible for their mode of antifungal action (Bakkali et al. 2008). Plant products have interaction with fungi. Natural fungitoxins are the substances of plant origin that cause a significant toxic or inhibitory action on the fungus growth at different concentration. Post-inhibitions are a group of fungitoxic vascular substances coexisting with the cytoplasmic-specific hydrolytic enzymes. Seven classes of these toxins are known which usually involved in the plant resistance to one or more phytopathogenic fungi. Some of these compounds are saponins, of sulfurated amino acid derivatives and several types of glycosides (Harborne 1987). The efficacy of many plant essential oils and their major compounds has displayed antifungal activity against a range of phytopathogenic fungi (Rai and Cecilia Carpinella 2006). The same compounds in the oils from different species of the same family may increase their probability of inhibiting particular microbes (Edris 2007).

Several studies have been published to confirm the effect of essential oil and their major compounds on pathogenic fungi. Reports from previous researches showed that the essential oils of *Thymus vulgaris* (Zambonelli et al. 2004; Lee et al. 2007; Abdel-Kader et al. 2011), *Salvia fruticosa* (Pitarokili et al. 2003), *Mentha piperita* (Zambonelli et al. 2004; Abdel-Kader et al. 2011), cinnamon, *Calocedrus macrolepis* var. *formosana* Florin (Chang et al. 2008), and *Monarda* spp. (Gwinn et al. 2010) have antifungal activity against *Rhizoctonia solani*. *Rhizoctonia solani* is an important necrotrophic pathogen, with a wide host range and little effective resistance in crop plants (Nikraftar et al. 2013(. In addition, antifungal activities of the essential oils of thyme (Abdel-Kader et al. 2011) against *Macrophomina phaseolina*; *Bunium persicum* against *Fusarium oxysporum* (Sekine et al. 2007); spearmint against *F. oxysporum* f. sp. *radiciscucumerinum* (Nosrati et al. 2011); cumin, thymus, lavandula, eucalyptus, rosemary, nigella, and dill against the *F. oxysporum* f. sp. *radicis-lycopersici* and *F. oxysporum* f. sp. *lycopersici* (Arici et al. 2013); *Syzygium aromaticum* against *F. moniliforme*, *F. oxysporum* (Rana et al. 2011), and *Fusarium* sp. (Borrego et al. 2012); *F. oxysporum*, *F. verticillioides*, and *F. avenaceum* (Ćosić et al. 2010); *Aspergillus* sp., *Mucor* sp., *Trichophyton rubrum*, and *Microsporum gypseum* (Rana et al. 2011); *Aspergillus niger*, *Aspergillus clavatus*, *Penicillium* sp. (Borrego et al. 2012), and *Botrytis cinerea* (Sirirat et al. 2009); and *Candida*, *Aspergillus*, and dermatophyte species (Pinto et al. 2009) were previously demonstrated.

Antifungal activities of *Mentha piperita* and *Echinophora platyloba* essential oils against phytopathogenic fungi, including *Fusarium oxysporum*, *Macrophomina phaseolina*, *Drechslera spicifera*, *Alternaria alternata*, *Curvularia fallax*, *Cytospora sacchari*, and *Colletotricbum tricbellum*, have been investigated (Moghaddam et al. 2013, 2015).

Differential inhibition of pathogens by essential oils may be due to their compositions that contribute to their biological activity. Volatile compounds of essential oils, mainly monoterpene and sesquiterpene hydrocarbons, are responsible for their biological activity. However the main components have good activity for the tested strains; minor compounds have very low activity (Tavares et al. 2010). Reports from previous researches showed that the main compounds are responsible for the antifungal properties of essential oil, for example:

Linalool (Koutsoudaki et al. 2005)
α-Pinene and β-pinene
p-Cymene
γ-Terpinene
β-Phellandrene
Spathulenol
δ-Cadinene
α-Cadinol (Dorman and Deans 2000; Chang et al. 2000, 2008; Chinou et al. 2004; Xianfei et al. 2007; Barra et al. 2010; Ho et al. 2011)
Citral in lemon grass oil (Paviani et al. 2006)
Eugenol
Iso-eugenol
Caryophyllene
Furfural
α-Pinene and eugenyl acetate in clove oil (Matan et al. 2006; Fichi et al. 2007; Ayoola et al. 2008; Rahimi et al. 2012)
Azadirachtin
Azadiradione
Fraxinellone
Nimbin salannin in neem oil (Silva and Fernandes 2010)
Menthol in peppermint oil (Bupesh et al. 2007)
Carvacrol
Thymol
γ-Terpinene and *p*-cymene in summer savory
Oregano and thyme (Lee et al. 2007; Dikbas et al. 2008; Stevic et al. 2014)
α-Phellandrene and α-pinene in *Wedelia prostrata* (İşcan et al. 2012)
Monoterpenes including *p*-cymene, isobornyl formate, citral, and phenolic compounds such as carvacrol and thymol in *Thymus vulgaris* and *Verbena officinalis* essential oils (Elshafie et al. 2015)
Dimethyl trisulfide, methyl-propyl-trisulfie, dietil-1,2,4-tritiolan,methyl-(1-propyl-trisulfideandmethyl-(1-propenyl)-trisulfide in *Allium cepa* (Kocic-Tanackov et al. 2012)
Linalool (Carson and Riley 1995) and geranyl acetate (José Goncalvesa et al. 2011) in *Coriandrum sativum*
β-Phellandrene, spathulenol, bicyclogermacrene, germacrene D, and 1,8-cineole in the *Mpumalanga specimens* (Moreno et al. 2009; Barra et al. 2010)

Results from different studies mark phenolic compounds as agents with the highest antifungal potential among different components of essential oils. Thyme oil and its major phenolic constituents thymol and carvacrol are clearly active against most fungal species tested (Kalemba and Kunicka 2003; Rai and Cecilia Carpinella 2006; Carmo et al. 2008; Gallucci et al. 2014). Diterpenoids isolated from *Nicotiana glutinosa*, isoterpenoids extracted from several species of *Lupinus*, and a sesquiterpenic lactone produced by *Chrysanthemum parthenium* are some examples of antifungal compounds produced by plant leaves as a reaction to the attack of a fungus (Harborne et al. 1976; Thara et al. 1984; Matos 2000). Moreover, antifungal activity of essential oils was significantly affected by the presence of biological precursor compounds such as γ-terpinene and *p*-cymene, as monoterpene hydrocarbons. Essential oils are rich in monoterpene hydrocarbons (α- and β-pinene, myrcene, cymene); monoterpenols (terpineol) and phenols (carvacrol, thymol, and other monoterpenes) are known to possess an important antifungal activity (Aggarwal et al. 2002; Matasyoh et al. 2007; Chang et al. 2008; Chutia et al. 2009; Tabassum and Vidyasagar 2013). These components showed toxic effects on the in vitro mycelium growth against several phytopathogenic fungi like *Penicillium*, *Fusarium*, and *Aspergillus* species and *Alternaria alternata*. Several other studies have confirmed the antifungal activity of eugenol against pathogens such as *A. ochraceus*, *F. graminearum*, *F. moniliforme*, *Penicillium citrinum*, *P. viridicatum*, *Trichophyton rubrum*, *T. mentagrophytes*, *C. tropicalis*, and *C. krusei*. (Kamatou et al. 2012). The presence of monoterpene hydrocarbons has been demonstrated to inhibit the mycelia growth of *F. oxysporum* (Singh et al. 2002) and *Candida* species, particularly

C. albicans (Filipowicz et al. 2003; Nissen et al. 2010; Rivas da Silva et al. 2012). Soković and Griensven (2006) observed antifungal activity of limonene and α-pinene against *Verticillium fungicola* and *Trichoderma harzianum* which are found in different amounts in different plant essential oils. Filtenborg et al. (1996) showed that the germination of *Penicillium digitatum* conidia was stimulated by certain combination of the volatiles such as limonene, β-pinene, sabiene, and β-myrcene. This might be due to the mechanism developed by some fungal pathogens using the secondary metabolites as a signal to initiate germination.

Antifungal activity of *Cymbopogon nardus* and *Corymbia citriodora* against *Aspergillus* spp., *Pyricularia* (*Magnaporthe*) *grisea*, and *Colletotrichum musae* was studied before where it was shown that the active constituent citronellal, as an aldehyde, was a potent inhibitor. Moreover this activity may be associated with synergistic interactions among the essential oil components such as geraniol, α-pinene, β-pinene, myrcene, and linalool (Nakahara et al. 2003).

Moreover, the sesquiterpenes, in particular (Z) caryophyllene, are known to possess more significant antifungal activity than monoterpene hydrocarbons such as limonene and β-myrcene (Chang et al. 2008). The highest MIC values were recorded for chamomile oil in which sesquiterpenes are dominant compounds (Stevic et al. 2014). The essential oil isolated from plantlets of *Thymus mastichina* L. ssp. mastichina was rich in 1,8 cineole (55.5 %) and linalool (24.5 %). The oil showed antifungal activity against eight pathogenic fungi of the genus *Fusarium* with MICs and MFCs ranging from 1500 to 2100 $\mu g\ mL^{-1}$ and from 2000 to 2400 $\mu g\ mL^{-1}$, respectively (Fraternale et al. 2003).

Monoterpene alcohols are one of the subgroups of oxygenated monoterpenes with antifungal activity which is highly active due to its good solubility in water and the presence of functional alcohol group. For example, high antifungal potential of rose oil can be explained by the dominance of monoterpene alcohols, citronellol, and geraniol. Essential oil of *Pelargonium graveolens* with high presence of citronellol (35 %) and geraniol (28.8 %) was found to be highly active against *Rhizoctonia solani* (Bouzenna and Krichen 2013). Furthermore, high antifungal potential of tea tree essential oil is the consequence of high content of terpinen-4-ol, monoterpene alcohol, which in previous studies showed good antifungal activity against *Aspergillus* and *Penicillium* (Hammer et al. 2003). Also, coriander oil exhibits moderate to good antifungal activity due to the domination of linalool and monoterpene alcohol. It is considered that monoterpene alcohols increase the permeability of the plasma membrane and inhibit process of respiration on mitochondrial membrane of fungi (Deba et al. 2008; Imelouane et al. 2009).

On the other hand, the high activity of anis oil was due to *trans*-anethole, as phenol ether, which is equally active in inhibiting the growth of fungi (Huang et al. 2010). Weaker antifungal potential of the basil oil can be attributed to the dominance of methyl chavicol which is a phenol ether (Tadtong et al. 2009).

In conclusion, the chemical composition of the main compounds and precursor could explain their strong antifungal activity (Stevic et al. 2014). Synergistic or antagonistic effects between major and minor oil components and their mutual interaction play an important role in the overall activity of the essential oils (Dikbas et al. 2008). Limonene and linalool acetate were predominant components in lavender and bergamot oils with low antifungal activity. So the good antifungal potential of bergamot and lavender oils is probably a result of the contributory activities of other compounds present in the oils (Hammer et al. 2003). Weak antifungal activity of essential oils of lemon, orange, and eucalyptus can be explained by the dominance of limonene which together with monoterpene acetates is the weakest inhibitor of fungal growth among the pure monoterpene compounds (Combrinck et al. 2011).

The constituents, structure, and functional groups of the oils play an important role in their antifungal activity that phenolic groups usually are being the most effective components (Holley and Patel 2005). Therefore, essential oils' antifungal action could be attributed to the presence of monoterpenes, sesquiterpenes, and oxygenated compounds such as alcohols, phenols, and aldehydes (Zhu et al. 2005). According to the results of preceding studies, monoterpene hydrocarbons, oxygenated monoterpenes, sesquiterpenes, and phenolic compounds are responsible for inhibiting the mycelia growth of various fungi. In the antimicrobial action of essential oil components, the lipophilic character of their hydrocarbon skeleton and the hydrophilic character of their functional groups are of fundamental importance. The activity rank of essential oil components is as follows: phenols > aldehydes > ketones > alcohols > ethers > hydrocarbons. The highest activity is reported for phenol thymol, carvacrol, and eugenol, which are explained by the acidic nature of the hydroxyl group, forming a hydrogen bond with an enzyme active center (Kalemba and Kunicka 2003). Therefore, essential oils with phenols as main compounds express the highest activity against microorganisms, and their activity spectrum is the broadest (El-Shiekh et al. 2012). In a study concerning *Candida* spp., several alkyl phenol thymol, methyl thymol, eugenol, methyl eugenol, anethole, estragole, and amphotericin, respectively, showed different MIC values (Fontenelle et al. 2011).

2.15 Effective Factors on Antifungal Activity of Essential Oils

As mentioned in previous section, the antifungal activity of essential oil is associated with phytochemical components. In addition to this, some other factors can influence their antifungal activity. Percentage inhibition of mycelia growth depends on various factors such as the antifungal activity method, day of observation, dosage, and examined fungal species. The antifungal activities of the essential oils depend on the concentration used (Bakkali et al. 2008). Minimum effective concentrations of the oil against fungal pathogens were different. Antifungal activity of essential oil is increased with rising in the concentration. Moreover, the day of observation and the screening methods which are used to determine antifungal activity can affect the results (Moghaddam et al. 2013, 2015).

The sensitivity of the various pathogens may depend on the morphological and physiological characteristics of the fungus hyphae. By comparing the percentage inhibition of mycelia growth for every fungus, a difference between the tolerances against the oil can be observed. This may be due to production of more enzymes by some fungus such as *Macrophomina phaseolina* which catalyzes the oxidation and thus inactivation of the added oil. Inhibition of the growth of these fungal pathogens may be due to emulsion causing damage to the cell wall and cell membrane to various degrees because of differential capacity to penetrate oil into the chitin-based cell walls of fungal hyphae (Moghaddam et al. 2013). The mode of action of the essential oil as antifungal agents may be due to inhibition of respiration and disrupting the permeability barriers of the cell membrane structures (Cox et al. 2000). Natural compounds act on the internal mechanisms of the fungus leading to malformation of important structures, cytoplasmic granulation, disorganization of cell contents, disruption of the plasma membrane, and inhibition of fungal enzymes, consequently inhibiting germination, germ tube elongation, and reduction or inhibition of the mycelia growth (Schwan-Estrada et al. 2000).

On the other hand, increased fungistatic potential of mixture of essential oils may be due to the combined activities of two or more components of the essential oils (Gõni et al. 2009; Begnami et al. 2010).

2.16 Results on Antifungal Activity of Essential Oil Obtained from Previous Researches

In previous sections, some important information were scripted about essential oils and their composition, antifungal activity, and mode of action. In recent years, many researchers investigated the effects of essential oils of different plant species on various pathogenic fungi. In Table 2.1,

Table 2.1 Antifungal activity of some essential oil recorded from various research projects on animal, human, plant, and food pathogenic species (2012–2015)

Family	Species	Plant part	Active compound	Pathogenic fungi	Method	100 % inhibitory concentration	MIC[a]	MFC	Reference
Apiaceae	*Angelica major*	Aerial parts	*cis*-β-Cymene (30.4 %), α- pinene (21.8 %)	*Cryptococcus neoformans*	Broth macrodilution	–	0.16 µLmL^{-1}	0.64 µLmL^{-1}	Cavaleiro et al. (2015)
				Candida albicans			≥10 µLmL^{-1}	> 10 µLmL^{-1}	
				Candida krusei			2.5–5[b] µLmL^{-1}	5 µLmL^{-1}	
				Candida tropicalis			≥10 µLmL^{-1}	>10 µLmL^{-1}	
				Candida parapsilosis			≥ 10 µLmL^{-1}	> 10 µLmL^{-1}	
				Aspergillus niger			5–10 µLmL^{-1}	> 10 µLmL^{-1}	
				Trichophyton verrucosum			1.25 µLmL^{-1}	1.25 µLmL^{-1}	
Lamiaceae	*Glechon spathulata*	Aerial parts	Bicyclogermacrene (17.1 %), β-caryophyllene (14.2 %)	*Trichophyton rubrum*	Broth microdilution	–	10–83 µgmL^{-1}	–	Venturi et al. (2015)
Rutaceae	*Glechon marifolia*	Leaves	β-Caryophyllene (32.2 %), bicyclogermacrene (16.5 %)	*Epidermophyton floccosum*	Agar diffusion method	–	83–500 µgmL^{-1}	–	Ibrahim et al. (2015)
	Aegle marmelos		α-Hellenderene (20.97 %), α-pinene (17.76 %), δ-carene (16.37 %), γ-cadinene (8.01 %), *trans*-2-hydroxycinnamic acid (6.85 %)	*Aspergillus niger*			50 µLmL^{-1}	–	
				Candida albicans			50 µLmL^{-1}	–	
Lamiaceae	*Lallemantia royleana*	Aerial parts	*trans*-Pinocarvyl acetate (26.0 %), pinocarvone (20.0 %), verbenone (7.1 %), *trans*-carveol (5.3 %), 3-thujen-2-one (5.1 %)	*Candida albicans*	Disk diffusion, microbroth dilution	–	3.1 µgmL^{-1}	–	Sharifi-Rad et al. (2015)
				Aspergillus niger			2.5 µgmL^{-1}		
Myrtaceae	*Corymbia citriodora*	Leaves	Citronellal (61.78 %)	*Pyricularia grisea*	Twofold dilution	–	200 µLL^{-1}	–	Aguiar et al. (2014)
				Aspergillus spp.			200 µLL^{-1}		
				Colletotrichum musae			50 µLL^{-1}		
Poaceae	*Cymbopogon nardus*		Citronellal (36.6 %)	*Pyricularia grisea*			100 µLL^{-1}		
				Aspergillus spp.			100 µLL^{-1}		
				Colletotrichum musae			50 µLL^{-1}		
Lamiaceae	*Mentha suaveolens*	Leaves	Piperitenone oxide (90%)	*Candida albicans*	Microbroth dilution	–	0.78 mgmL^{-1}	–	Stringaro et al. (2014)

Myrtaceae	*Eugenia caryophylla*	Flower bud	Eugenol (80.0 %), β-caryophyllene (8.3 %), eugenol acetate (6.7 %)	*Candida albicans*	Broth microdilution	–	0.5 μLmL^{-1}	1 μLmL^{-1}	Nyegue et al. (2014)
				Tricophyton rubrum			0.25 μLmL^{-1}	0.125 μLmL^{-1}	
				T. violaceum			0.25 μLmL^{-1}	0.125 μLmL^{-1}	
				T. soudanensis			0.25 μLmL^{-1}	0.125 μLmL^{-1}	
Lamiaceae	*Mentha piperita*	Aerial parts	Piperitone (67.5 %), menthol (10.0 %), β-phellandrene (5.8 %)	*Candida albicans*			NI[b]	–	
Rutaceae	*Citrus reticulata*	Branches	Dimethyl anthranilate (34.7 %), γ-terpinene (33.6 %), limonene (11.2 %)	*Trichophyton mentagrophytes*	Agar well diffusion	–	0.98 μgmL^{-1}	–	Eldahshan (2015)
				Candida albicans			7.8 μgmL^{-1}	–	
				Aspergillus fumigatus			0.12 μgmL^{-1}	–	
		Leaves layers heavier	Dimethyl anthranilate (65.3 %), γ-terpinene (19.8 %)	*Trichophyton mentagrophytes*			0.24 μgmL^{-1}	–	
				Candida albicans			3.9 μgmL^{-1}	–	
				Aspergillus fumigatus			0.12 μgmL^{-1}	–	
		Leaves lighter layers	Dimethyl anthranilate (60.6 %) , γ-terpinene (22.8 %), limonene (5.3 %)	*Trichophyton mentagrophytes*			7.8 μgmL^{-1}	–	
				Candida albicans			15.63 μgmL^{-1}	–	
				Aspergillus fumigatus			3.9 μgmL^{-1}	–	
Apiaceae	*Eryngium tricuspidatum*	Aerial parts	α-Bisabolol (32.6 %), α-curcumene (6.5 %)	*Candida albicans*	Micro-well determination	–	4.6 μgmL^{-1}	–	Merghache et al. (2014)

(continued)

Table 2.1 (continued)

Family	Species	Plant part	Active compound	Pathogenic fungi	Method	100 % inhibitory concentration	MIC[a]	MFC	Reference
Lamiaceae	*Ocimum americanum*	Leaves	1,8-Cineole (25.9 %), (Z)-methyl cinnamate (29.4 %)	*Candida albicans*	Broth microdilution	–	5000 μgmL^{-1}	5000 μgmL^{-1}	Vieira et al. (2014)
				Candida tropicalis			5000 μgmL^{-1}	NI	
				Candida parapsilosis			2500 μgmL^{-1}	5000	
				Candida glabrata			NI	NI	
				Candida krusei			NI	NI	
				Candida parapsilosis			NI	NI	
				C. albicans			5000 μgmL^{-1}	5000 μgmL^{-1}	
	Ocimum basilicum var. *purpurascens*	Leaves	Linalool (41.5 %), α-muurolol (11.8 %)	*Candida albicans*			NI	NI	
				Candida tropicalis			NI	NI	
				Candida parapsilosis			5000 μgmL^{-1}	NI	
				Candida glabrata			NI	NI	
				Candida krusei			NI	NI	
				Candida parapsilosis (America Type Culture Collection)			NI	NI	
				C. albicans			5000 μgmL^{-1}	NI	
Verbenaceae	*Vitex negundo*	Seeds	*n*-Hexadecanoic acid (17.68 %), eudesm-4(14)-en-11-ol (12.39 %), caryophyllene oxide (10.79 %)	*Candida albicans*	Broth microdilution	–	4.0 μgmL^{-1}	–	Hong-Wei et al. (2014)
				Cryptococcus neoformans		–	64 μgmL^{-1}	–	
				Microsporum gypseum		–	64 μgmL^{-1}	–	
				Candida tropicalis		–	64 μgmL^{-1}	–	
				Trichophyton rubrum		–	32 μgmL^{-1}	–	
				Fonsecaea compacta		–	>64 μgmL^{-1}	–	
				Aspergillus fumigatus		–	>64 μgmL^{-1}	–	
				Candida parapsilosis		–	>64 μgmL^{-1}	–	

Lamiaceae	*Ocimum basilicum* var. *minimum*	Leaves	Linalool (44 %), 1,8 cineole (15.5 %)	*Candida albicans*	Broth microdilution	–	NI	NI	Vieira et al. (2014)
				Candida tropicalis			NI	NI	
				Candida parapsilosis			NI	NI	
				Candida glabrata			NI	NI	
				Candida krusei			NI	NI	
				Candida parapsilosis			NI	NI	
				C. albicans			NI	NI	
	Ocimum micranthum	Leaves	Eugenol (64.11 %), β-caryophyllene (14.3 %)	*Candida albicans*			1250 μgmL^{-1}	1250 μgmL^{-1}	
				Candida tropicalis			1250 μgmL^{-1}	1250 μgmL^{-1}	
				Candida parapsilosis			312.5 μgmL^{-1}	625 μgmL^{-1}	
				Candida glabrata			625 μgmL^{-1}	625 μgmL^{-1}	
				Candida krusei			1250 μgmL^{-1}	1250 μgmL^{-1}	
				Candida parapsilosis (America Type Culture Collection)			1250 μgmL^{-1}	1250 μgmL^{-1}	
				C. albicans			625 μgmL^{-1}	625 μgmL^{-1}	
Poaceae	*Canarium schweinfurthii*	Resins	Limonene (42.7), α-terpineol (34.4), *p*-cymene (9.8 %)	*Aspergillus flavus*	Disk diffusion	–	1800 μLL^{-1}	–	Ambindei et al. (2014)
				Aspergillus niger		–	2800 μLL^{-1}	–	
				Aspergillus fumigatus		–	1300 μLL^{-1}	–	
Burseraceae	*Aucoumea klaineana*	Resins	α-Phellandrene (30.9 %), α-pinene (29.3 %), *p*-cymene (9.2 %)	*Aspergillus flavus*		–	>5000 μLL^{-1}	–	
				Aspergillus niger		–	>5000 μLL^{-1}	–	
				Aspergillus fumigatus		–	>5000 μLL^{-1}	–	
Verbenaceae	*Aloysia gratissima*	Leaves	1,8 Cineole (13.7 %), germacrene-D (13.4 %), β-caryophyllene (12.7 %), β-pinene (11.7 %)	*Aspergillus flavus*	Broth microdilution	–	>1000 μgmL^{-1}	–	Santos et al. (2015)
				Aspergillus fumigates		–	>1000 μgmL^{-1}	–	
				Rhizopus sp.		–	>1000 μgmL^{-1}	–	
				Microsporum canis		–	>1000 μgmL^{-1}	–	
				Epidermophyton floccosum		–	1000 μgmL^{-1}	–	
				Microsporum gypseum		–	1000 μgmL^{-1}	–	
				Trichophyton rubrum		–	1000 μgmL^{-1}	–	
				Cryptococcus neoformans		–	1000 μgmL^{-1}	–	
				Candida albicans		–	1000 μgmL^{-1}	–	
				Trichophyton mentagrophytes		–	500 μgmL^{-1}	–	

(continued)

Table 2.1 (continued)

Family	Species	Plant part	Active compound	Pathogenic fungi	Method	100 % inhibitory concentration	MIC[a]	MFC	Reference
Apiaceae	*Coriandrum sativum*	Leaves	Decanal (19.09 %), *trans*-2-decenal (17.54 %), 2-decen-1-ol (12.33 %), cyclodecane (12.15 %), cis-2-dodecenal (10.72 %)	*Candida albicans*	Agar dilution	–	15.6 μgmL^{-1}	31.2 μgmL^{-1}	Freires et al. (2014)
				C. tropicalis		–	31.2 μgmL^{-1}	62.5 μgmL^{-1}	
				C. krusei		–	15.6 μgmL^{-1}	31.2 μgmL^{-1}	
				C. dubliniensis		–	31.2 μgmL^{-1}	62.5 μgmL^{-1}	
				C. rugosa		–	15.6 μgmL^{-1}	31.2 μgmL^{-1}	
Lamiaceae	*Thymbra spicata*	Aerial parts	Carvacrol (82.68 %), cymene (7.58 %), γ-terpinene (6.04 %)	*Saprolegnia parasitica*	Disk diffusion	–	50 μLmL^{-1}	–	Gormez and Diler (2014)
	Origanum onites		Carvacrol (82.24 %), thymol (13.04 %)			–	10 μLmL^{-1}	–	
	Satureja thymbra		Carvacrol (52.38 %), thymol (26.45 %), caryophyllene (7.28 %)			–	100 μLmL^{-1}	–	
Myrtaceae	*Myrtus communis*	Leaves and Flowers	α-Pinene (50.8 and 33.6 %), 1,8 cineole (21.9 and 13.3 %), linalool (2.7 and 14.8 %)	*Candida albicans*	Broth macrodilution	–	2.5 mgmL^{-1}	2.5 mgmL^{-1}	Bouzabata et al. (2015)
				C. tropicalis		–	2.5 mgmL^{-1}	2.5 mgmL^{-1}	
				C. krusei		–	2.5 mgmL^{-1}	2.5 mgmL^{-1}	
				C. guilliermondii		–	2.5 mgmL^{-1}	2.5 mgmL^{-1}	
				C. parapsilosis		–	2.5 mgmL^{-1}	5 mgmL^{-1}	
				Aspergillus flavus		–	5 mgmL^{-1}	>10 mgmL^{-1}	
				Aspergillus niger		–	2.5 mgmL^{-1}	>10 mgmL^{-1}	
				Aspergillus fumigatus		–	2.5 mgmL^{-1}	>10 mgmL^{-1}	
				Microsporum canis		–	0.64 mgmL^{-1}	0.64 mgmL^{-1}	
				Cryptococcus neoformans		–	0.64 mgmL^{-1}	0.64–1.25 mg mL^{-1}	
				Epidermophyton floccosum		–	0.64 mgmL^{-1}	0.64 mgmL^{-1}	
				Trichophyton rubrum		–	0.64 mgmL^{-1}	0.64 mgmL^{-1}	
				Microsporum gypseum		–	1.25 mgmL^{-1}	1.25 mgmL^{-1}	
				Trichophyton mentagrophytes		–	1.25 mgmL^{-1}	1.25 mgmL^{-1}	
				Trichophyton mentagrophytes var. *interdigitale*		–	1.25 mgmL^{-1}	1.25 mgmL^{-1}	
				Trichophyton verrucosum		–	1.25 mgmL^{-1}	1.25 mgmL^{-1}	

Asteraceae	*Artemisia abrotanum*	Aerial parts	Eucalyptol (25.27 %)	*Sclerotinia clerotiorum*	Disk diffusion	–	1200 µLL^{-1}	–	Badea and Delian (2014)
	Artemisia santonica		α-Thujone (44.80 %)			–	2400 µLL^{-1}	–	
	A. pontica		Eucalyptol (62.41 %)			–	2400 $^{-1}$	–	
	A. annua		Artemisia ketone (31.31 %)			–	2400 µLL^{-1}	–	
	A. austriaca		Camphor (35.75 %)			–	2400 µLL^{-1}	–	
	A. dracunculus		Sabinene (51.06 %)			–	2400 µLL^{-1}	–	
	A. lerchiana		Eucalyptol (44.00 %)			–	2400 µLL^{-1}	–	
	A. scoparia		Capillene (66.20 %)			–	2400 µLL^{-1}	–	
	A. vulgaris		Germacrene-D (27.69 %)			–	2400 µLL^{-1}	–	
	A. dracunculus var. *pilosa*		Borneol (28.29 %)			–	2400 µLL^{-1}	–	
Lamiaceae	*Ocimum selloi*	Leaves	Linalool (16.8 %), anethole (52.2 %)	*Candida albicans*	Broth microdilution	–	1250 µgmL^{-1}	2500 $^{-1}$	Vieira et al. (2014)
				Candida tropicalis			1250 µgmL^{-1}	1250 µgmL^{-1}	
				Candida parapsilosis			312.5 µgmL^{-1}	625 µgmL^{-1}	
				Candida glabrata			1250 µgmL^{-1}	5000 µgmLl^{-1}	
				Candida krusei			1250 µgmL^{-1}	1250 µgmL^{-1}	
				Candida parapsilosis			625 µgmL^{-1}	2500 µgmL^{-1}	
				C. albicans			1250 µgmL^{-1}	1250 µgmL^{-1}	
Lamiaceae	*Ocimum canum*	Aerial parts	Camphor (33.2 %), limonene (5.6 %), camphene (6.2 %), β-selinene (10.3 %), α-selinene (8.1 %), β-caryophyllene (5.7 %)	*Rhizoctonia solani*	Poisoned food	–	1000 µLL^{-1}	–	Pragadheesh et al. (2013)
				Choanephora cucurbitarum			900 µLL^{-1}		
	Ocimum kilimandscharicum	Aerial parts	Camphor (66.5 %), limonene (6.6 %), camphene (5 %)	*Rhizoctonia solani*			1200 µLL^{-1}		
				Choanephora cucurbitarum			730 µLL^{-1}		
Ranunculaceae	*Nigella sativa*	Seed	Thymoquinone (42.4 %), *p*-cymene (14.1 %), carvacrol (10.3 %), longifolene (6.1 %), 4-terpineol (5.1 %)	*Trichophyton mentagrophytes*	Disk diffusion	–	4 mgmL^{-1}	–	Mahmoudvand et al. (2014)
Myrtaceae	*Eucalyptus erythrocorys*	Leaves	1,8 cineole (54.8 %), viridiflorol (8.97 %), α-pinene (7.84 %), ledol (5.32 %), acide octadecanoique (5.32 %)	*Bipolaris sorikiniana*	Disk diffusion	–	>4 µLmL^{-1}	–	Ben Ghnaya et al. (2013)
				Botrytis cinerea			>4 µLmL^{-1}		
				Fusarium solani			>4 µLmL^{-1}		
				F. avenaceum			>4 µLmL^{-1}		
				F. oxysporum			>4 µLmL^{-1}		

(continued)

Table 2.1 (continued)

Family	Species	Plant part	Active compound	Pathogenic fungi	Method	100 % inhibitory concentration	MIC[a]	MFC	Reference
Cupressaceae	*Juniperus phoenicea*	Berries	α-Pinene (84.2 %)	*Candida albicans*	Broth macrodilution	–	5 μLmL^{-1}	–	Abu-Darwish et al. (2013)
				C. tropicalis		–	10 μLmL^{-1}	–	
				C. krusei		–	5 μLmL^{-1}	–	
				C. guilliermondii		–	1.25 μLmL^{-1}	–	
				C. parapsilosis		–	5 μLmL^{-1}	–	
				Cryptococcus neoformans		–	0.64–1.25 μLmL^{-1}	–	
				Trichophyton mentagrophytes		–	1.25 μLmL^{-1}	–	
				Microsporum canis		–	1.25 μLmL^{-1}		
				M. gypseum		–	1.25 μLmL^{-1}	–	
				Trichophyton rubrum		–	0.64 μLmL^{-1}	–	
				Epidermophyton floccosum		–	0.64 μLmL^{-1}	–	
				T. mentagrophytes var. *interdigitale*		–	2.5 μLmL^{-1}	–	
				T. verrucosum		–	2.5 μLmL^{-1}	–	
				Aspergillus niger		–	≥10 μLmL^{-1}	–	
				A. flavus		–	≥10 μLmL^{-1}	–	
				A. fumigatus		–	>2.5 μLmL^{-1}	–	
Lamiaceae	*Origanum vulgare*	Leaves	Carvacrol (20.04 %), *p*-cymene (6.62 %), caryophyllene (5.5 %), α –terpinene (5.4 %)	*Candida albicans*	Agar well diffusion	–	0.6 μLmL^{-1}	1.25 μLmL^{-1}	Abdulmajeed (2015)
				C. tropicalis		–	0.6 μLmL^{-1}	1.25 μLmL^{-1}	
				C. parapsilosis		–	1.25 μLmL^{-1}	2.5 μLmL^{-1}	
				C. glabrata		–	1.25 μLmL^{-1}	2.5 μLmL^{-1}	

Apiaceae	*Oenanthe crocata*	Aerial parts	*trans*-β-Ocimene (31.3 %), sabinene (29.0 %) , *cis*-β-ocimene (12.3 %)	*Cryptococcus neoformans*	Macrodilution broth	–	0.64–1.25 μLmL^{-1}	–	Valente et al. (2013)
				Candida albicans		–	0.64–1.25 μLmL^{-1}	–	
				Trichophyton verrucosum		–	0.64–1.25 μLmL^{-1}	–	
				Candida guilliermondii		–	0.64 μLmL^{-1}	–	
				Candida krusei		–	1.25 μLmL^{-1}	–	
				Candida parapsilosis		–	1.25 μLmL^{-1}	–	
				Candida tropicalis		–	1.25 μLmL^{-1}	–	
				Aspergillus fumigatus		–	1.25 μLmL^{-1}	–	
				Aspergillus niger		–	1.25 μLmL^{-1}	–	
				Epidermophyton floccosum		–	0.08 μLmL^{-1}	–	
				Microsporum canis		–	0.08 μLmL^{-1}	–	
				Microsporum gypseum		–	0.08 μLmL^{-1}	–	
				Trichophyton rubrum		–	0.08 μLmL^{-1}	–	
				Trichophyton mentagrophytes		–	0.16 μLmL^{-1}	–	
				Trichophyton mentagrophytes var. *interdigitale*		–	0.16 μLmL^{-1}	–	
				Aspergillus flavus		–	2.5 μLmL^{-1}	–	
Ranunculaceae	*Nigella sativa*	Seed	Thymoquinone (42.4 %), *p*-cymene (14.1 %), carvacrol (10.3 %), longifolene (6.1 %) ,4-Terpineol (5.1 %)	*Trichophyton mentagrophytes*	Disk diffusion	–	4 mgmL^{-1}	–	Mahmoudvand et al. (2014)
				Microsporum canis		–	4 mgmL^{-1}	–	
				Microsporum gypseum		–	4 $^{-1}$	–	

(continued)

Table 2.1 (continued)

Family	Species	Plant part	Active compound	Pathogenic fungi	Method	100 % inhibitory concentration	MIC[a]	MFC	Reference
Lamiaceae	*Lavandula multifida*	Aerial parts	Carvacrol (42.8–41.5 %), *cis*-β-ocimene (27.4–27.0 %)	*Candida albicans*	Broth macrodilution	–	0.32 μLmL^{-1}	0.64 μLmL^{-1}	Zuzarte et al. (2012)
				Candida tropicalis		–	0.32 μLmL^{-1}	0.64 μLmL^{-1}	
				Candida parapsilosis		–	0.32 μLmL^{-1}	0.64 μLmL^{-1}	
				Aspergillus fumigatus		–	0.64 μLmL^{-1}	0.64 μLmL^{-1}	
				Candida krusei		–	0.64 μLmL^{-1}	0.64 μLmL^{-1}	
				Candida guilliermondii		–	0.32 μLmL^{-1}	0.32 μLmL^{-1}	
				Cryptococcus neoformans		–	0.16 μLmL^{-1}	0.32 μLmL^{-1}	
				Trichophyton mentagrophytes		–	0.16 μLmL^{-1}	0.32 μLmL^{-1}	
				Trichophyton mentagrophytes var. interdigitale		–	0.16 μLmL^{-1}	0.32 μLmL^{-1}	
				Trichophyton rubrum		–	0.16 μLmL^{-1}	0.32 μLmL^{-1}	
				Trichophyton verrucosum		–	0.16 μLmL^{-1}	0.32 μLmL^{-1}	
				Microsporum canis		–	0.16 μLmL^{-1}	0.32 μLmL^{-1}	
				Microsporum gypseum		–	0.16 μLmL^{-1}	0.32 μLmL^{-1}	
				Epidermophyton floccosum		–	0.16 μLmL^{-1}	0.32 μLmL^{-1}	
				Aspergillus niger		–	0.32 μLmL^{-1}	1.25 μLmL^{-1}	
				Aspergillus flavus		–	0.64 μLmL^{-1}	1.25 μLmL^{-1}	
Lamiaceae	*Origanum vulgare* ssp. *vulgare*	Aerial parts	–	*Aspergillus niger*	Well diffusion	–	3.125 mgmL^{-1}	3.125 mgmL^{-1}	Yotova and Ignatova-Ivanova (2015)
				Penicillium claviforme		–	3.125 mgmL^{-1}	3.125 mgmL^{-1}	
				Saccharomyces cerevisiae		–	12.5 mgmL^{-1}	12.5 mgmL^{-1}	
				Candida albicans		–	6.25 mgmL^{-1}	3.125 mgmL^{-1}	
				Candida glabrata		–	12.5 mgmL^{-1}	6.25 mgmL^{-1}	

Lamiaceae	*Monarda didyma*	–	Thymoquinone (36.02 %), thymol (16.7 %), vinyl amyl carbinol (6.54 %), methyl thymol (7.86 %)	*Aspergillus flavus*	Broth dilution	–	6.0 %v/v	–	Mickiene et al. (2014)
				Aspergillus flavus	Vapor phase	–	2.0 %v/v	–	
				Aspergillus niger	Broth dilution	–	6.0 %v/v	–	
				Aspergillus niger	Vapor phase	–	2.0 %v/v	–	
				Aspergillus ochraceus	Broth dilution	–	4.0 %v/v	–	
				Aspergillus ochraceus	Vapor phase	–	0.5 %v/v	–	
				Penicillium chrysogenum	Broth dilution	–	11 %v/v	–	
				Penicillium chrysogenum	Vapor phase	–	8.0 %v/v	–	
				Penicillium commune	Broth dilution	–	11.0 %v/v	–	
				Penicillium commune	Vapor phase	–	8.0 %v/v	–	
				Penicillium citrinum	Broth dilution	–	11.0 %v/v	–	
				Penicillium citrinum	Vapor phase	–	8.0 %v/v	–	
				Cladosporium cladosporioides	Broth dilution	–	9.0%v/v	–	
				Cladosporium cladosporioides	Vapor phase	–	5.0 %v/v	–	
				Alternaria alternata	Broth dilution	–	1.0 %v/v	–	
				Alternaria alternata	Vapor phase	–	0.05 %v/v	–	
				Rhizopus microsporus	Broth dilution	–	5.0 %v/v	–	
				Rhizopus microsporus	Vapor phase	–	1.5 %v/v	–	
				Trichoderma viride	Broth dilution	–	10.0 %v/v	–	
				Trichoderma viride	Vapor phase	–	5.0 %v/v	–	

(continued)

Table 2.1 (continued)

Family	Species	Plant part	Active compound	Pathogenic fungi	Method	100 % inhibitory concentration	MIC[a]	MFC	Reference
Lamiaceae	*Lamium album*	–	Tetradecanol –n (20.77 %), benzoate-isopentyl (15.24 %), squalene (9.82 %), phytone (7.25 %), undecane -n (5.25 %)	*Aspergillus flavus*	Broth dilution	–	ND	–	Mickiene et al. (2014)
				Aspergillus flavus	Vapor phase	–	ND	–	
				Aspergillus niger	Broth dilution	–	ND	–	
				Aspergillus niger	Vapor phase	–	ND	–	
				Aspergillus ochraceus	Broth dilution	–	20.0 %v/v	–	
				Aspergillus ochraceus	Vapor phase	–	18.0 %v/v	–	
				Penicillium chrysogenum	Broth dilution	–	ND	–	
				Penicillium chrysogenum	Vapor phase	–	ND	–	
				Penicillium commune	Broth dilution	–	ND	–	
				Penicillium commune	Vapor phase	–	ND	–	
				Penicillium citrinum	Broth dilution	–	25.0 %v/v	–	
				Penicillium citrinum	Vapor phase	–	23.0 %v/v	–	
				Cladosporium cladosporioides	Broth dilution	–	25.0 %v/v	–	
				Cladosporium cladosporioides	Vapor phase	–	10.0 %v/v	–	
				Alternaria alternata	Broth dilution	–	15.0 %v/v	–	
				Alternaria alternata	Vapor phase	–	10.0 %v/v	–	
				Rhizopus microsporus	Broth dilution	–	15.0 %v/v	–	
				Rhizopus microsporus	Vapor phase	–	13.0 %v/v	–	
				Trichoderma viride	Broth dilution	–	25.0 %v/v	–	
				Trichoderma viride	Vapor phase	–	20.0 %v/v	–	
Poaceae	*Canarium schweinfurthii*	Resins	Limonene (42.7), α-terpineol (34.4), *p*-cymene (9.8 %)	*Aspergillus flavus*	Disk diffusion	–	1800 μLL^{-1}	–	Ambindei et al. (2014)
				Aspergillus niger		–	2800 μLL^{-1}	–	
				Aspergillus fumigatus		–	1300 μLL^{-1}	–	
Burseraceae	*Aucoumea klaineana*	Resins	α-Phellandrene (30.9 %), α-pinene (29.3 %), *p*-cymene (9.2 %)	*Aspergillus flavus*		–	>5000 μLL^{-1}	–	

Apiaceae	*Myrrhis odorata*	–	Nerolidol –(E) (27.13 %), anethole –(Z) (15.89 %), germacrene –(D) (6.46 %), caryophyllene –E (6.36 %), β-bisabolene (5.07 %), neophytadiene (5.28 %)	*Aspergillus flavus*	Broth dilution	–	7.0 %v/v	–	Mickiene et al. (2014)
				Aspergillus flavus	Vapor phase	–	5.0 %v/v	–	
				Aspergillus niger	Broth dilution	–	8.0 %v/v	–	
				Aspergillus niger	Vapor phase	–	7.0 %v/v	–	
				Penicillium chrysogenum	Broth dilution	–	13.0 %v/v	–	
				Penicillium chrysogenum	Vapor phase	–	8.0 %v/v	–	
				Penicillium commune	Broth dilution	–	15.0 %v/v	–	
				Penicillium commune	Vapor phase	–	10.0 %v/v	–	
				Penicillium citrinum	Broth dilution	–	15.0 %v/v	–	
				Penicillium commune	Vapor phase	–	10.0 %v/v	–	
				Cladosporium cladosporioides	Broth dilution	–	15.0 %v/v	–	
				Cladosporium cladosporioides	Vapor phase	–	8.0 %v/v	–	
				Alternaria alternata	Broth dilution	–	4.0 %v/v	–	
				Alternaria alternata	Vapor phase	–	2.0 %v/v	–	
				Rhizopus microsporus	Broth dilution	–	8.0 %v/v	–	
				Rhizopus microsporus	Vapor phase	–	5.0 %v/v	–	
				Trichoderma viride	Broth dilution	–	14.0 %v/v	–	
				Trichoderma viride	Vapor phase	–	10.0 %v/v	–	
Aizoaceae	*Mesembryanthemum edule*	Leaves	Benzoic acid (15.68 %), phytol (2-hexadecan-1-ol,3,7,11,15-tetramethyl) (12.41 %), tetradecamethylcycloheptasiloxane (13.57 %), 2-pentadecanone,6,10,14-trimethyl (9.28 %)	*Candida albicans*	Agar dilution	–	0.02 $mgmL^{-1}$	–	Omoruyi et al. (2014)
				Candida krusei		–	0.04 $mgmL^{-1}$	–	
				Candida rugosa		–	0.08 $mgmL^{-1}$	–	
				Candida glabrata		–	0.31 $mgmL^{-1}$	–	
				Cryptococcus neoformans		–	0.08 $mgmL^{-1}$	–	

(continued)

Table 2.1 (continued)

Family	Species	Plant part	Active compound	Pathogenic fungi	Method	100 % inhibitory concentration	MIC[a]	MFC	Reference
Apiaceae	*Echinophora platyloba*	Seeds	*p*-Cymene (22.15 %), α-pinene (18.52 %), β-phellandrene (14.40 %), α-phellandrene (9.69 %)	*Alternaria alternate*	Agar dilution	>800 μLL^{-1}	–	–	Moghaddam et al. (2015)
				Curvularia fallax		200 μLL^{-1}	–	–	
				Macrophomina phaseolina		200 μLL^{-1}	–	–	
				Fusarium oxysporum		400 μLL^{-1}	–	–	
				Cytospora sacchari		400 μLL^{-1}	–	–	
				Colletotricbum tricbellum		>800 μLL^{-1}	–	–	
				Alternaria alternate	Disk diffusion	>800 μLL^{-1}	–	–	
				Curvularia fallax		100 μLL^{-1}	–	–	
				Macrophomina phaseolina		100 μLL^{-1}	–	–	
				Fusarium oxysporum		100 μLL^{-1}	–	–	
				Cytospora sacchari		200 μLL^{-1}	–	–	
				Colletotricbum tricbellum		800 μLL^{-1}	–	–	
Asteraceae	*Helichrysum splendidum*	Aerial parts	α-Muurolol (3.43–7.09 %), germacrene-D (3.41–10.1 %), δ-cadinene (9.32–16.9 %), α-cadinene (2.27–7.58 %), τ-cadinol (3.61–9.01 %), α-cadinol (5.84–16.8 %)	*Alternaria alternata*	Agar dilution	–	500 μLL^{-1}	–	Mashigo et al. (2015)
				Colletotrichum gloeosporioides		–	500 μLL^{-1}	–	
				Fusarium oxysporum		–	>500 μLL^{-1}	–	
				Lasiodiplodia theobromae		–	>500 μLL^{-1}	–	
				Penicillium digitatum		–	>500 μLL^{-1}	–	
				Penicillium expansum		–	>500 μLL^{-1}	–	
				Penicillium italicum		–	>500 μLL^{-1}	–	

Liliaceae	*Allium cepa*	–	–	*Fusarium oxysporum* f. sp. *melonis*	Agar dilution	500 μLL^{-1}	–	–	Elgorban et al. (2015)
	Nigella sativa	–	–	*Fusarium oxysporum* f. sp. *melonis*		>500 μLL^{-1}	–	–	
	Eucalyptus globulus	–	–	*Fusarium oxysporum* f. sp. *melonis*		>500 μLL^{-1}	–	–	
	Allium cepa	–	–	*Fusarium solani*		500 μLL^{-1}	–	–	
	Nigella sativa	–	–	*Fusarium solani*		>500 μLL^{-1}	–	–	
	Eucalyptus globulus	–	–	*Fusarium solani*		250 μLL^{-1}	–	–	
	Allium cepa	–	–	*Sclerotinia sclerotiorum*		>500 μLL^{-1}	–	–	
	Nigella sativa	–	–	*Sclerotinia sclerotiorum*		>500 μLL^{-1}	–	–	
	Eucalyptus globulus	–	–	*Sclerotinia sclerotiorum*		>500 μLL^{-1}	–	–	
Apiaceae	*Coriandrum sativum*	Fruits	Linalool (66.29 %), γ-terpinene (5.26)	*Candida albicans*	Microdilution	–	0.39 μLmL^{-1}	–	Sourmaghi et al. (2015)
Asteraceae	*Artemisia annua*	Upper third of the foliage	Camphor (up to 48 %), germacrene-D (up to 18.9 %), artemisia ketone (up to 68 %), 1,8 cineole (up to 51.5 %)	*Candida albicans*	Agar disk diffusion	–	2 mgmL^{-1}	–	Bilia et al. (2014)
				Saccharomyces cerevisiae		–	2 mgmL^{-1}	–	
				Aspergillus fumigatus		–	5 mgmL^{-1}	–	
Rubiaceae	*Galium tricornutum* subsp. *longipedanculatum*	–	Eucalyptol (20.8 %), L-camphor (13.9 %), L-borneol (11.6 %), myrtenol (5 %)	*Candida albicans*	–	–	0.27 mgmL^{-1}	–	Jan et al. (2015)
				Aspergillus flavus		–	0.18 mgmL^{-1}	–	
				Microsporum canis		–	110.8 mgmL^{-1}	–	
				Fusarium solani		–	200 mgmL^{-1}	–	
				Candida glabrata		–	200 mgmL^{-1}	–	
Myrtaceae	*Syzygium aromaticum*	Flower buds	–	*Fusarium oxysporum* f. sp *radicis-lycopersici*	Agar diffusion	–	1 μLmL^{-1}	–	Hamini-Kadar et al. (2014)
				F. redolens		–	1 μLmL^{-1}	–	
				F. commune		–	0.5 μLmL^{-1}	–	
Rutaceae	*Ruta chalepensis*	Leaves	Undecan-2-one (49.08 %), nonan-2-one (33.15 %)	*Fusarium proliferatum*	Disk agar diffusion	–	15 μLmL^{-1}	–	Bouajaj et al. (2014)
				F. pseudograminearum		–	10 μLmL^{-1}	–	
				F. culmorum		–	5 μLmL^{-1}	–	
				F. graminearum		–	20 μLmL^{-1}	–	
				F. polyphialidicum		–	>20 μLmL^{-1}	–	

(continued)

Table 2.1 (continued)

Family	Species	Plant part	Active compound	Pathogenic fungi	Method	100 % inhibitory concentration	MIC[a]	MFC	Reference
Rutaceae	*Zanthoxylum bungeanum*	Seeds	α-Pinene (19.59 %)	*Fusarium sulphureum*	Agar dilution	–	400 μLL^{-1}	–	XingDong and HuaLi (2014)
Lamiaceae	*Salvia officinalis*	Aerial parts	α-Thujone (24.92 %), camphor (15.30 %),1,8-cineole (11.53 %), α-pinene (6.36 %), D-limonene (5.90 %), *trans*-caryophyllene (5.07 %)	*Penicillium verrucosum*	Macrodilution broth	–	62 μLmL^{-1}	125 μLmL^{-1}	Ozcakmak et al. (2012)
Piperaceae	*Piper divaricatum*	Aerial parts	Methyleugenol (75.0 %), eugenol (10.0 %)	*Fusarium solani* f. sp. *piperis*	Agar dilution	–	>2.50 mgmL^{-1}	–	Silva et al. (2014)
Lamiaceae	*Origanum vulgare*	Aerial parts	Carvacrol (64.06 %), linalool (17.56 %)	*Epicoccum nigrum*	Microatmosphere, macrodilution	–	0.1 μLmL^{-1}	0.1 μLmL^{-1}	Stupar et al. (2014)
				Aspergillus niger		–	1.25 μLmL^{-1}	1.5 μLmL^{-1}	
				Aspergillus ochraceus		–	0.1 μLmL^{-1}	0.75 μLmL^{-1}	
				Bipolaris spicifera		–	0.25 μLmL^{-1}	1.0 μLmL^{-1}	
				Penicillium sp.		–	1.0 μLmL^{-1}	1.75 μLmL^{-1}	
				Trichoderma viride		–	1.75 μLmL^{-1}	2.0 μLmL^{-1}	
				Trichoderma viride	Microdilution	–	1.75 μLmL^{-1}	2.0 μLmL^{-1}	
				Aspergillus niger		–	1.25 μLmL^{-1}	1.5 μLmL^{-1}	
				Aspergillus ochraceus		–	0.1 μLmL^{-1}	0.75μLmL^{-1}	
				Bipolaris spicifera		–	0.25 μLmL^{-1}	1.0 μLmL^{-1}	
				Penicillium sp.		–	1.0 μLmL^{-1}	1.75 μLmL^{-1}	
				Trichoderma viride		–	0.5 μLmL^{-1}	0.5 μLmL^{-1}	
				Aspergillus niger		–	0.25 μLmL^{-1}	2.0 μLmL^{-1}	
				Aspergillus ochraceus		–	0.2 μLmL^{-1}	2.0 μLmL^{-1}	
				Bipolaris spicifera		–	0.2 μLmL^{-1}	1.0 μLmL^{-1}	
				Penicillium sp.		–	0.2 μLmL^{-1}	1.0 μLmL^{-1}	

Lamiaceae	*Rosmarinus officinalis*	Aerial parts	1,8-Cineole (44.28 %), camphor (12.54 %), α-pinene (11.62 %), β-pinene (8.26 %)	*Epicoccum nigrum*	Microatmosphere, Macrodilution	–	10.0 μLmL^{-1}	10.0 μLmL^{-1}	Stupar et al. (2014)
				Trichoderma viride		–	100 μLmL^{-1}	100 μLmL^{-1}	
				Aspergillus niger		–	30.0 μLmL^{-1}	80.0 μLmL^{-1}	
				Aspergillus ochraceus		–	80.0 μLmL^{-1}	80.0 μLmL^{-1}	
				Bipolaris spicifera		–	100.0 μLmL^{-1}	100.0 μLmL^{-1}	
				Penicillium sp.		–	80.0 μLmL^{-1}	100.0 μLmL^{-1}	
				Trichoderma viride	Microdilution	–	80.0 μLmL^{-1}	100.0 μLmL^{-1}	
				Aspergillus niger		–	100.0 μLmL^{-1}	100.0 μLmL^{-1}	
				Aspergillus ochraceus		–	30.0 μLmL^{-1}	100.0 μLmL^{-1}	
				Bipolaris spicifera		–	100.0 μLmL^{-1}	100.0 μLmL^{-1}	
				Penicillium sp.		–	50.0 μLmL^{-1}	100.0 μLmL^{-1}	
Lamiaceae	*Lavandula angustifolia*	Aerial parts	Linalool (37.61 %), Linalool acetate (34.86 %)	*Epicoccum nigrum*	Microatmosphere, Macrodilution	–	20.0 μLmL^{-1}	40.0 μLmL^{-1}	Stupar et al. (2014)
				Trichoderma viride		–	80.0 μLmL^{-1}	100.0 μLmL^{-1}	
				Aspergillus niger		–	60.0 μLmL^{-1}	60.0 μLmL^{-1}	
				Aspergillus ochraceus		–	60.0 μLmL^{-1}	100.0 μLmL^{-1}	
				Bipolaris spicifera		–	55.0 μLmL^{-1}	100.0 μLmL^{-1}	
				Penicillium sp.		–	40.0 μLmL^{-1}	40.0 μLmL^{-1}	
				Trichoderma viride	Microdilution	–	75.0 μLmL^{-1}	75.0 μLmL^{-1}	
				Aspergillus niger		–	100.0 μLmL^{-1}	100.0 μLmL^{-1}	
				Aspergillus ochraceus		–	50.0 μLmL^{-1}	75.0 μLmL^{-1}	
				Bipolaris spicifera		–	100.0 μLmL^{-1}	100.0 μLmL^{-1}	
				Penicillium sp.		–	25.0 μLmL^{-1}	25.0 μLmL^{-1}	
Poaceae	*Cymbopogon citratus*	Fresh whole plan	Geranial (39.53 %), Neral (33.31 %), Myrcene (11.41 %)	*Aspergillus flavus*	Paper disc diffusion inhibition test	–	118 $mgmL^{-1}$	–	Shaaban et al. (2013)
				Aspergillus parasiticus		–	59 $mgmL^{-1}$	–	
				Aspergillus ochraceus		–	59 $mgmL^{-1}$	–	
				Aspergillus niger		–	15 $mgmL^{-1}$	–	
				Aspergillus fumigatus		–	59 $mgmL^{-1}$	–	
Lamiaceae	*Salvia aucheri* var. *mesatlantica*	Aerial parts	Camphor (49.0 %), 1,8-Cineole (9.5 %), Viridiflorol (8.8 %), Camphene (7.8 %)	*Alternaria* sp.	Poison food	>2 μLmL^{-1}	–	–	Laghchimi et al. (2014)
				Penicillium expansum		>2 μLmL^{-1}	–	–	
				Rhizopus stolonifer		>2 μLmL^{-1}	–	–	

(continued)

Table 2.1 (continued)

Family	Species	Plant part	Active compound	Pathogenic fungi	Method	100 % inhibitory concentration	MIC[a]	MFC	Reference
Asteraceae	*Erigeron karwinskianus*	Aerial parts	*trans*-2-cis-8-Matricariaester (25.4 %), germacrene-D (13.1 %), α-muurolene (6.4 %), β-elemene (6.0 %)	*Fusarium oxysporum*	Agar dilution	–	1000.0 $mgmL^{-1}$	–	Kumar et al. (2014)
				Rhizoctonia solani		–	NI	–	
				Sclerotinia sclerotiorum		–	NI	–	
				Helminthosporium maydis		–	1000.0 $mgmL^{-1}$	–	
				Alternaria solani		–	3000.0 $mgmL^{-1}$	–	
	Erigeron mucronatus		*trans*-2-cis-8-Matricariaester (62.1 %), *cis*-lachnophyllum ester (21.2 %)	*Fusarium oxysporum*		–	1000.0 $mgmL^{-1}$	–	
				Rhizoctonia solani		–	NI	–	
				Sclerotinia sclerotiorum		–	NI	–	
				Helminthosporium maydis		–	1000.0 $mgmL^{-1}$	–	
				Alternaria solani		–	3000.0 $mgmL^{-1}$	–	
	Erigeron annuus		*cis*-Lachnophyllum ester (68.1 %), germacrene-D (10.4 %)	*Fusarium oxysporum*		–	1000.0 $mgmL^{-1}$	–	
				Rhizoctonia solani		–	NI	–	
				Sclerotinia sclerotiorum		–	NI	–	
				Helminthosporium maydis		–	1000.0 $mgmL^{-1}$	–	
				Alternaria solani		–	3000.0 $mgmL^{-1}$	–	
Lamiaceae	*Mentha piperita*	Leaves	–	*Rhizoctonia solani*	Agar dilution	–	850 μLL^{-1}	1000 μLL^{-1}	Khaledi et al. (2015)
				Macrophomina phaseolina		–	950 μLL^{-1}	NI	
Apiaceae	*Bunium persicum*	Seeds	–	*Rhizoctonia solani*		–	1200 μLL^{-1}	1500 μLL^{-1}	
				Macrophomina phaseolina		–	975 μLL^{-1}	NI	
Lamiaceae	*Thymus vulgaris*	Leaves	–	*Rhizoctonia solani*		–	1100 μLL^{-1}	1500 μLL^{-1}	
				Macrophomina phaseolina		–	1150 μLL^{-1}	1500 μLL^{-1}	

Lamiaceae	*Thymus vulgaris*	Aerial parts	–	*Rhizopus oryzae*	Poisoned food	–	350 ppm/100 mL	650 ppm/100 mL	Mohammadi et al. (2013)
				Botrytis cinerea		–	150 ppm/100 mL	250 ppm/100 mL	
				Penicillium expansum		–	150 ppm/100 mL	350 ppm/100 mL	
				Rhizopus stolonifer		–	250 ppm/100 mL	450 ppm/100 mL	
	Thymus migricus	Aerial parts	–	*Rhizopus oryzae*		–	100 ppm/100 mL	350 ppm/100 mL	
				Botrytis cinerea		–	50 ppm/100 mL	100 ppm/100 mL	
				Penicillium expansum		–	50 ppm/100 mL	150 ppm/100 mL	
				Rhizopus stolonifer		–	100 ppm/100 mL	250 ppm/100 mL	
	Thymus koschyanus	Aerial parts	–	*Rhizopus oryzae*		–	100 ppm/100 mL	200 ppm/100 mL	
				Botrytis cinerea		–	50 ppm/100 mL	100 ppm/100 mL	
				Penicillium expansum		–	50 ppm/100 mL	150 ppm/100 mL	
				Rhizopus stolonifer		–	100 ppm/100 mL	150 ppm/100 mL	
	Mentha spicata	Aerial parts	–	*Rhizopus oryzae*		–	350 ppm/100 mL	1000 ppm/100 mL	
				Botrytis cinerea		–	100 ppm/100 mL	350 ppm/100 mL	
				Penicillium expansum		–	150 ppm/100 mL	450 ppm/100 mL	
				Rhizopus stolonifer		–	150 ppm/100 mL	650 ppm/100 mL	

(continued)

Table 2.1 (continued)

Family	Species	Plant part	Active compound	Pathogenic fungi	Method	100 % inhibitory concentration	MIC[a]	MFC	Reference
	Mentha pulegium	Aerial parts	–	*Rhizopus oryzae*		–	150 ppm/100 mL	350 ppm/100 mL	
				Botrytis cinerea		–	50 ppm/100 mL	100 ppm/100 mL	
				Penicillium expansum		–	50 ppm/100 mL	150 ppm/100 mL	
				Rhizopus stolonifer		–	100 ppm/100 mL	250 ppm/100 mL	
	Satureja mutica	Aerial parts	–	*Rhizopus oryzae*		–	250 ppm/100 mL	450 ppm/100 mL	
				Botrytis cinerea		–	50 ppm/100 mL	150 ppm/100 mL	
				Penicillium expansum		–	100 ppm/100 mL	250 ppm/100 mL	
				Rhizopus stolonifer		–	150 ppm/100 mL	350 ppm/100 mL	
Lamiaceae	*Melissa officinalis*	Aerial parts	1,2-Propanediol (31.26 %), 2-propanol 1,1 oxybis (23.30 %), dihyromycrenol (10.99 %), benzene 1,1-oxybis (6.20 %)	*Penicillium verrucosum*	Macrodilution broth	–	125 $\mu L mL^{-1}$	250 $\mu L mL^{-1}$	Ozcakmak et al. (2012)
Cupressaceae	*Juniperus communis*	Aerial parts	α.-Cedrol (19.44 %), 3,4A,7,7,10A-pentamethyl-3- vinyldodecahydro-1H-benzo (F)chromene (7.86 %)	*Rhizoctonia solani*	Mycelia radial growth inhibition	>1 $mg mL^{-1}$	–	–	Abbassy and Marei (2013)
Lamiaceae	*Thymus vulgaris*	Flowering tops and leaves	Thymol (45.35 %)	*Rhizopus stolonifer*	Agar dilution	>1 $mg mL^{-1}$	–	–	Abbassy and Marei (2013)
				Rhizoctonia solani		>1 $mg mL^{-1}$	–	–	
				Rhizopus stolonifer		>1 $mg mL^{-1}$	–	–	
Pinaceae	*Pinus halepensis*	Spines	Caryophyllene-oxide (48.15 %), thumbergol (8.34 %), humulene-oxide (6.67 %), phenethyl-valerate (5.79 %)	*Aspergillus niger*	Disk diffusion	>15 $\mu L mL^{-1}$	–	–	Abi-Ayad et al. (2011)
				A. flavus		>15 $\mu L mL^{-1}$	–	–	
				Fusarium oxysporum		>15 $\mu L mL^{-1}$	–	–	
				Rhizopus stolonifer		>15 $\mu L mL^{-1}$	–	–	

Lamiaceae	*Mentha piperita*	Leaves	Menthone (30.63 %), menthol (25.16 %), menthofuran (6.47 %), β-phellandrene (5.59 %)	*Drechslera spicifera*	Agar dilution	–	800 μLL^{-1}		Moghaddam et al. (2013)
				Fusarium oxysporum f. sp. *ciceris*		–	1600 μLL^{-1}	–	
				Macrophomina phaseolina		–	>1600 μLL^{-1}		
Lamiaceae	*Rosmarinus officinalis*	Leaves	1,8 Cineole (52.2 %), camphor (15.2 %), α-pinene (12.4 %)	*Fusarium verticillioides*	Agar dilution	–	150 μgmL^{-1}	150 μgmL^{-1}	Bomfim et al. (2014)
Compositae	*Inula britannica*	Flowers	–	*Alternaria alternata*	–	>1 $mgmL^{-1}$	–	–	Zhao et al. (2013)
				A. solani		>1 $mgmL^{-1}$	–	–	
				Botrytis cinerea		>1 $mgmL^{-1}$	–	–	
				Cercosporium personata		>1 $mgmL^{-1}$	–	–	
				Colletotrichum dematium		>1 $mgmL^{-1}$	–	–	
				C. orbiculare		>1 $mgmL^{-1}$	–	–	
				Fusarium graminearum		>1 $mgmL^{-1}$	–	–	
				F. monilifome		>1 $mgmL^{-1}$	–	–	
				Gaeumannomyces graminis var. tritici		1 $mgmL^{-1}$	–	–	
				Gibberella zeae		>1 $mgmL^{-1}$	–	–	
				Gloeosporium fructigenum		>1 $mgmL^{-1}$	–	–	
				Phytophthora capsici		>1 $mgmL^{-1}$	–	–	
				Pyricularia grisea		>1 $mgmL^{-1}$	–	–	
				Rhizoctonia cerealis		>1 $mgmL^{-1}$	–	–	
				Sclerotinia sclerotiorum		>1 $mgmL^{-1}$	–	–	
				Verticillium dahliae		>1 $mgmL^{-1}$	–	–	

(continued)

Table 2.1 (continued)

Family	Species	Plant part	Active compound	Pathogenic fungi	Method	100 % inhibitory concentration	MIC[a]	MFC	Reference
Apiaceae	*Angelica koreana*	Dried roots	Sabinene (31.85 %)	*Aspergillus flavus*	Broth dilution	–	1000–2000 μgmL^{-1}	–	JungHyun and SeungWon (2014)
				Aspergillus fumigatus		–	2000 μgmL^{-1}	–	
				Aspergillus niger		–	1000–2000 μgmL^{-1}	–	
				Aspergillus terreus		–	1000 μgmL^{-1}	–	
				Aspergillus versicolor		–	250 μgmL^{-1}	–	
				Trichophyton mentagrophytes		–	250 μgmL^{-1}	–	
				Trichophyton rubrum		–	500 μgmL^{-1}	–	
				Trichophyton tonsurans		–	250–500 μgmL^{-1}	–	

[a]>, ≥: 100 % inhibitory concentration was not reported in tested concentration
[b]Different ranges are related to various fungal strains

the summary of preceding results is reported which can help to design and manufacture new natural fungicidal products.

References

Abbassy MA, Marei GI (2013) Antifungal and chemical composition of essential oils of Juniperus communis L. and Thymus vulgaris L. against two phytopathogenic fungi. J Appl Sci Res 9(8):4584–4588

Abdel-Kader MM, El-Mougy NS, Lashin SM (2011) Essential oils and *Trichoderma harzianum* as an integrated control measure against faba bean root rot pathogens. J Plant Prot Res 51:306–313

Abdulmajeed SM (2015) Detection of oral candida species and investigation from anti-candida activity of *Origanum vulgare* L. essential oil. Middle-East J Sci Res 23(1):18–25

Abi-Ayad M, Abi-Ayad FZ, Lazzouni HA, Rebiahi SA, Ziani-Cherif C, Bessiere (2011) Chemical composition and antifungal activity of Aleppo pine essential oil. J Med Plants Res 5(22):5433–5436

Abou-Jawdah Y, Sobh H, Salameh A (2002) Antimycotic activities of selected plant flora, growing wild in Lebanon, against phytopathogenic fungi. J Agric Food Chem 50(11):3208–3213

Abraham S, Prakasam V (2000) Efficacy of botanicals on post harvest pathogens of carrot. J Mycol Plant Pathol 30:257

Abu-Darwish MS, Gonçalves MJ, Cabral C, Cavaleiro C, Salgueiro L (2013) Chemical composition and antifungal activity of essential oil from *Juniperus phoenicea* subsp. *Phoenicea* berries from Jordan. Acta Aliment (Budapest) 42(4):504–511

Adebayo O, Dang T, Bélanger A, Khanizadeh S (2013) Antifungal studies of selected essential oils and a commercial formulation against *Botrytis cinerea*. J Food Res 2(1):217–226

Agarwal M, Walia S, Dhingra S, Khambay BPS (2001) Insect growth inhibition antifeedant and antifungal activity of compounds isolated/derived from *Zingiber officinale* Roscoe (ginger) rhizomes. Pest Manag Sci 57:289–300

Aggarwal KK, Khanuja SPS, Ahmad A, Kumar TRS, Gupta VK, Kumar S (2002) Antimicrobial activity profiles of the two enantiomers of limonene and carvone isolated from the oils of *Mentha spicata* and *Anethum sowa*. Flavour Frag J 17:59–63

Agnese AM, Perez C, Cabrera JL (2001) *Adesmia aegiceras*: antimicrobial activity and chemical study. Phytomedicine 8(5):389–394

Aguiar RWS, Ootani MA, Ascencio SD, Ferreira TPS, Santos MMdos, Santos GRdos (2014) Fumigant antifungal activity of *Corymbia citriodora* and *Cymbopogon nardus* essential oils and citronellal against three fungal species. Sci World J 2014: Article ID 492138: 1–8: http://dx.doi.org/10.1155/2014/492138

Ahmad I, Aqil F, Owais M (2006) Modern phytomedicine. Wiley-VCH velarg GmbH & Co. KGaA, Gemany

Alvarez-Castellans PP, Bishop CD, Pascual-villalobos MJ (2001) Antifungal activity of the essential oil of flower heads of garland chrysanthemum (*Chrysanthemum coronarium*) against agricultural pathogens. Phytochemistry 57:99–102

Ambindei WA, Ngouné LT, Sameza ML, Sonwa ET, Nguimatsia F, Dongmo PMJ (2014) Antifungal activities against some *Aspergillus* species of the essential oils of *Canarium schweinfurthii* and *Aucoumea klaineana* growing in Cameroon. Int J Curr Microbiol Appl Sci 3(5):691–701

Amusa NA, Okechukwu RU, Akinfenwa B (2007) Reactions of cowpea to infection by *Macrophomina phaseolina* isolates from leguminous plants in Nigeria. Afr J Agric Res 2:73–75

Arici SE, Bozat G, Akbulut I (2013) Investigation of potential biological control of *Fusarium oxysporum* f.sp. *radicislycopersici* and *F. oxysporum* f.sp. *lycopersici* by essential oils, plant extract and chemical elicitors *in vitro*. Pak J Bot 45:2119–2124

Ayoola GA, Lawore FM, Adelowotan T, Aibinu IE, Adenipekun E, Coker HAB, Odugbemi TO (2008) Chemical analysis and antimicrobial activity of the essential oil of *Syzygium aromaticum* (clove). Afr J Microbiol Res 2:162–166

Badea ML, Delian E (2014) In vitro antifungal activity of the essential oils from *Artemisia* spp. L. on Sclerotinia sclerotiorum. Rom Biotechnol Lett 19(3):9345–9352

Bailey JA, O'Connell RJ, Pring RJ, Nash C (1992) Infection strategies of *Colletotrichum* species. In: Bailey JA, Jeger MJ (eds) *Colletotrichum*: biology, pathology and control. CAB International, Wallingford, pp 88–120

Bakkali F, Averbeck S, Averbeck D, Idaomar M (2008) Biological effects of essential oils – a review. Food Chem Toxicol 46:446–475

Balbi-Peoa MI, Becker A, Stangarlin JR, Franzener G, Lopes MC, Schwan-Estrada KRF (2006) Control of *Alternaria solani* in tomato by *Curcuma longa* extracts and curcumin – I. *In vitro* evaluation. Fitopatol Bras 31:310–314

Banihashemi Z, Abivardi C (2011) Evaluation of fungicidal and fungistatic activity of plant essential oils towards plant pathogenic and saprophytic fungi. Phytopathol Mediterr 50:245–256

Bansal RK, Gupta RK (2000) Evaluation of plant extracts against *Fusarium oxysporum* wilt pathogen of fenugreek. Indian Phytopathol 53:107–108

Barra A, Coroneo V, Dessi S, Cabras P, Angion A (2010) Chemical variability, antifungal and antioxidant activity of *Eucalyptus camaldulensis* essential oil from Sardinia. Nat Prod Commun 5:329–335

Baser KHC, Buchbauer G (2010) Handbook of essential oils science, technology and applications. CRC Press, Taylor & Francis Group, Boca Raton

Begnami AF, Duarte MCT, Furletti V, Rehder VLG (2010) Antimicrobial potential of *Coriandrum sativum* L. against different *Candida* species in vitro. Food Chem 118:74–77

Ben Ghnaya A, Hanana M, Amri I, Balti H, Gargouri S, Jamoussi B, Hamrouni L (2013) Chemical composition of *Eucalyptus erythrocorys* essential oils and evaluation of their herbicidal and antifungal activities. J Pest Sci 86(3):571–577

Bilia AR, Santomauro F, Sacco C, Bergonzi MC, Donato R (2014) Essential oil of *Artemisia annua* L.: an extraordinary component with numerous antimicrobial properties. Evid Based Complement Alternat Med 2014: Article ID 159819 1–7: http://dx.doi.org/10.1155/2014/159819

Blaschke T, Bjornsson T (1995) Pharmacokinetics and pharmacoepidemiology. Sci Am 8:1–14

Bomfim NDAS, Nakassugi LP, Oliveira JFP, Kohiyama CY, Mossini SAG, Grespan R, Nerilo SB, Mallmann CA, Filho BAA, Machinski M Jr (2014) Antifungal activity and inhibition of fumonisin production by Rosmarinus officinalis L. essential oil in Fusarium verticillioides (Sacc.) Nirenberg. Food Chem 166:330–336

Borrego S, Valdés O, Vivar I, Lavin P, Guiamet P, Battistoni P, Gomez de Saravia S, Borges P (2012) Essential oils of plants as biocides against microorganisms isolated from Cuban and Argentine documentary heritage. ISRN Microbiol 2012:1–7

Bouajaj S, Romane A, Benyamna A, Amri I, Hanana M, Hamrouni L, Romdhane M (2014) Essential oil composition, phytotoxic and antifungal activities of *Ruta chalepensis* L. leaves from High Atlas Mountains (Morocco). Nat Prod Res 28(21):1910–1914

Bouzabata A, Cabral C, Gonçalves MJ, Cruz MT, Bighelli A, Cavaleiro C, Casanova J, Tomi F, Salgueiro L (2015) *Myrtus communis* L. as source of a bioactive and safe essential oil. Food Chem Toxicol 75:166–172

Bouzenna H, Krichen L (2013) *Pelargonium graveolens* L'Her. and *Artemisia arborescens* L. essential oils: chemical composition, antifungal activity against *Rhizoctonia solani* and insecticidal activity against *Rhyzopertha dominica*. Nat Prod Res 27(9):841–846

Bowles EJ (2003) The chemistry of aromatherapeutic oils, 3rd edn. Griffin Press, South Australia

Bryant B, Knights K, Salerno E (2003) Pharmacology for health professionals. Mosby, Elsevier (Australia) Pty Ltd, Marrickville, p 2

Bu'Lock JD (1965) The biosynthesis of natural products. McGraw-Hill, New York

Bupesh G, Amutha C, Nandagopal S, Ganeshkumar A, Sureshkumar P, Saravana Murali K (2007) Antibacterial activity of *Mentha piperita* L. (peppermint) from leaf extracts – a medicinal plant. Acta Agric Slovenica 89:73–79

Calderone RA (2002) Introduction and historical perspectives. *Candida* and *Candidiasis*. ASM Press, Washington, DC, pp 15–25

Carmo ES, de Oliveira Lima E, de Souza EL (2008) The potential of *Origanum vulgare* L (Lamiaceae) essential oil in inhibiting the growth of some food-related *Aspergillus* species. Braz J Microbiol 39:362–367

Carson CF, Riley TV (1995) Antimicrobial activity of the major components of the essential oil of *Melaleuca alternifolia*. J Appl Bacteriol 78:264–269

Cavaleiro C, Salgueiro L, Gonçalves M-J, Hrimpeng K, Pinto J, Pinto E (2015) Antifungal activity of the essential oil of Angelica major against *Candida*, *Cryptococcus*, *Aspergillus* and dermatophyte species. J Nat Med 69:241–248

Cavanagh HMA (2007) Antifungal activity of the volatile phase of essential oils: a brief review. Nat Prod Commun 2:1297–1302

Chang ST, Wang SY, Wu SY, Chen PF (2000) Comparison of the antifungal activity of cadinane skeletal sesquiterpenoids from Taiwania (*Taiwania cryptomerioides* Hayata) heartwood. Holzforschung 54:241–245

Chang HT, Cheng YH, Wu CL, Chang ST, Chang TT, Su YC (2008) Antifungal activity of essential oil and its constituents from *Calocedrus macrolepis* var. formosana florin leaf against plant pathogenic fungi. Bioresour Technol 99:66–70

Chinou IB, Bougatsos C, Perdetzoglou D (2004) Chemical composition and antimicrobial activities of *Helichrysum amorginum* cultivated in Greece. J Essent Oil Res 16:243–245

Chutia M, Deka Bhuyan P, Pathak MG, Sarma TC, Boruah P (2009) Antifungal activity and chemical composition of *Citrus reticulata* Blanco essential oil against phytopathogens from Northeast India. LWT–Food Sci Technol 42:777–780

Combrinck S, Regnier T, Kamatou G (2011) In vitro activity of eighteen essential oils and some major components against common postharvest fungal pathogens of fruit. Ind Crops Prod 33:344–349

Ćosić J, Vrandečić K, Postić J, Jurković D, Ravlić M (2010) In vitro antifungal activity of essential oils on growth of phytopathogenic fungi. Poljoprivreda 16:25–28

Cox SD, Mann CM, Markham JL (2000) The mode of antimicrobial action of the essential oil of *Melaleuca alternifolia* (tea tree oil). J Appl Microbial 88:170–175

Croteau R (1987) Biosynthesis and catabolism of monoterpenoids. Chem Rev 87:929–954

Czaika VA (2013) Effective treatment of tinea corporis due to Trichophyton mentagrophytes with combined isoconazole nitrate and diflucortolone valerate therapy. Mycoses 56(1):30–32

Da Silva Bomfim N, Polis Nakassugi L, Faggion Pinheiro Oliveira J, Yumie Kohiyama C, Aparecida Galerani Mossini S, Grespan R, Botiao Nerilo S, Augusto Mallmann C, Alves Abreu Filho B, Machinski M Jr

(2015) Antifungal activity and inhibition of fumonisin production by *Rosmarinus officinalis* L. essential oil in *Fusarium verticillioides* (Sacc.) Nirenberg. Food Chem 166:330–336

Dagenais TRT, Keller NP (2009) Pathogenesis of *Aspergillus fumigatus* in invasive Aspergillosis. Clin Microbiol Rev 22(3):447–465

De Billerbeck VG, Roques CG, Bessiére J-M, Fonvieille J-L, Dargent R (2001) Effects of *Cymbopogon nardus* (L.) W. Watson essential oil on the growth and morphogenesis of *Aspergillus niger*. Can J Microbiol 47 (1):9–17

Deba F, Xuan T, Yasuda M, Tawata S (2008) Chemical composition and antioxidant, antibacterial and antifungal activities of the essential oils from *Bidens pilosa* Linn. var. Radiata. Food Control 19:346–352

Dikbas N, Kotan R, Dadasoglu F, Sahin F (2008) Control of *Aspergillus flavus* with essential oil and methanol extract of *Satureja hortensis*. Int J Food Microbiol 124:179–182

Dorman HJD, Deans SG (2000) Antimicrobial agents from plants: antibacterial activity of plant volatile oils. J Appl Microbiol 88:308–316

Dwivedi SK, Singh KP (1999) Fungitoxicity of some higher plant products against *Macrophomina phaseolina* (Tassi) Goid. Flavour Frag J 13 (6):397–399

Edris AE (2007) Pharmaceutical and therapeutic potentials of essential oils and their individual volatile constituents: a review. Phytother Res 21:308–323

Eldahshan OA (2015) Comparison of chemical and antimicrobial studies of Egyptian mandarin leaves and green branches volatile oil. Eur J Med Plants 5 (3):248–254

Elgorban AM, Bahkali AH, El-Metwally MA, Elsheshtawi M, Abdel-Wahab MA (2015) In vitro antifungal activity of some plant essential oils. Int J Pharmacol 11(1):56–61

Elshafie HS, Mancini E, Camele I, De Martino L, DeFeo V (2015) In vivo antifungal activity of two essential oils from Mediterranean plants against postharvest brown rot disease of peach fruit. Ind Crops Prod 66:11–15

El-Shiekh YWA, Nour El-Din AHM, Shaymaa AA, Karam EL-Din AZ (2012) Antifungal activity of some naturally occurring compounds against economically important phytopathogenic fungi. Nat Sci 10:114–123

Fichi G, Flamini G, Giovanelli F, Otranto D, Perrucci S (2007) Efficacy of an essential oil of *Eugenia caryophyllata* against *Psoroptes cuniculi*. Exp Parasitol 115:168–172

Filipowicz N, Kamiński M, Kurlenda J, Asztemborska M, Ochocka JR (2003) Antibacterial and antifungal activity of juniper berry oil and its selected components. Phytother Res 17:227–231

Filtenborg O, Frisvad JC, Thrane U (1996) Moulds in food spoilage. Int J Food Microbiol 33:85–102

Fontenelle RO, Morais SM, Brito EH, Brilhante RS, Cordeiro RA, Lima YC, Brasil NV, Monteiro AJ, Sidrim JJ, Rocha MF (2011) Alkylphenol activity against *Candida* spp. and *Microsporum canis*: a focus on the antifungal activity of thymol, eugenol and O-methyl derivatives. Molecules 16:6422–6431

Fraternale D, Giamperi L, Ricci D (2003) Chemical composition and antifungal activity of essential oil obtained from in vitro plants of *Thymus mastichina* L. J Essent Oil Res 15:278–281

Freires IA, Murata RM, Furletti VF, Sartoratto A, Mde Alencar S, Figueira GM, AdeO Rodrigues J, Duarte MCT, Rosalen PL (2014) *Coriandrum sativum* L. (Coriander) essential oil: antifungal activity and mode of action on *Candida* spp., and molecular targets affected in human whole-genome expression. PLoS One 9(6):e99086

Friedman M, Henika PR, Levin CE, Mandrell RE (2004) Antibacterial activities of plant essential oils and their components against *Escherichia coli* O157:H7 and *Salmonella enterica* in apple juice. J Agric Food Chem 52:6042–6048

Gachitua SA, Valiente C, Katy Diaz P, Hernandez JC, Uribe MM, Sanfeutes EV (2009) Identification and biological characterization of isolates with activity inhibitive against *Macrophomina phaseolina* (Tassi) Goid. Chilean J Agric Res 69:526–533

Gallucci MN, Carezzano ME, de Las M, Oliva M, Demo MS, Pizzolitto RP, Zunino MP, Zygadlo JA, Dambolena JS (2014) In vitro activity of natural phenolic compounds against fluconazole-resistant *Candida* species. A quantitative structure-activity relationship analysis. J Appl Microbiol 116 (4):795–804

Gõni P, Lopez P, Sanchez C, Gomez-Lus R, Becerril R, Nerin C (2009) Antimicrobial activity in the vapour phase of a combination of cinnamon and clove essential oils. Food Chem 116(4):982–989

Gormez O, Diler O (2014) *In vitro* antifungal activity of essential oils from *Thymbra*, *Origanum*, *Satureja* species and some pure compounds on the fish pathogenic fungus, *Saprolegnia parasitica*. Aquac Res 45 (7):1196–1201

Gustot T, Maillart E, Bocci M, Surin R, Trépo E, Degré D, Lucidi V, Taccone FS, Delforge ML, Vincent JL, Donckier V, Jacobs F, Moreno C (2014) Invasive aspergillosis in patients with severe alcoholic hepatitis. J Hepatol 60(2):267–274

Gwilt R (1994) Pharmacokinetics. In: Craig C, Stitzel R (eds) Modern pharmacology. Little, Brown & Co., Boston, pp 55–64

Gwinn KD, Ownley BH, Greene SE, Clark MM, Taylor CL, Springfield TN, Trently DJ, Green JF, Reed A, Hamilton SL (2010) Role of essential oils in control of *Rhizoctonia* damping-off in tomato with bioactive monarda herbage. Phytopathology 100:493–501

Hamini-Kadar N, Hamdane F, Boutoutaou R, Kihal M, Henni JE (2014) Antifungal activity of clove

(*Syzygium aromaticum* L.) essential oil against phytopathogenic fungi of tomato (*Solanum lycopersicum* L) in Algeria. J Exp Biol Agric Sci (JEBAS) 2 (5):447–454

Hammer K, Carson C, Riley T (2003) Antifungal activity of the components of *Melaleuca alternifolia* (teatree) oil. J Appl Microbiol 95(4):853–860

Harborne BJ (1987) Natural fungitoxins. Biologically active natural products. Proc Phytochem Soc Eur 27:195–211

Harborne JB, Ingham JL, King L, Payne M (1976) Isopentenyl-isoflavoneluteone as a preinfeccional antifungal agent in the genus *Lupinus*. Phytochemistry 15:1485–1487

Harris JP, Mantle PG (2001) Biosynthesis of ochratoxins by *Aspergillus ochraceus*. Phytochemistry 58:709–716

Harvey RB, Edrington TS, Kubena LF, Rottinghaus GE, Turk JR, Genovese KJ, Nisbet DJ (2001) Toxicity of moniliformin from *Fusarium fujikuroi* culture material to growing barrows. J Food Protect 64:1780–1784

Ho CL, Liao PC, Wang EI, Su YC (2011) Composition and antifungal activities of the leaf essential oil of *Neolitsea parvigemma* from Taiwan. Nat Prod Commun 6:1357–1360

Holley RA, Patel D (2005) Improvement in shelf-life and safety of perishable foods by plant essential oils and smoke antimicrobials. J Food Microbiol 22:273–292

Hong-Wei A, Yue-Xin K, Yang C, Cheng-Jian Z (2014) Antifungal properties and chemical analysis of essential oil from *Vitex negundo* seeds. Br J Pharm Res 4 (5):541–548

Hryncewicz-Gwóźdź A, Kalinowska K, Plomer-Niezgoda E, Maj J, Kołodziej T, Czarnecka A (2010) Tinea capitis and tinea cutis glabrae with severe inflammation due to Trichophyton tonsurans. Mikol Lek 17(1):57–60

Huang Y, Jianglin Z, Ligang Z, Jihua W, Youwen G, Xujun C, Zejian G, Qi W, Weibo J (2010) Antifungal activity of the essential oil of *Illicium verum* fruit and its main component trans-anethole. Molecules 15:7558–7569

Huang X, Zhang N, Yong X, Yang X, Shen Q (2012) Biocontrol of *Rhizoctonia solani* damping-off disease in cucumber with *Bacillus pumilus* SQR-N43. Microbiol Res 167:135–143

Hussein HS, Brasel JM (2001) Toxicity, metabolism, and impact of mycotoxins on humans and animals. Toxicology 167:101–134

Ibrahim NA, El-Sakhawy FS, Mohammed MMD, Farid MA, AbdelWahed NAM, Deabes DAH (2015) Chemical composition, antimicrobial and antifungal activities of essential oils of the leaves of *Aegle marmelos* (L.) Correa growing in Egypt. J Appl Pharm Sci (JAPS) 5(2):1–5

Imelouane B, Amhamdi H, Wathelet P, Ankit M, Khedid K, El Bachiri A (2009) Chemical composition of the essential oil of thyme (*Thymus vulgaris*) from Eastern Morocco. Int J Agric Biol 11:205–208

International School of Aromatherapy (1993) A safety guide on the use of essential oils. Natural by Nature Oils Ltd, London

İşcan G, Kirimer N, Demirci F, Demirci B, Noma Y, Başer KH (2012) Biotransformation of (-)-(R)-α-phellandrene: antimicrobial activity of its major metabolite. J Chem Biodivers 9:1525–1532

Jan AK, Anis I, Shah MR (2015) Chemical composition and antifungal activity of the essential oil of *Galium tricornutum* subsp. *longipedanculatum* from Pakistan. Chem Nat Compd 51(1):164–165

Javaid A, Naqvi SF (2012) Evaluation of antifungal potential of *Cenchrus pennisetiformis* for the management of *Macrophomina phaseolina*. World Acad Sci Eng Technol 69:32–835

Jiang L (2011) Comparison of disk diffusion, agar dilution, and broth microdilution for antimicrobial susceptibility testing of five chitosans. Ph.D. dissertation, Louisiana State University, United States

Jing L, Lei Z, Zhou Z, Li L, Xie R, Xi W, Guan Y, Sumner LW (2014) Antifungal activity of citrus essential oils. J Agric Food Chem 62:3011–3033

José Goncalvesa M, Teresa Cruzb M, Cristina Tavaresc A, Cavaleiroa C, Celeste Lopesb M, Canhotoc J, Salgueiroa L (2011) Composition and biological activity of the essential oil from Thapsia minor, a new source of geranyl acetate. Ind Crop Prod 35:166–171

JungHyun R, SeungWon S (2014) Antifungal and antioxidant activities of the essential oil from *Angelica koreana* Nakai. Evid Based Complement Alternat Med 2014: Article ID 398503: 1–7: http://dx.doi.org/10.1155/2014/398503

Kalemba D, Kunicka A (2003) Antibacterial and antifungal properties of essential oils. Curr Med Chem 10:813–829

Kamanzi Atindehou K, Koné M, Terreaux C, Traore D, Hostettmann K, Dosso M (2002) Evaluation of the antimicrobial potential of medicinal plants from the Ivory Coast. Phytother Res 16(5):497–502

Kamatou GP, Vermaak I, Viljoen AM (2012) Eugenol—from the remote maluku islands to the international market place: a review of a remarkable and versatile molecule. Molecules 17:6953–6981

Khaledi N, Taheri P, Tarighi S (2015) Antifungal activity of various essential oils against *Rhizoctonia solani* and *Macrophomina phaseolina* as major bean pathogens. J Appl Microbiol 118(3):704–717

Kocic-tanackov S, Dimic G, Levic J, Tanackov I, Tepic A, Vujicic B, Gvozdanovic-varga J (2012) Effects of onion (*Allium cepa* L.) an garlic (*Allium sativum* L.) essential oils on the *Aspergillus versicolor* growth and sterigmatocystin production. J Food Sci 77:278–284

Koutsoudaki C, Krsek M, Rodger A (2005) Chemical composition and antibacterial activity of the essential oil and the gum of *Pistacia lentiscus* var. chia. J Agric Food Chem 53:7681–7685

Kumar V, Mathela CS, Tewari G, Singh D, Tewari AK, Bisht KS (2014) Chemical composition and antifungal activity of essential oils from three Himalayan *Erigeron* species. LWT Food Sci Technol 56(2):278–283

Kurtzman CP, Piškur J (2006) Taxonomy and phylogenetic diversity among the yeasts. Springer, Berlin, pp 29–46

Laghchimi A, Znini M, Majidi L, Paolini J, Desjobert JM, Costa J (2014) Liquid and vapour-phase antifungal activities of essential oil of *Salvia aucheri* Boiss. var. *mesatlantica* Maire. (endemic from Morocco) against fungi commonly causing deterioration of apple. Der Pharm Chem 6(1):370–378

Lahlou M (2004) Methods to study the phytochemistry and bioactivity of essential oils. Phytother Res 18:435–448

Lang D (1998) Europe's medicinal and aromatic plants: their use trade and conservation. Traffic International Goldaming, Surrey, p 119

Lang G, Buchbauer G (2012) A review on recent research results (2008–2010) on essential oils as antimicrobials and antifungals. A review. Flavour Frag J 27:13–39

Latgé J-P, Steinbach W (2009) Aspergillus fumigatus and aspergillosis. ASM Press, Washington, DC

Lee H-C, Cheng S-S, Chang S-T (2005) Antifungal property of the essential oils and their constituents from *Cinnamomum osmophloeum* leaf against tree pathogenic fungi. J Sci Food Agric 85(12):2047–2053

Lee SO, Choi GJ, Jang KS, Kim JC (2007) Antifungal activity of five plant essential oils as fumigant against postharvest and soilborne plant pathogenic fungi. Plant Pathol J 23:97–102

Leyngold I, Olivi A, Ishii M, Blitz A, Burger P, Subramanian PS, Gallia G (2014) Acute chiasmal abscess resulting from perineural extension of invasive sino-orbital aspergillosis in an immunocompetent patient. World Neurosurg 81(1):203–206

Lichter A, Zutchy Y, Sonego L, Dvir O, Kaplunov T, Sarig P, Ben-Arie R (2002) Ethanol controls postharvest decay of table grapes. Postharvest Biol Technol 24:301–308

Liu WT, Chu CL (2002) Thymol and acetic acid vapors reduce postharvest brown Rot of apricots and plums. HortSci 37:151–156

Mahmoud YAG, Aly AA, Omar MR, Ismail AWA (2006) Variation in sensitivity among some isolates of *Macrophomina phaseolina* isolated from cotton roots to flutolanil fungicide. Mycobiology 34:99–103

Mahmoudvand H, Sepahvand A, Jahanbakhsh S, Ezatpour B, Mousavi SAA (2014) Evaluation of antifungal activities of the essential oil and various extracts of *Nigella sativa* and its main component, thymoquinone against pathogenic dermatophyte strains. J Mycol Méd 24(4):155–161

Mann J, Davidson RS, Hobbs JB, Banthorpe DV, Harbourne JB (1994) Natural products: their chemistry and biological significance. Longman, London

Mapelli ET, Borghi E, Cerri A, Sciota R, Morace G, Menni S (2012) Tinea corporis due to *Trichophyton rubrum* in a woman and Tinea capitis in her 15-day-old baby: molecular evidence of vertical transmission. Mycopathologia 173(2–3):135–138

Mashigo M, Combrinck S, Regnier T, du Plooy W, Augustyn W, Mokgalaka N (2015) Chemical variations, trichome structure and antifungal activities of essential oils of *Helichrysum splendidum* from South Africa. S Afr J Bot 96:78–84

Matan N, Rimkeeree H, Mawson AJ, Chompreeda P, Haruthaithanasan V, Parker M (2006) Antimicrobial activity of cinnamon and clove oils under modified atmosphere conditions. Intl J Food Microbiol 107 (2):180–185

Matasyoh JC, Kiplimo JJ, Karubiu NM, Hailstorks TP (2007) Chemical composition and antimicrobial activity of essential oil of *Tarchonanthus camphorates*. Food Chem 101:1183–1187

Matos OC (2000) Uso de substâncias naturais de origem vegetal com actividade biológica na protecc-aõ das culturas agrícolas. Agron Lusitânica 48(2):1–44

Mayek-Pérez N, Lopez-Castaoeda C, Lopez-Salinas E, Cumpion-Gutiérrez J, Joaquan-Torres IC, Padilla-Ramdrez JS, Acosta-Gallegos JA (2003) Effect of *Macrophomina phaseolina* (Tassi) Goid. on grain yield of common beans (*Phaseolus vulgaris* L.) and its relationship with yield stability parameters. Rev Mex Fitopatol 21:168–175

Mc Guinness H (2007) Aromatherapy therapy basics, 2nd edn. Hodder Arnold, India

Merghache D, Boucherit-Otmani Z, Merghache S, Chikhi I, Selles C, Boucherit K (2014) Chemical composition, antibacterial, antifungal and antioxidant activities of Algerian *Eryngium tricuspidatum* L. essential oil. Nat Prod Res 28(11):795–807

Messgo-Moumene S, Li Y, Bachir K, Houmani Z, Bouznad Z, Chemat F (2015) Antifungal power of citrus essential oils against potato late blight causative agent. J Essent Oil Res 27(2):169–176

Mickiene R, Bakutis B, Maruška A, Ragazinskiene O, Kaškoniene V (2014) Effect of the volatile secondary metabolites of *Monarda didyma* L., *Lamium album* L. and *Myrrhis odorata* L. plants against micromycetes of indoor environments of animals. Vet Zootech Lith 68:48–54

Mihail JD, Taylor SJ (1995) Interpreting of variability among isolates of *Macrophomina phaseolina* in pathogenicity, pycnidium production, and chlorate utilization. Can J Bot 73:1596–1603

Moghaddam M, Pourbaige M, Kourosh Tabar H, Farhadi N, Ahmadi Hosseini SM (2013) Composition and antifungal activity of peppermint (*Mentha piperita*) essential oil from Iran. J Essent Oil Bearing Plant 16:506–512

Moghaddam M, Taheri P, Ghasemi Pirbalouti A, Mehdizadeh L (2015) Chemical composition and antifungal activity of essential oil from the seed of Echinophora platyloba DC. against phytopathogens fungi by two different screening methods. LWT -ood Sci Technol 61:536–542

Mohammadi P, Lotfi N, Naseri L, Etebarian HR (2013) Antifungal activities of essential oils from some Iranian medicinal plants against various post harvest moulds. J Med Plants Res 7(23):1699–1708

Molenberg D, Deleuran M, Sommerlund M (2010) Connubial tinea gladiatorum due to *Trichophyton mentagrophytes*. Mycoses 53(6):533–534

Moreno HP, Lima MEL, Caruzo MBR, Torres DSC, Young MCM (2009) Chemical composition and antimicrobial activity of the essential oil from *Croton heterocalyx* Baill. (Euphorbiaceae s.s.) leaves. J Essent Oil Res 21:190–192

Nakahara K, Alzoreky NS, Yoshihashi T, Nguyen HTT, Trakoontivakorn G (2003) Chemical composition and antifungal activity of essential oil from Cymbopogon nardus (citronella grass). Jpn Agric Res Q 37(4):249–252

Nguefack J, Wulff GE, Dongmo JBL, Fouelefack FR, Fotio D, Mbo J, Torp J (2013) Effect of plant extracts and an essential oil on the control of brown spot disease, tillering, number of panicles and yield increase in rice. Eur J Plant Pathol 137:871–882

Nikraftar F, Taheri P, Falahati Rastegar M, Tarighi S (2013) Tomato partial resistance to *Rhizoctonia solani* involves antioxidative defense mechanisms. Physiol Mol Plant Pathol 81:74–83

Nissen L, Zatta A, Stefanini I, Grandi S, Sgorbati B, Biavati B, Monti A (2010) Characterization and antimicrobial activity of essential oils of industrial hemp varieties (*Cannabis sativa* L.). Fitoterapia 81:413–419

Nosrati S, Esmailzadeh-Hosseini SA, Sarpeleh A, Soflaei Shahrbabak M, Soflaei Shahrbabak Y (2011) Antifungal activity of spearmint (*Mentha spicata L.*) essential oil on *Fusarium oxysporum* f. sp. *radicis-cucumerinum* the causal agent of stem and crown rot of greenhouse cucumber in Yazd, Iran. In: International Conference on Environmental and Agricultural Engineering, Chengdu, China held on pp 52–56

Nyegue MA, Ndoyé-Foe FMC, Riwom SE, Hockmeni TC, Etoa FX, Menut C (2014) Chemical composition of essential oils of *Eugenia caryophylla* and *Mentha piperita* and their *in vitro* antifungal activities on six human pathogenic fungi. Afr J Tradit Complement Altern Med 11(6):40–46

Omoruyi BE, Bradley G, Afolayan AJ (2014) Chemical composition profiling and antifungal activity of the essential oil and plant extracts of *Mesembryanthemum edule* (L.) bolus leaves. Afr J Tradit Complement Altern Med 11(4):19–30

Ozcakmak S, Dervisoglu M, Yilmaz A (2012) Antifungal activity of lemon balm and sage essential oils on the growth of ochratoxigenic *Penicillium verrucosum*. Afr J Microbiol Res 6(12):3079–3084

Pauli A, Schilcher H (2010) In vitro antimicrobial activities of essential oils monographed in the European pharmacopoeia. In: Baser KHC, Buchbauer G (eds) Handbook of essential oils: science, technology, and applications, 6th edn. Taylor & Francis, Boca Raton, pp 353–547

Paviani L, Pergher SBC, Dariva C (2006) Application of molecular sieves in the fractionation of lemongrass oil from high-pressure carbon dioxide extraction. Braz J Chem Eng 23:219–225

Pfaller MA, Diekema DJ (2004) Rare and emerging opportunistic fungal pathogens: concern for resistance beyond *Candida albicans* and *Aspergillus fumigatus*. J Clin Microbiol 42:4419–4431

Pinto E, Vale-Silva L, Cavaleiro C, Salgueiro L (2009) Antifungal activity of the clove essential oil from *Syzygium aromaticum* on *Candida*, *Aspergillus* and dermatophyte species. J Med Microbiol 58:1454–1462

Pitarokili D, Tzakou O, Loukis A, Harvala C (2003) Volatile metabolites from *Salvia fruticosa* as antifungal agents in soilborne pathogens. J Agric Food Chem 51:3294–3301

Plodpai P, Chuenchitt S, Petcharat V, Chakthong S, Voravuthikunchai SP (2013) Anti-*Rhizoctonia solani* activity by *Desmos chinensis* extracts and its mechanism of action. Crop Prot 43:65–71

Pragadheesh VS, Saroj A, Yadav A, Samad A, Chanotiya CS (2013) Compositions, enantiomer characterization and antifungal activity of two *Ocimum* essential oils. Ind Crops Prod 50:333–337

Price S, Price L (1999) Aromatherapy for health professionals. Churchill Livingstone, London

Rahimi AA, Ashnagar A, Nikoei H (2012) Isolation and characterization of 4-allyl-2-methoxyphenol (eugenol) from clove buds marketed in Tehran city of Iran. Int J Chem Tech Res 4:105–108

Rai M, Cecilia Carpinella M (2006) Advances in phytomedicine series, vol 3. Naturally occurring boioactive compounds. Elsevier, Amsterdam, pp 1–502

Rana BK, Singh UP, Taneja V (1997) Antifungal activity and kinetics of inhibition by essential oil isolated from *Aegle marmelos*. J Ethnopharmacol 57:29–34

Rana IS, Rana AS, Rajak RC (2011) Evaluation of antifungal activity in essential oil of the *Syzygium aromaticum* (L.) by extraction, purification and analysis of its main component eugenol. Braz J Microbiol 42:1269–1277

Rao GP, Sharma SR, Singh PK (2000) Fungitoxic and insect repellent efficacy of limonene against sugarcane pests. J Essent Oil Bearing Plant 3:157–163

Rawal P, Thakore BBL (2000) Effect of plant extracts on *Fusarium* rot of sponge gourd. J Mycol Plant Pathol 30:278

Rawlins MD (1989) Clinical pharmacology of the skin. In: Turner P (ed) Recent advances in clinical and pharmacology and toxicology. Churchill Livingstone, Edinburgh, pp 121–135

Rivas da Silva AC, Lopes PM, Barros de Azevedo MM, Costa DC, Alviano CS, Alviano DS (2012) Biological activities of α-pinene and β-pinene enantiomers. Molecules 17:6305–6316

Romagnoli C, Bruni R, Andreotti E, Rai MK, Vicentini CB, Mares D (2005) Chemical characterization and antifungal activity of essential oil of capitula from wild Indian *Tagetes patula* L. Protoplasma 225:57–65

Santos TG, Laemmle J, Rebelo RA, Dalmarco EM, Cruz AB, Schmit AP, Cruz RCB, Zeni ALB (2015) Chemical composition and antimicrobial activity of *Aloysia gratissima* (Verbenaceae) leaf essential oil. J Essen Oil Res 27(2):125–130

Scallan E, Hoekstra RM, Angulo FJ, Tauxe RV, Widdowson MA, Roy SL, Jones JL, Griffin PM (2011) Foodborne illness acquired in the United States-major pathogens. Emerg Infect Dis 17:7–15

Schadler DL, George AA (2006) Synthesis and bioassay of a volatile fungistatic agent. The Plant Health Instructor

Schwan-Estrada KRF, Stangarlin JR, Cruz MEDS (2000) Uso de extratos vegetais no controle de fungos fitopatogênicos. Floresta 30:129–137

Seema M, Devaki NS (2010) Effect of some essential oils on *Rhizoctonia solani* Kuhn infecting flue-cured virginia tobacco. J Biopest 3:563–566

Sekine T, Sugano M, Majid A, Fujii Y (2007) Antifungal effects of volatile compounds from Black Zira (*Bunium persicum*) and other spices and herbs. J Chem Ecol 33:2123–2132

Shaaban HA, Ramadan MM, Amer MM, El-Sideek LE, Osman F (2013) Chemical composition of *Cymbopogon citratus* essential oil and antifungal activity against some species of mycotoxigenic *Aspergillus* fungi. J Appl Sci Res 9(11):5770–5779

Sharifi-Rad J, Hoseini-Alfatemi SM, Sharifi-Rad M, Setzer WN (2015) Chemical composition, antifungal and antibacterial activities of essential oil from *Lallemantia royleana* (Benth. in Wall.) Benth. J Food Saf 35(1):19–25

Sharma N, Tripathi A (2008) Effects of *Citrus sinensis* (L.) osbeck epicarp essential oil on growth and morphogenesis of *Aspergillus niger* (L.) Van Tieghem. Microbiol Res 163(3):337–344

Silva NCC, Fernandes JA (2010) Biological properties of medicinal plants: a review of their antimicrobial activity. J Venom Anim Toxins 16:402–413

Silva S, Negri M, Henriques M, Oliveira R, Williams DW, Azeredo J (2012) *Candida glabrata*, *Candida parapsilosis* and *Candida tropicalis*: biology, epidemiology, pathogenicity and antifungal resistance. FEMS Microbiol Rev 36:288–305

Silva JKR, Silva JRA, Nascimento SB, da Luz SFM, Meireles EN, Alves CN, Ramos AR, Maia JGS (2014) Antifungal activity and computational study of constituents from *Piper divaricatum* essential oil against *Fusarium* infection in black pepper. Molecules 19(11):1792–17942

Singh UP, Sarma BK, Mishra PK, Ray AB (2000) Antifungal activity of venenatine an indole alkaloid isolated from *Alstonia venenata*. Folia-Microbiol 45:173–176

Singh G, Singh OP, Maurya S (2002) Chemical and biocidal investigations on essential oils of some Indian *Curcuma* species. Prog Cryst Growth Charact Mater 45:75–81

Sirirat S, Wimolpun R, Sanit S (2009) Antifungal activity of essential oils derived from some medicinal plants against grey mould (*Botrytis cinerea*). As J Food Ag-Ind Special Issue, S229–S233.

Soković M, Griensven LJLD (2006) Antimicrobial activity of essential oils and their components against the three major pathogens of cultivated button mushroom *Agaricus bisporus*. Eur J Plant Pathol 116:211–224

Sourmaghi MHS, Kiaee G, Golfakhrabadi F, Jamalifar H, Khanavi M (2015) Comparison of essential oil composition and antimicrobial activity of *Coriandrum sativum* L. extracted by hydrodistillation and microwave-assisted hydrodistillation. J Food Sci Technol 52(4):2452–2457

Soylu EM, Kurt S, Soylu S (2010) *In vitro* and *in vivo* antifungal activities of the essential oils of various plants against tomato grey mould disease agent *Botrytis cinerea*. Int J Food Microbiol 143:183–189

Sreedevi B, Charitha Devi M, Saigopal DVR (2011) Isolation and screening of effective *Trichoderma* spp. against the root rot pathogen *Macrophomina phaseolina*. J Agric Technol 7:623–635

Stević T, Berić T, Savikin K, Soković M, Godevac D, Dimkić I, Stanković S (2014) Antifungal activity of selected essential oils against fungi isolated from medicinal plant. Ind Crops Prod 55:116–122

Stringaro A, Vavala E, Colone M, Pepi F, Mignogna G, Garzoli S, Cecchetti S, Ragno R, Angiolella L (2014) Effects of *Mentha suaveolens* essential oil alone or in combination with other drugs in *Candida albicans*. Evid Based Complement Alternat Med 2014: Article ID 125904: 1–9: http://dx.doi.org/10.1155/2014/125904

Stupar M, Grbić ML, Dzamić A, Unković N, Ristić M, Jelikić A, Vukojević J (2014) Antifungal activity of selected essential oils and biocide benzalkonium chloride against the fungi isolated from cultural heritage objects. S Afr J Bot 93:118–124

Su G, Suh SO, Schneider RW, Russin JS (2001) Host specialization in the charcoal rot fungus, *Macrophomina phaseolina*. Phytopathology 91:120–126

Tabassum N, Vidyasagar GM (2013) Antifungal investigations on plant essential oils. A review. Int J Pharm Pharma Sci 5:19–28

Tadtong S, Wannakhot P, Poolsawat W, Athicomkulchai S, Ruangrunsi N (2009) Antimicrobial activity of essential oil from *Etlingera punicea* rhizome. J Health Res 23(2):77–79

Talas-Ogras T, Ipekci Z, Bajrovic K, Gozukirmizi N (2005) Antibacterial activity of seed proteins of *Robinia pseudoacacia*. Fitoterapia 76(1):67–72

Tavares AC, Gonçalves MJ, Cruz MT, Cavaleiro C, Lopes MC, Canhoto J, Salgueiro LR (2010) Essential oils from *Distichoselinum tenuifolium*: chemical composition, cytotoxicity, antifungal and anti-inflammatory properties. J Ethnopharmacol 130:593–598

Tejeswini MG, Sowmya HV, Swarnalatha SP, Negi PS (2014) Antifungal activity of essential oils and their combinations in *in vitro* and *in vivo* conditions. Arch Phytopathol Plant Protect 47(5):564–570

Thara S, Ingham JL, Nakahara S, Mizutani J, Harborne JB (1984) Fungitoxic dihydrofuranoisoflavones and related compounds in white lupine, *Lupinus albus*. Phytochemistry 23:1889–1900

Thind TS, Dhaliwal HS, Arora JK, Mohan C (2005) Role of plant products in ecofriendly management of plant disease. In: Rai MK, Deshmukh SK (eds) Fungi: diversity and biotechnology. Scientific Publishers (India), Jodhpur, pp 403–404

Thompson DP (1989) Fungitoxic activity of essential oil components on food storage fungi. Mycologia 81:151–153

Tian J, Ban X, Zeng H, He J, Huang B, Youwei W (2011) Chemical composition and antifungal activity of essential oil from *Cicuta virosa* L. var. latisecta Celak. Int J Food Microbiol 145:464–470

Tian J, Huang B, Luo X, Zeng H, Ban X, He J, Wang Y (2012) The control of *Aspergillus flavus* with *Cinnamomum jensenianum* Hand.-Mazz essential oil and its potential use as a food preservative. Food Chem 130:520–527

Tisserand R, Balacs T (1995) Essential oil safety: a guide for health professionals. Churchill Livingstone, Edinburgh, p 199

Torres P, Avila JG, Romo de Vivar A, Garcı a AM, Marín JC, Aranda E, Céspedes CL (2003) Antioxidant and insect growth regulatory activities of stilbenes and extracts from *Yucca periculosa*. Phytochemistry 64:463–473

Tsao R, Zhou T (2000) Antifungal activity of monoterpenoids against postharvest pathogens *Botrytis cinerea* and *Monilinia fructicola*. J Essential Oil Res 12:113–121

Valente J, Zuzarte M, Gonçalves MJ, Lopes MC, Cavaleirom C, Salgueiro L, Cruz MT (2013) Antifungal, antioxidant and anti-inflammatory activities of *Oenanthe crocata* L. essential oil. Food Chem Toxicol 62:349–354

Varela NP, Friendship R, Dewey C, Valdivieso A (2008) Comparison of agar dilution and E-test for antimicrobial susceptibility testing of *Campylobacter coli* isolates recovered from 80 Ontario swine farms. Can J Vet Res 72:168–174

Venturi CR, Danielli LJ, Klein F, Apel MA, Montanha JA, Bordignon SAL, Roehe PM, Fuentefria AM, Henriques AT (2015) Chemical analysis and *in vitro* antiviral and antifungal activities of essential oils from *Glechon spathulata* and *Glechon marifolia*. Pharm Biol 53(5):682–688

Veraldi S, Chiaratti A, Harak H (2010) Onychomycosis caused by *Aspergillus versicolor*. Mycoses 53 (4):363–365

Vieira PRN, Morais SM, Bezerra FHQ, Ferreira PAT, Oliveira ÍR, Silva MGV (2014) Chemical composition and antifungal activity of essential oils from *Ocimum* species. Ind Crops Prod 55:267–271

Walton NJ, Brown DE (eds) (1999) Chemicals from plants: perspectives on plant secondary product. Imperial College Press and World Scientific, London

WHO (2005) National policy on traditional medicine and regulation of herbal medicines. WHO Press, Geneva

Woodfolk JA (2005) Allergy and dermathophytes. Clin Microbiol Rev 18(1):30–43

Wu L, Sun P, Chang Y (2013) Extensive deep dermatophytosis cause by *Trichophyton rubrum* in a patient with liver cirrhosis and chronic renal failure. Mycopathologia 176:457–462

Xianfei X, Xiaoqiang C, Shunying Z, Guolin Z (2007) Chemical composition and antimicrobial activity of essential oils of *Chaenomeles speciosa* from China. Food Chem 100:1312–1315

XingDong L, HuaLi X (2014) Antifungal activity of the essential oil of *Zanthoxylum bungeanum* and its major constituent on *Fusarium sulphureum* and dry rot of potato tubers. Phytoparasitica 42(4):509–517

Yotova I, Ignatova-Ivanova T (2015) *In vitro* study of antifungal activity of oregano (*Origanum vulgare*). Int J Curr Microbiol App Sci 4(3):321–326

Zambonelli A, D'aulerio AZ, Severi A, Benvenuti S, Maggi L, Bianchi A (2004) Chemical composition and fungicidal activity of commercial essential oils of *Thymus vulgaris* L. J Essent Oil Res 16:69–74

Zhang M, Su X, Sun W-K, Chen F, Xu X-Y, Shi Y (2014) Efficacy of the combination of voriconazole and caspofungin in experimental pulmonary Aspergillosis by different *Aspergillus* species. Mycopathologia 77 (1–2):11–18

Zhao T, Fei G, Lin Z, TianYou S (2013) Essential oil from *Inula britannica* extraction with SF-CO_2 and its antifungal activity. J Integr Agr 12(10):1791–1798

Zhu SY, Yang Y, Yu HD, Ying Y, Zou GL (2005) Chemical composition and antimicrobial activity of the essential oils of *Chrysanthemum indicum*. J Ethnopharmacol 96(1–2):151–158

Zuzarte M, Vale-Silva L, Gonçalves MJ, Cavaleiro C, Vaz S, Canhoto J, Pinto E, Salgueiro L (2012) Antifungal activity of phenolic-rich *Lavandula multifida* L. essential oil. Eur J Clin Microbiol Infect Dis 31:1359–1366

Antifungal Peptides with Potential Against Pathogenic Fungi 3

Camila G. Freitas and Octávio L. Franco

Abstract

Systemic fungal infections have increased over time due to the rise in the at-risk population, which includes immunocompromised patients, those submitted to organ transplantation or undergoing chemotherapy. Clinically available antifungals are limited since some of them have important side effects, being toxic to the host cells, and some can quell filamentous fungi, but their activity against pathogenic yeasts is not killing but controlling their multiplication. Antimicrobial peptides are multifunctional molecules expressed by several microorganisms or synthetized by different techniques. They can play a central role in infection and inflammation. Some of their other effects include chemotactic and immunomodulating activities and wound repair. Antimicrobial peptides (AMPs) can be isolated from a large variety of microorganisms, such as plants, vertebrates, insects, bacteria, and fungi. They are classified into categories according to their amino acid composition, size, and conformational structures, and naturally occurring peptides can be synthetized. Solid-phase peptide synthesis allows the use of nonproteinogenic amino acids and permits changes in structural and physicochemical properties. In these terms, peptide engineering is a useful tool to adjust features such as net charge, surface hydrophobicity, and polarity, and it may also optimize activity and overcome the limitations inherent to natural peptides. AMPs have potential applications in antifungal therapeutics in human health, and recent uses of synthetic AMPs against fungal infections are discussed in this article.

C.G. Freitas
Instituto Federal de Brasília, Brasilia, Brazil

Centro de Análises Proteômicas e Bioquímicas, Pós-Graduação em Ciências Genômicas e Biotecnologia, Universidade Católica de Brasília, Brasilia, Brazil

O.L. Franco (✉)
Centro de Análises Proteômicas e Bioquímicas, Pós-Graduação em Ciências Genômicas e Biotecnologia, Universidade Católica de Brasília, Brasilia, Brazil

S-Inova Biotech, Pos-Graduação em Biotecnologia, Universidade Catolica Dom Bosco, Campo Grande MS, Brazil
e-mail: ocfranco@gmail.com

A. Basak et al. (eds.), *Recent Trends in Antifungal Agents and Antifungal Therapy*,
DOI 10.1007/978-81-322-2782-3_3

Abbreviations

ABP-dHC	Antimicrobial peptide drury *Hyphantria cunea*
AIDS	Acquired immune deficiency syndrome
AMPs	Antimicrobial peptides
ATP	Adenosine triphosphate
BMAP	Bovine myeloid antimicrobial peptides
DNA	Desoxyribonucleic acid
EDMC	Electrostatically driven Monte Carlo
FDA	US Food and Drug Administration
GPI	Glycosylphosphatidylinositol
HIV	Human immunodeficiency virus
hLF	Human lactoferrin
HNP	Human neutrophils peptides
HP	Human defensins
MDR	Multidrug-resistant
MPTP	Mitochondrial permeability forming transition pores
PI	Propidium iodate
PMAP	Porcine myeloid antimicrobial peptide
SMAP	Sheep myeloid antimicrobial peptide

3.1 Introduction

Pathogenic fungal infections are the seventh most common cause of infection-related deaths in the United States, and the fourth cause of nosocomial infection is due to the fungal pathogen *Candida albicans* (McNeil et al. 2001; Fisher et al. 2012; Wisplinghoff et al. 2004). Systemic mycoses can be classified according to whether the causative agent is a systemic fungal pathogen (*Coccidioides immitis*, *Histoplasma capsulatum*, and *Paracoccidioides brasiliensis*) or one of the increasing number of opportunistic fungal pathogens, including *C. albicans*, *Cryptococcus neoformans* var. *grubii* (Maurya et al. 2011; Rodriguez-Cerdeira et al. 2014), and several other ones. Those infections have become more frequent among the increasingly large population of individuals with severe immune deficiencies, including those with human immunodeficiency virus (HIV) and acquired immune deficiency syndrome (AIDS) (Martinez and Temesgen 2006; Marukutira et al. 2014).

The treatments of those systemic fungal infections are primarily based on itraconazole, fluconazole, or amphotericin B (Rodriguez-Cerdeira et al. 2014; Kahn et al. 2014). The triazole group enhances the specificity for fungal cytochrome P450 target, and the extra methyl group in fluconazole enhances the hydrophobic interactions at the active site, but they present limitations of toxicity or bioavailability problems that affect their potential use as systemic agents (Fukuoka et al. 2003; Ostrosky-Zeichner et al. 2010).

Amphotericin B is used for the treatment of many types of invasive fungal infections and binds to ergosterol to form membrane pores, which leads to leakage of intracellular constituents (Gabrielska et al. 2006). Despite its undeniable antifungal activity, amphotericin B has many side effects, such as nephrotoxicity (Fanos and Cataldi 2001; Wong-Beringer et al. 1998). These side effects occur mainly because drug targets in fungi are homologues of some molecular sites in humans. In addition to this limitation, the development of antimicrobial resistance due to the use of broad-spectrum antifungal drugs and their limited number for clinical purposes is a concern for public health. Considering these facts, research into new molecules with future potential therapeutic application is extremely important. Antimicrobial peptides have potential use as antifungal agents, killing the microorganism directly, or immunomodulating the host immune system response (Maurya et al. 2011; Zhai et al. 2010; Lakshminarayanan et al. 2014; Wong et al. 2013; Steinstraesser et al. 2011; Lim et al. 2015).

3.2 Antimicrobial Peptides

Antimicrobial peptides (AMPs) are naturally occurring molecules that play an important role in the first line of defense against microbial threats. They can be isolated from organisms as diverse as humans, plants, insects, and even other microorganisms like bacteria (Zasloff 2002). They are produced due to an exposure to infecting microorganisms and act in order to kill or to slow the growth of invading microorganisms and to aid allied mechanisms of natural and adaptive immunity (Fox 2013; Brogden 2005).

AMPs have a broad spectrum of activity against bacteria, fungi, enveloped viruses, parasites, and even cancerous cells. They can act directly on microorganism membranes or other nonspecific cell targets, which is an advantage in avoiding the development of microbial resistance by gene mutation, as it might happen when drugs have specific proteins as targets (Peschel and Sahl 2006). Moreover, those peptides can be extremely variable in length, amino acid composition, and structure (Nguyen et al. 2011). According to their predominant secondary structure, AMPs are divided into four categories: (a) α-helical, (b) β-sheet, (c) mixed α-helix/β-sheet, and (d) extended. The net positive charge of cationic peptides (+2 or +9) mediates their selective activity against microorganisms' cells that carry a negative net charge due to arginine and lysine residues. Those peptides also have approximately 50 % of hydrophobic amino acids that facilitate interactions with the fatty acyl chains (Steckbeck et al. 2014; Hancock and Patrzykat 2002; Garibotto et al. 2010; Hancock and Rozek 2002).

Some of the AMP mechanisms of action involve different membrane interactions. Membrane disruption can occur through the formation of toroidal pores, composed of loosely associated peptides with interdigitating phospholipid head groups among them. Those peptides are at a critical threshold concentration (Brogden 2005). Differently, in the "barrel-stave model," the peptides are not associated with the lipid head groups, but their hydrophobic regions align with the lipid core region of the bilayer, and the hydrophilic peptide regions form the interior region of the pore (Yang et al. 2001). Another type of membrane interaction is through peptide accumulation on the bilayer surface, since they are electronically attracted to the anionic phospholipid head groups at numerous sites, thus covering the surface of the membrane in a carpet-like manner. At high peptide concentrations, these surface-oriented peptides may act like detergents, leading to the formation of micelles (Shai 1999; Ladokhin and White 2001).

The mechanism of action for the antifungal activity of peptides is generally more complex and often involves entry of the peptide into the cell. It occurs mainly because of the fungal cell architecture, briefly described since they are targets for antifungal drugs (Yu et al. 2014). The cell wall has been shown to be primarily composed of chitin, glucans, and glycoproteins, all of them being covalently cross-linked together. The glycoproteins presented in the cell wall are extensively modified with both N- and O-linked carbohydrates and, in many instances, contain a glycosylphosphatidylinositol (GPI) anchor as well (Bowman and Free 2006). The β-glucan network consists largely of (1–3)-β-glucans with (1–6)-β-branches. In yeast (1–6)-β-glucans are also present. The cell wall is also composed of glycosylated proteins that can be decorated with mannose, galactose, glucose, and uronic acid residues. Chitin and (1-3)-β-glucan layer are a target for a wide range of antifungal molecules that can affect their synthesis and lead to a growth inhibitory effect (Fontaine et al. 2000; Schoffelmeer et al. 1999; Theis and Stahl 2004). Figure 3.1 is a schematic representation of some possible peptide membrane interactions and the basic fungi cell wall structure. Fungal plasma membranes are composed of three main lipids including phospholipids, sphingolipids, and sterols, mainly ergosterol (differently from mammalian cells, composed of cholesterol). The difference in sterol content has been exploited in the mechanism of antifungal drugs such as amphotericin B and azoles (van der Weerden et al. 2013; Kaminski 2014). The mechanisms through which AMPs have antimicrobial activity involve not only membrane physiology interference or disruptions (as described by the proposed models of barrel-stave, carpet, or toroidal pores). They might also interact with protein targets associated with the membrane or by intracellular

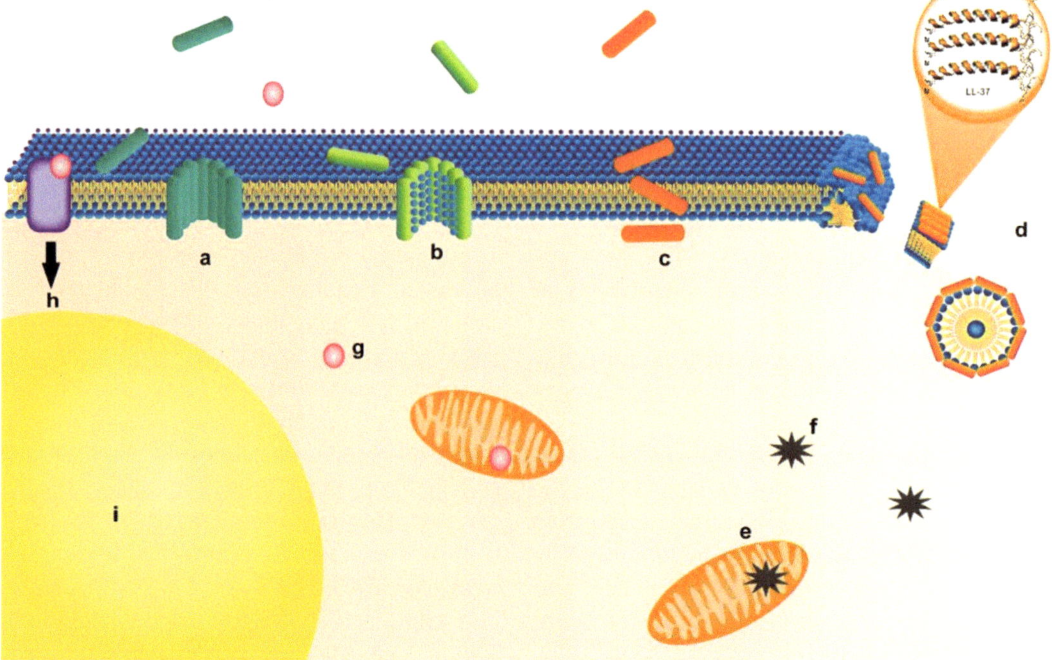

Fig. 3.1 Peptide interaction with the fungal membrane. Schematic representation of the fungal cell wall, composed of outer protein layer with carbohydrate residues (*dark purple*). Peptides with hydrophilic head group and hydrophobic acyl side chain regions (*dark green*, *light green*, and *orange*) interacting with fungal plasma membrane as barrel-stave pore (**a**), toroidal pore (**b**), or by membrane translocation (**c**) and membrane disruption in a carpet-like manner (**d**), as highlighted for LL-37. These interactions depend on the peptide, its concentration, and lipid composition of the membrane. Some peptides such as histatins, β-defensins, lactoferricins, RK-31, KS-30, and hLF (1-11) can reach internal targets such as the mitochondria (**e**). Histatin 5 also leads to the generation of (**f**) reactive oxygen species – ROS (*black spindles*). The non-lytic release of ATP (*pink*) (**g**) by HNP-1, HNP-2, HNP-3a, histatin-5, hLF(1–11), hLF(21–31), and B4010 might activate cell death pathway and (**h**) induce G1 phase arrest of the nucleus (**i**)

targets, including DNA and protein synthesis, protein folding, enzymatic activity, and cell wall synthesis, which may confound the generation of resistance development (Lakshminarayanan et al. 2014; Brogden 2005; Jenssen et al. 2006; Hale and Hancock 2007; Hancock et al. 2012; Yount et al. 2006). As an example, histatins bind to a receptor in the fungal cell membrane, enter the cytoplasm, and induce the non-lytic loss of ATP from actively respiring cells. Their action can also disrupt the cell cycle and lead to the generation of reactive oxygen species (Kavanagh and Dowd 2004; De Smet and Contreras 2005) (Table 3.1).

The isolation and characterization of a natural peptide is a long and laborious process that can hinder the clinical use of AMPs. A new approach is the design of synthetic sequences, which are the result of optimizing sequence and chemical characteristics that are common to many types of AMPs (pharmacophoric patron). Ideally, an antifungal peptide agent should be as short as possible, and therefore the de novo peptide design approaches help to minimize costs production and can help to overcome the low in vivo activity, the labile nature of peptides, and potential toxicity (Steckbeck et al. 2014; Garibotto et al. 2010).

Table 3.1 Antifungal peptides

Antifungal peptides				
Origin	Name	Species	Effective	References
Insect	Alo-3	*Acrocinus longimanus*	*Candida albicans*, *C. albicans* ATCC 90030, *C. glabrata*, C. glabrata ATCC 36082	van der Weerden et al. (2013) and Barbault et al. (2003)
	Termicin	*Pseudocanthotermes spiniger*	*C. albicans*, *Cryptococcus neoformans*	Da Silva et al. (2003) and Lamberty et al. (2001)
	Holotricin-3	*Holotrichia diomphalia*	*C. albicans*	Lee et al. (1995)
	Tenecin-3	*Tenebrio molitor*	*C. albicans*	Kim et al. (1998)
	Cecropin A	*Hyalophora cecropia*, *Drosophila*	*Aspergillus fumigatus*	Steiner et al. (1988) and De Lucca et al. (1997 2000)
	ABP-dHC-cecropin	*Hyphantria cunea*	*C. albicans*, *Neurospora crassa*, *Rhyzopus*, *Fusarium*, *Alternaria*, *Mucor*	Zhang et al. (2015)
	Rondonin	*Acanthoscurria rondoniae*	*C. albicans*, *C. krusei*, *C. glabrata*, *C. parapsilosis*, *C. tropicalis*, *C. guilliermondii*	Riciluca et al. (2012)
Amphibian	Brevenin-1BYa	*Rana boylii*	*C. albicans*, *Staphylococcus aureus*	Conlon et al. (2003), Yeaman and Yount (2003), and Pal et al. (2006)
	Brevenin-1Pa, brevenin-1Pb, brevenin-1Pc	*Rana pipiens*	*C. albicans*, *S. aureus*, *Escherichia coli*	Goraya et al. (2000) and Marenah et al. (2004)
	Temporin A	*Rana temporaria*	*C. albicans*	Mangoni et al. (2000)
Mammalian	Indolicidin	Cytoplasmic granules of neutrophils	*C. albicans*, *C. neoformans*, *S. aureus*, *E. coli*	Selsted et al. (1992), Lee et al. (2003) and Hsu et al. (2005)
	BMAP-28	Bovine myeloid antimicrobial peptide	*C. albicans*, mammalian tumor cells	Risso et al. (2002)
	SMAP-29	Sheep myeloid antimicrobial peptide	*C. albicans*, *Pseudomonas aeruginosa*	Lee et al. (2002a), Shin et al. (2001) and Dawson and Liu (2011)
	PMAP-23	Porcine myeloid antimicrobial peptide	*C. albicans*	Park et al. (2002b) and Lee et al. (2001, 2002b)
	Protegrin-1	Porcine cathelicidin	*C. albicans*, *C. neoformans*	Dawson and Liu (2010)
	LL-37	Human secretions	*C. albicans*, Gram-positive and Gram-negative bacteria	den Hertog et al. (2005), Durr et al. (2006), and Oudhoff et al. (2010)
	HNP1, HNP-2, HNP3a	Human neutrophils	*C. albicans*, *C. neoformans*, *Coccidioides immitis*, *Rhizopus oryzae*, *A. fumigatus*	Ganz (2003, 2005), Raj and Dentino (2002), and Lehrer et al. (1988)
	Histatin 5	Human salivary peptide	*C. albicans*, *C. neoformans*, *A. fumigatus*	Xu et al. (1999), Tsai and Bobek (1997), Situ et al. (2000), and Helmerhorst et al. (2001)
	hLF(1–11), hLF (21–31)	Human lactoferrin	*C. albicans*	Lupetti et al. (2000) and Viejo-Diaz et al. (2004)

(continued)

Table 3.1 (continued)

Antifungal peptides				
Origin	Name	Species	Effective	References
Synthetic	B4010	Originated from a secondary peptide derived from human β-defensin3	*C. albicans*	Eckert (2011) and Nguyen et al. (2010)
	Penetratin 1	Cell-penetrate peptide	*C. albicans*, *C. neoformans*	Milletti (2012) and Masman et al. (2009)

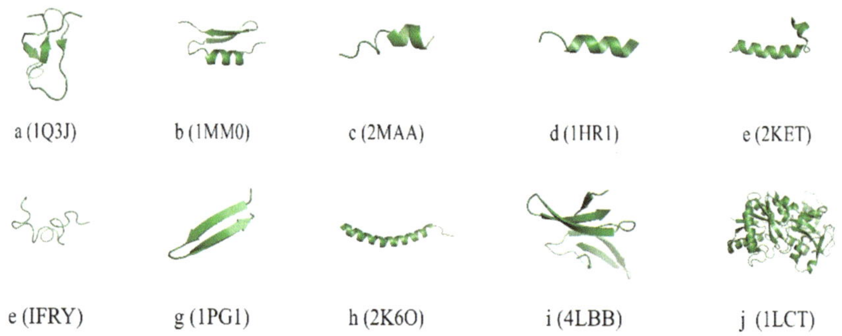

Fig. 3.2 Structures of antifungal peptides. Schematic representation of antifungal peptide based on nuclear magnetic resonance. Each PDB code is in parenthesis. Alo-3 (**a**), termicin (**b**), temporin-1 (**c**), indolicidin (**d**), BMAP-27(**e**), SMAP-29 (**f**), protegrin-1 (**g**), LL-37 (**h**), HNP1 (**i**), and (hLF) (**j**). All images were done using The Pymol molecular graphic system, v1.7.4

*De novo*peptide design includes high-throughput combinatorial library screening, structure base modeling, predictive algorithms, and introduction of non-coded modifications to conventional peptide chemistry (Nguyen et al. 2011; Blondelle and Lohner 2010). The great importance of designed peptides is that many specific properties, such as hydrophobicity, hydrophobic length, or nature of flanking residues, can be systematically varied. Ideally, artificial transmembrane peptides should serve as mimics for transmembrane segments of membrane proteins (Holt and Killian 2010). It is possible to synthetize shorter peptides structurally related to another known peptide, exploring the influence of amino acid substitutions and deletions on its antifungal activity. Linear peptides are flexible, and their possible tridimensional conformations are therefore very complex to determine. It is necessary to use tools to perform conformational analysis for these structures (Garibotto et al. 2010). Figure 3.2 shows structures of some antifungal peptides deposited in RSCB Protein Data Bank website by their authors, including its PDB code (Barbault et al. 2003; Da Silva et al. 2003; Saravanan et al. 2013; Friedrich et al. 2001; Yang et al. 2009; Tack et al. 2002; Fahrner et al. 1996; Wang 2008; Zhao et al. 2013; Day et al. 1993).

3.3 Insect Antifungal Peptides

Insect peptide Alo-3 was isolated from the coleopteran *Acrocinus longimanus*. This peptide contains six cysteine residues, forming three disulfide bridges and an antiparallel β-sheet with a long flexible loop connecting the first strand to the second strand and a series of turns. Alo-3 belongs to the knottin-type family of proteins with a cysteine-stabilized, "knotted" topology, defined by two parallel disulfide bonds, threaded by a third one. It has no negatively charged residues and displays a cationic pole on its surface that may contribute to its antifungal activity (van der Weerden et al. 2013). Barbault and coworkers (2003) tested two other homologous peptides, Alo-1 and Alo-2, with sequence identity above 80 %, but Alo-3 was the most effective against *Candida glabrata* and *C. albicans*, both tested not only against clinical isolates of those pathogens but also against ATCC strains (ATCC90030 and ATCC 36082, respectively). Another peptide derived from insects is termicin, isolated from the fungus-growing termite *Pseudocanthotermes spiniger* (heterometabole insect, Isoptera). Termicin is a cysteine-rich antifungal peptide with a α-helical segment and two antiparallel β-sheets forming a "cysteine αβ motif," also found in antibacterial and antifungal defensins and from plants. Termicin showed activity against *C. albicans* and *C. neoformans*, but was inactive against *C. glabrata* (Da Silva et al. 2003; Lamberty et al. 2001).

Likewise, *C. albicans* growth was inhibited by holotricin-3 and tenecin-3 peptides, isolated from the hemolymph of the coleopteran insect *Holotrichia diomphalia* and from the larvae of *Tenebrio molitor*, respectively (Kim et al. 1998; Lee et al. 1995). Tenecin-3 had a better candidacidal effect, and its uptake and internalization by the cell are essential for its antimicrobial activity. This indicates an inner target involvement in the process of killing, since membrane permeabilization and calcein release were not observed. Uptake of tenecin-3 was inhibited at low temperature (0 °C) and by the presence of the oxidative phosphorylation inhibitor, sodium azide (Kim et al. 2001). Chae and coworkers (2012) isolated a new 14 kDa peptide, named tenecin-4, which was effective against *Escherichia coli* but not against *Bacillus subtilis* or *C. albicans*.

An interesting group of insect peptides is the cecropins, which were originally isolated from the cecropin moth (*Hyalophora cecropia*) and have been found in insects like *Drosophila* (van der Weerden et al. 2013). Cecropins are basic 35–39 amino acid residue peptides that can fold into two amphipathic α-helices, separated by a more flexible hinge. Their mode of action against bacteria is based on the formation of either voltage-dependent ion channels or general disruption of the membrane by a "carpet-like" mechanism (Steiner et al. 1988). Cecropin A, a 37 amino acid residue peptide, is complexed with lipopolysaccharide and in germinating cells of *Aspergillus fumigatus* induces death, whereas binding and cell death were not observed with non-germinating hyphae (De Lucca et al. 1997). De Lucca and coworkers (2000) have proposed the mode of action of this peptide as involving disruption of the plasma membrane. The ABP-dHC-cecropin A (antimicrobial peptide drury *Hyphantria cunea*), a highly cationic peptide isolated from the fat bodies of drury moths (*H. cunea*), has shown a strong antifungal activity against both *C. albicans* and *Neurospora crassa* as well as *Rhyzopus*, *Fusarium*, *Alternaria*, and *Mucor* species (Zhang et al. 2015).

Riciluca and coworkers (2012) have recently isolated a peptide named rondonin from the spider *Acanthoscurria rondoniae*. This peptide shows a molecular mass of 1236.77 Da and activity against *C. albicans*, *C. krusei*, *C. glabrata*, *C. parapsilosis*, *C. tropicalis*, and *C. guilliermondii*. Otherwise, no deleterious activities against human erythrocytes or Gram-positive and Gram-negative strains were observed.

3.4 Amphibian Antifungal Peptides

Amphibians inhabit environments that provide a great challenge for their immunity, providing valuable information about prospective functional molecules. In this context, peptides have been isolated and described. Some of them, like brevinins and temporins, were effective against human fungal pathogens (Xu and Lai 2015).

Brevinins consist of two families named brevinin-1 (24 residues) and brevinin-2 (33–34 residues), and they were first described in 1992 by Morikawa and coworkers (Morikawa et al. 1992), who isolated these peptides from the skin of the Japanese frog *Rana brevipoda porsa*, demonstrating microbicidal activity against a wide range of Gram-positive, Gram-negative bacteria and pathogenic fungi strains. Up to now about 350 types of brevinins have been discovered, sharing common features like linearity, amphipathicity, and cationicity, and some of them have a C-terminal disulfide-bridge cyclic heptapeptide, called a rana box (Novkovic et al. 2012; Savelyeva et al. 2014). Brevinin-1 exists as a random coil in aqueous solution, but adopts an amphipathic α-helical structure in a hydrophobic membrane-mimetic environment such as 50 % trifluoroethanol. The brevinin-1 peptides generally comprise an N-terminal hydrophobic region, a proline containing a hinge region in the central portion, and a C-terminal disulfide-bonded loop (Kwon et al. 1998). The α-helical structure leads to perturbation of the phospholipid bilayer of target membranes in the "barrel-stave" and "carpet-like" models (Savelyeva et al. 2014)

Conlon and coworkers (2003) isolated the peptide brevinin-1BYa (FLPILASLAAKFGPKLFCL VTKKC) from the norepinephrine-stimulated skin secretions from the foothill yellow-legged frog *Rana boylii*. This peptide was potent against *C. albicans* and *Staphylococcus aureus*, but its therapeutic potential is limited due to its strong hemolytic activity. The research group has also substituted amino acid in the original molecule, which leads to brevenin-1BYb and brevenin-1BYc peptides with fourfold and tenfold reduction against *C. albicans*, respectively. The change in the cationic residues can be the explanation of the observed result, since this global net charge reduction affects the initial binding to the negatively charged phospholipids in the microorganisms' cell membranes (Yeaman and Yount 2003). Another study evaluated the antimicrobial activity of brevinin-1BYa and investigated the growth inhibitory activity of a synthetic replicate of this peptide: [Ser^{18}, Ser^{24}] brevinin-1BYa (FLPILASL AAKFGPKLFSLVTKKS). The group observed an eightfold reduced hemolytic activity compared to the native peptide and suggested that this reduction arises from destabilization of the α-helix in the C-terminal region of the peptide associated with replacement of the cysteine bridge. Antimicrobial activities against *C. albicans* and Gram-negative bacteria were reduced. In contrast, substituting the two cysteines for serines abolished the antifungal activity (Park et al. 2002a; Pal et al. 2006). Peptides isolated from *Rana pipiens*, such as brevinin-1Pa, brevinin-1Pb, and brevinin-1Pc, were also effective against *C. albicans*, *S. aureus*, and *Escherichia coli* (Goraya et al. 2000). In addition, brevinins are able to stimulate insulin release, which causes hypoglycemia in frogs attacking predators. This property can also be explored for the treatment of patients with type 2 diabetes (Marenah et al. 2004; Abdel-Wahab et al. 2010).

Another class of amphibian peptides is the temporins. They were initially identified in 1996 in skin secretion of the European red frog *Rana temporaria* (Simmaco et al. 1996), but they can be isolated from several other frog species as well as from wasp venom (Rollins-Smith et al. 2003). They are a group of linear short peptides (10–14 amino acid residues) with a net cationic charge and an amidated C-terminus. In an apolar environment, temporins showed a marked propensity to adopt an amphipathic α-helical structure. Temporins are not as basic as other cationic peptides, although in the most potent, temporin A, the one basic residue is essential for activity. Temporins A and B have been reported to be active against *C. albicans* and Gram-positive and Gram-negative bacteria. These peptides permeate both artificial and biological membranes, but they do not lyse human erythrocytes, which suggests there are additional factors involved in the mechanism of action on different cell types (van der Weerden et al. 2013; Mangoni et al. 2000; Wade et al. 2000; Hujakka et al. 2001; Carotenuto et al. 2008).

3.5 Mammalian Antifungal Cathelicidins

Cathelicidins are peptides of approximately 100 amino acid residues, and their sequences are related to cathelin, a cystatin-like protein. They are commonly found in humans and other species such as sheep, pigs, horses, cattle, chickens, rabbits, and some species of fishes, being usually stored in the secretory granules of neutrophils and macrophages. They can also be released extracellularly upon leukocyte activation (Zanetti 2005; Kosciuczuk et al. 2012). The term cathelicidins was first proposed in 1995 to acknowledge the evolutionary relationship of the novel protein family to cathelin, and it is used to denote holoproteins that contain a cathelin-like sequence and a cationic antimicrobial domain (Zanetti et al. 1995). The first is a conserved N-terminal sequence ("cathelin" domain, the cathepsin L inhibitor), and the second is a C-terminal antimicrobial domain of varied sequence and length (both interspecies and intraspecies), which express their activity after they have been cleaved from the holoprotein (Gennaro and Zanetti 2000). In mammals, cathelicidins were first identified in bone marrow myeloid cell, and therefore they are also named "myeloid antimicrobial peptides" (MAP) (Zanetti 2005).

Among the cathelicidin peptides, some present antifungal activity and this is stronger against yeast than against filamentous fungi (Benincasa et al. 2006). Indolicidin is a tryptophan-rich bovine cathelicidin peptide of 13 amino acid residues (ILPWKWPWWPWRR-NH2), purified from the cytoplasmic granules of neutrophils and found in bone marrow cells as a 144-long amino acid precursor (Selsted et al. 1992; Del Sal et al. 1992). This peptide showed activity against fungi *C. albicans* and *C. neoformans*, as well toward bacteria *S. aureus* and *E. coli* (Benincasa et al. 2006). The fungicidal activity involves membrane disruption, DNA binding and topoisomerase 1 inhibition. In the first situation, the peptide interacts with the lipid bilayers in a salt- and energy-dependent manner. The DNA interaction occurs by DNA synthesis inhibitors binding DNA or proteins involved in the process. Indolicidins may also interact in other biosynthesis pathways or in cell cycle signal transduction (Lee et al. 2003; Hsu et al. 2005).

Other bovine cathelicidins with fungicidal activity are bovine myeloid antimicrobial peptides (BMAP) of 27 and 38 amino acid residues, BMAP-27 and BMAP-28, respectively. BMAP-28 is toxic for mammalian tumor cells, inducing their apoptosis, and it was also demonstrated that it induces mitochondrial permeability, forming transition pores (MPTP), resulting in the release of cytochrome *c*. The cytotoxic activity has been related to the structural features of the peptide, which consists of a cationic N-terminal sequence predicted to assume an amphipathic α-helical conformation (residues 1–18) and a C-terminal hydrophobic tail (residues 19–27). This hydrophobic tail is responsible for the peptide activity, since its analogue, BMAP28 (1–18), which comprises the 18 N-terminal residues, showed a reduction in MPTP effect. BMAP28 cytotoxicity requires an active metabolism of the target cells (Risso et al. 2002).

SMAP-29 is cathelicidin-like peptide derived from myeloid sheep with α-helical structure in a hydrophobic environment, and its C-terminal hydrophobic domain has a strong membrane permeability (Chen et al. 2011; Skerlavaj et al. 1999). This peptide concentrates on the plasma membrane of treated cells and causes propidium iodide (PI) uptake, provided by the cells that are metabolically active. Lee and coworkers (2002a) suggest that membrane disruption by SMAP-29 occurs via pore formation, due to a direct interaction with the lipid bilayers and irregularly disrupted fungal membranes in an energy- and salt-dependent manner. SMAP-29, however, is strongly hemolytic against human erythrocytes. A variant of SMAP-29, $[K^{22,25,27}]$-SMAP-29, is effective against bacterial and fungal cells in physiological salt concentrations and was not injurious to eukaryotic cells, such as human erythrocytes (Shin et al. 2001; Dawson and Liu 2011).

Isolated from porcine myeloid, peptide PMAP-23 was identified by cDNA cloning and is 23 residues long, cationic (+7), and amphipathic. In a hydrophobic environment, it forms two α-helices joined by a flexible region when membrane-bound (Roversi et al. 2014; Park et al. 2002b). This peptide is capable of binding to plasma membrane of *C. albicans* protoplasts, indicating that an interaction with the cell wall is not a requirement for the inhibitory activity of this peptide, which also did not show hemolytic activity (Park et al. 2002b; Lee et al. 2001). Lee and coworkers (2002b) designed several analogs of PMAP-23, with amino acid substitutions in order to increase the net hydrophobicity by Trp (W)-substitution at positions 10, 13, or 14 on the hydrophilic face of the peptide. In *C. albicans* the P6 analog peptide exerted its fungicidal effect on the blastoconidia by disrupting the mycelial forms, causing significant morphological changes. Meanwhile, P6 also displayed about fourfold greater antitumor activity than the parent PMAP-23.

Porcine cathelicidins, such as protegrin-1 (PG-1), showed activity against clinical isolates of fungi, including those resistant to conventional medicines used in human therapy (Benincasa et al. 2006). These peptides and their variants have a rather rigid antiparallel β-sheet (β-hairpin) structure that is stabilized by two intramolecular disulfide bonds. The linear peptide forms have been reported to be considerably less active than the native form, being sensitive to physiological salt concentrations (Dawson and Liu 2010). PG-1, BMAP-27, BMAP-28, SMAP-29, and indolicidin showed deleterious activity against a number of nosocomial yeast strains, mainly *Candida* spp. and *C. neoformans* (Benincasa et al. 2006).

The most famous prototype of cathelicidin group is human cathelicidin LL-37, which attains a helical structure when bounded to cell wall and plasma membrane of treated cells and has the protein hCAP-18 as its precursor. LL-37 is secreted in human sweat and further processed into RK-31 and KS-30, more active peptides that retain their activity even in high salt conditions. Moreover, they are able to enter the cytoplasm of *C. albicans* cells, suggesting that their increased activity may result from interaction with intracellular targets. In contrast, LL-37 could not be detected in the cytoplasm (den Hertog et al. 2005, 2006; Lopez-Garcia et al. 2005). LL-37 showed a pH-dependent activity against *C. albicans*, disrupting its membrane and allowing leakage of proteins of up to 40 kDa into the medium (den Hertog et al. 2005). This peptide discriminates against phospholipid monolayers containing negatively charged lipids, as SMPA-29 does. However, experiments comparing these two peptides demonstrated that the LL-37 peptide had a more potent effect than the SMAP-29, suggesting that its interaction with monolayers involves other factors, such as hydrophobicity, size, and charge distribution, although SMAP-29 has a higher net positive charge (+10), and it would be expected to be more attracted to the negatively charged lipid monolayers (Neville et al. 2010). When compared to histatin-5, LL-37 induced higher morphological defects, but the efflux of nucleotides is similar in comparable candidacidal concentrations, suggesting that the loss of nucleotides plays an important role in the killing process (den Hertog et al. 2005). LL-37 has been found to have additional activities, such as regulating the inflammatory response and chemo-attracting cells of the adaptive immune system to wound or infection sites, helping to neutralize the microorganism and promoting re-epithelization and wound closure (Durr et al. 2006; Oudhoff et al. 2010).

3.6 Mammalian Antifungal Defensins

Mammalian defensins are a large group of peptides with an important role in the host's immune system. They can be divided into the α- and β-defensins based on their structural characteristics and cysteine spacing pattern (van der Weerden et al. 2013). Both defensins were first identified as antimicrobial compounds involved in innate immunity. In humans, α- and β-defensins are expressed mainly in different sites: the α-defensins are mostly expressed in neutrophils (known as human neutrophil peptides (HNP), or human defensins (HP) when expressed in natural killer cells) and the β-defensins are secreted by the epithelial cells of the skin and mucosae and known as HβD (Suarez-Carmona et al. 2014). Apart from the antimicrobial activities, the defensins appeared as modulators of the adaptive immune system and angiogenesis, key mediators of wound healing and determinant players in male fertility (Ganz 2003; Oppenheim et al. 2003). The α-defensins comprise 29–35 amino acid peptides that share six conserved cysteine residues with three disulfide bonds. Their structure is formed by an amphipathic and antiparallel β-sheet. They also have a β-hairpin loop containing cationic charged molecules. The β-defensins are longer than their α-counterparts (34–42 residues in length) and are triple-stranded with an antiparallel β-sheet as well as a short α-helix (De Smet and Contreras 2005; Selsted and Ouellette 2005). The human defensins present activity against a wide range of microorganisms, including fungal pathogens. HNP 1–3 are identical apart from one N-terminal amino acid, which makes HNP 3 completely inactive against *C. albicans*, while HNP 1–2 is candidacidal. HNP 4 is also toxic to *C. albicans* cells, and the mechanism of action of those peptides on fungal cells has been proposed to be by membrane permeabilization (Ganz and Lehrer 1995; Ganz 2005). NP 1 seems to work in the same way, but differently from HNP, and it was shown to be dependent on the metabolic activity of the target cells. HNP-1 causes *C. albicans* to release ATP, just as histatin 5 does, but in contrast, it did not seem to lyse cells. NP-1, NP2, and NP3a were highly effective against *C. albicans*, and NP-1 was effective against *C. neoformans*, *Coccidioides immitis*, and hyphae and germinating conidia of *Rhizopus oryzae* and *Aspergillus fumigatus* (van der Weerden et al. 2013; Raj and Dentino 2002; Lehrer et al. 1988).

Moreover, β-defensins, HβD 2 and HβD 3, are potent inhibitors of *C. albicans*. Exposure to this microorganism and to *Trichophyton rubrum* and *A. fumigatus* stimulates HβD 2 expression. *A. fumigatus* also induces the expression of HβD 9. The mechanisms of action of these peptides are not well known, but some requirements seem to be necessary for their activities, such as metabolically active cells and a low concentration, since they have a strong positive net charge, which in high concentrations of cations may decrease efficacy (Dhople et al. 2006; Joly et al. 2004; Liu et al. 2002; Alekseeva et al. 2009). Another antifungal peptide is Novexatin, a cyclic and highly cationic peptide (1,093 daltons), arginine rich, based on the human α and β defensins. It is currently in phase 1/2 of clinical trials for use against fungal infections of the toenails; NovaBiotics (Aberdeen, UK) is the company responsible for its development (Fox 2013).

3.7 Mammalian Antifungal Histatins

The histatin family consists of 12 members of histidine-rich peptides from which histatins 1 and 3 (the full-length proteins and gene products) and histatin 5 (a cleavage product of histatin 3) are the main ones and constitute 70–80 % of the total amount (Xu et al. 1999). The other nine members are proteolytic cleavage products of these peptides (Fitzgerald et al. 2003). Although they are named due to a high number of histidine residues, other amino acids including lysines (Lys^5 and Lys^{13}) rather than histidines have key importance for fungicidal activity (Kumar et al. 2011; Rothstein et al. 2001). Studies demonstrate that histatins have a number of biological activities in vitro, such as the maintenance of tooth surface integrity, histamine release induction, and potentiation of rabbit chondrocyte growth (Hay 1975; Oppenheim et al. 1986; Sugiyama et al. 1985; Murakami et al. 1994). The histatins are encoded by two closely related genes (HIS1 and HIS2), with histatin 1 and histatin 3 as primary products of HISI and HIS2, respectively (Sabatini and Azen 1989).

Histatin 5 was obtained from histatin 3 (Raj et al. 1998) and has the strongest antimicrobial activity against pathogenic fungi *C. albicans*, *C. neoformans*, and *A. fumigatus* (Kavanagh and Dowd 2004). This histatin is 24-amino acid residues long with seven histidines, four arginines, and three lysines, and its fungicidal activity resides in a region of 11–24 residues at the C-terminal, referred to as the functional domain or dh-5 (Driscoll et al. 1995). In a non-aqueous environment, the peptide adopts a α-helical conformation, and, like histatin 3, in an aqueous environment it adopts a random coil structure (van der Weerden et al. 2013; Helmerhorst et al. 1999a; Tsai and Bobek 1997). The ability to form a α-helix was thought to be important for the mode of action of histatin 5, but Situ and coworkers (2000) demonstrated that one variant of histatin 5 (3P) with reduced ability to form α-helix had an antifungal ability comparable to that of histatin 5. Its antifungal mechanisms against *C. albicans* involve binding to a specific receptor, translocation across the membrane, and interaction with internal targets such as mitochondria and non-lytic release of cellular ATP (Fitzgerald et al. 2003; Helmerhorst et al. 1999b; Koshlukova et al. 1999, 2000). Histatins do not display lytic activities to lipid membranes, measured by release and dequenching of the fluorescent dye calcein (Edgerton et al. 1998).

Histatin interaction with the cell wall and its uptake by the cell are two independent events, since the fungal cell wall binding itself does not result in uptake of histidines. Li and coworkers (2006) demonstrated that the heat shock protein Ssa2p is the binding site for histatin 5 measured by yeast two-hybrid analyses, whereas Ssa1p appears to have a lesser role in histatin 5 toxicity. This heat shock protein 70 (Ssap 1/2) is located in the fungal cell envelope. These interactions, however, can be prevented in the presence of Ca^{+}, which presumably disrupts the interaction between histatin 5 and Ssa2p (Li et al. 2006). Internalization of histatin 5 occurs by translocation, and its uptake is dependent on two polyamine transporters, Dur3 and Dur31 (which usually function in spermidine uptake), since *C. albicans* showed a reduced intracellular transport of histatin 5 upon growth in a medium rich in spermidine, implicating polyamine transporters in uptake of this peptide (van der Weerden et al. 2013; Kumar et al. 2011).

The mitochondrion's primeval functionality is as an energy-generating organelle, although the molecular machinery in charge of its role shows a broad divergence among different phyla (Tielens et al. 2002). It has been proposed that a negatively charged phospholipid in the mitochondrion membrane, called cardiolipin, attracts histatin 5 toward the mitochondrion. This interaction causes ATP release into the cytoplasm, and it was shown by the evidence of unchanged levels of ATP when cells were treated with respiration uncouplers, such as azide or cyanide, and then exposed to histatin 5, indicating that respiration is essential for histatin activity to occur (Helmerhorst et al. 1999a; Koshlukova et al. 1999). Gyurko and coworkers (2000) also demonstrated that fungal cells incapable of respiration (respiratory-deficient or petite mutants) are resistant to the action of histatins. The ATP efflux occurs via ATP-binding cassette (ABC) proteins, and extracellular ATP activates a purigenic-like receptor that triggers cell death (Koshlukova et al. 2000). Another ATP release consequence is the effect on the regulation of cell volume homeostasis, which can halt cells at the G1 phase, disrupting the cell cycle and leading to cell death that is not in a programmed pathway (Wunder et al. 2004). Histatin 5 is responsible for the generation of reactive oxygen species (ROS), which is measured using an oxygen radical sensitive probe (dihydroethidium), and it is one of the components responsible for the disruption of cell organelle structure or function (Baev et al. 2001). Another important factor of histatin-5 is that it retards the transition of *C. albicans* from the blastopore to the hyphal stage of growth, a process that may assist in arresting tissue penetration by the fungus (Helmerhorst et al. 1997).

3.8 Mammalian Lactoferrin-Derived Antifungal Peptides

Lactoferrin (Lf) is a multifunctional protein (80 kDa), a member of the transferrin family of non-heme iron-binding glycoproteins that is expressed and secreted by granular epithelial cells and secreted into mucosal fluids that bathe the body surface; it is found in the secondary granules of neutrophils during the myelocyte stage of maturation (Ward et al. 2005; Levay and Viljoen 1995). It was first isolated from bovine milk and later identified in mice, pigs, and humans (van der Weerden et al. 2013). On the mucosal surface, this peptide works as a component of the first line of host defense, being released from neutrophils during infection, inflammation, tumor development, and iron overload (Levay and Viljoen 1995; Legrand et al. 2008). It also acts in the regulation of iron homeostasis, cellular growth, and differentiation and protection against cancer development and metastasis (Ward et al. 2005; Shimamura et al. 2004).

Proteolytic cleavages of lactoferrin revealed some peptides (lactoferricins) with better antifungal activity than that of the whole protein, such as the first and second cationic domains derived from human lactoferrin (hLF) hLF (1–11) and hLF (21–31), respectively (Lupetti et al. 2000). A study using those synthetic peptides revealed a dose-dependent release of ATP by *C. albicans* upon exposure to hLF (1–11). The same study demonstrated that a metabolic active cell is necessary for the hLF (1–11) mode of action, since cells incubated with sodium azide had a reduced candidacidal activity and a lower PI uptake. The use of the fluorescent dye rhodamine 123 showed an accumulation inside the mitochondria and later was released into the cytoplasm when cells were treated with hLF (1–11), which indicates that the peptide triggers the energized mitochondrion (Lupetti et al. 2000). This ATP efflux was also observed in a short synthetic peptide following the N-terminal amino sequence of bovine lactoferrin (peptide 2 or Pep2). Pep 2 activated pertussis toxin-insensitive and cholera toxin-sensitive G-protein and activated signals downstream through phosphatidylinositol 3-kinase to protein kinase C. This indicates that Pep 2 induced ATP efflux mediated by G-protein activation (Tanida et al. 2006; Helmerhorst et al. 2002). Another study showed that cell wall interaction and therefore membrane binding with hLF are not the major mode of action, thus demonstrating a slight efflux of K^+ from *C. albicans* cells, but this did not allow Na^+ release or membrane disruption (Viejo-Diaz et al. 2004). Some other peptides, derived both from bovines (LfcinB), which comprise the region spanning 17–40 residues, and from humans (LfcinH), dissipated the proton gradient across the plasma membrane. However, they did not seem to act by nonspecific permeabilization of the membrane as they did not cause calcein release from artificial liposomes (Nguyen et al. 2005; Viejo-Diaz et al. 2003).

3.9 Synthetic Antifungal Peptides

Natural peptides can have their activity enhanced by peptide engineering, which can help to develop novel peptides with desirable biological properties (Ryu et al. 2014). The synthetic peptide B4010 originated from a 10-residue peptide (RGRKVVRRKK), which in turn is a synthetic analogue of human β-defensin-3 (HβD-3). Its properties have been previously reported by Lakshminarayanan and coworkers (2014) as showing potent activity against *Pseudomonas aeruginosa*, but poor activity against fungi (Bai et al. 2009).

That research group also extended the previous analysis to a higher order of covalently linked peptides. B4010 is a tetravalent synthetic peptide, which carries four copies of the sequence RGRKVVRR through a branched lysine core. This strategy of developing multivalent peptides by assembling multiple copies of monomeric peptides around a core molecule is an alternative to circumventing the drawbacks of antimicrobial peptides, such as the loss of antimicrobial properties in a physiological concentration of salts and polyanionic polymers (Knappe et al. 2010; Eckert 2011).

B4010 presented deleterious activity against *C. albicans* strains (Lakshminarayanan et al. 2014). The researchers also tested another peptide by linking two copies of the sequence through a branched lysine, B2088, and observed a substantial decrease in MIC (minimal inhibitory concentration) values when compared to the monomer or linear retrodimer peptide (RGRKVVRRKKRRVVVKRGR). The MIC values of B4010 (0.37 μM) for two clinical isolates of *C. albicans* were also lower when compared to the MIC values for amphotericin B (1.4 μM) and natamycin (15 μM). This last one is the only US FDA-approved antifungal for ophthalmic applications (Lakshminarayanan et al. 2014; Arora et al. 2011), which is also nonhemolytic and nontoxic to mice when administrated by intravenous (100 $mg.kg^{-1}$) or intraperitoneal (200 $mg.kg^{-1}$) routes and had no affinity for cell wall polysaccharides. It was proposed that its mode of action includes a rapid dissipation of membrane potential and release of vital ions and ATP when challenged by *C. albicans*, and some studies suggest that the first arginine is important for mediating peptide-bilayer interactions (Lakshminarayanan et al. 2014; Nguyen et al. 2010).

Cell-penetrating peptides (CPPs) are part of a group of synthetic peptides with up to 30 peptides and are able to enter cells in an energy-independent manner, translocating across the membrane (Milletti 2012). Penetratin 1 is a 16 amino acid long CPP from the third helix of the Antennapedia homeodomain of *Drosophila*. It can be classified as a cationic amphipathic peptide, and its proposed mechanism of action is by "inverted micelle" pathway. This peptide was synthetized and had its fungicidal activity evaluated by Masman and coworkers (2009). This research group tested not only penetratin 1 against *C. albicans* and *C. neoformans* but also tested other peptide sequences, among them a trapeptide (RQKK-NH_2), identified as 8 in their study. Both peptides, 1 and 8, displayed the most potent inhibitory effect against those pathogenic fungi (Garibotto et al. 2010; Masman et al. 2009).

3.10 Prospects for Clinical Use of Antimicrobial Peptides

The development of new drugs is a remarkable challenge. The last new class of antibiotics, the lipopeptide daptomycin, was introduced in 2003, more than 40 years after the introduction of fluoroquinolones, the last antibiotics used to treat multidrug-resistant (MDR) organisms such as *Klebsiella* and *Acinetobacter* species. Peptides are a class of molecules with potential for therapeutic use against fungal infections (Steckbeck et al. 2014), but their use as novel drugs has to overcome some therapeutic difficulties, such as their poor chemical and physical stability and short circulating plasma half-life. Additionally, those molecules are antagonized by physiological concentrations of salt and polyanionic polymers (mucins, DNA, and glycosaminoglycans) and by the action of proteolytic enzymes, thus limiting their therapeutic potential (Lakshminarayanan et al. 2014; Knappe et al. 2010; Otvos and Wade 2014). Another difficulty is the expensive manufacturing of peptides when compared to small-molecule drugs. Companies are scrambling for financial support both from federal programs and from corporate partners through the later and most expensive stages of clinical evaluations (Fox 2013).

AMPs have only been tested in clinical trials relatively recently, and to date, with the exception of gramicidin for topical administrations, none has received US Food and Drug Administration (FDA) approval. Kaspar and Reichert (Kaspar and Reichert 2013) highlight that in 2012 the first marketing approvals were made for six peptides (lucinactant, peginesatide, pasireotide, carfilzomib, linaclotide, and teduglutide) in the United States and European Union, which only disapproved peginesatide due to safety issues. Another example of a peptide that was undergoing clinical trials is pexiganan, a synthetic analog of the AMP magainin. Pexiganan had impressive results in early phase I and II clinical trials to treat diabetic ulcers, but its performance was not superior when compared to traditional antibiotics used in treating feet ulcers (Moore 2003). At present, approximately 140 peptide therapeutics are being evaluated in clinical trials (Fosgerau and Hoffmann 2015).

In spite of all the challenges, antimicrobial peptides have the potential for development as novel antifungal agents, because they have rapid action and a broad spectrum, being active against species that are multiple resistant to currently used antimycotics. Moreover, peptides are selective and efficacious signaling molecules with specific targets, properties that are important features for drug safety, and they show diminished side effects for the patient (Matsuzaki

2009; Wang et al. 2010; Takahashi et al. 2010). Natural or designed peptides can avoid cell toxicity and hemolytic activity and other undesirable features (Fjell et al. 2012). Some chemical strategies that have been used are sugar molecules incorporated in the peptide N-terminus, which may improve tissue penetration, and the conjugation to passive and active transport enhancers in order to increase oral bioavailability (Charlton et al. 2008; Gupta et al. 2013; Sachdeva et al. 2013). In addition, peptides play important roles in innate immune response and wound healing, additional features for their antimicrobial activity (Pena et al. 2013; Steinstraesser et al. 2012).

In summary, peptides are promising molecules with potential as future therapeutics for fungal diseases. Research is imperative to understand their mode of action and specific targets against fungi and to improve their pharmacological properties to meet clinical therapeutics and drug manufacturing needs.

References

Abdel-Wahab YH, Patterson S, Flatt PR, Conlon JM (2010) Brevinin-2-related peptide and its [D4K] analogue stimulate insulin release in vitro and improve glucose tolerance in mice fed a high fat diet. Hormone Metabol Res Horm Stoffwechselforschung Horm Metab 42(9):652–656. doi:10.1055/s-0030-1254126

Alekseeva L, Huet D, Femenia F, Mouyna I, Abdelouahab M, Cagna A et al (2009) Inducible expression of beta defensins by human respiratory epithelial cells exposed to Aspergillus fumigatus organisms. BMC Microbiol 9:33. doi:10.1186/1471-2180-9-33

Arora R, Gupta D, Goyal J, Kaur R (2011) Voriconazole versus natamycin as primary treatment in fungal corneal ulcers. Clin Experiment Ophthalmol 39 (5):434–440. doi:10.1111/j.1442-9071.2010.02473.x

Baev D, Li X, Edgerton M (2001) Genetically engineered human salivary histatin genes are functional in Candida albicans: development of a new system for studying histatin candidacidal activity. Microbiology 147 (Pt 12):3323–3334

Bai Y, Liu S, Jiang P, Zhou L, Li J, Tang C et al (2009) Structure-dependent charge density as a determinant of antimicrobial activity of peptide analogues of defensin. Biochemistry 48(30):7229–7239. doi:10.1021/bi900670d

Barbault F, Landon C, Guenneugues M, Meyer JP, Schott V, Dimarcq JL et al (2003) Solution structure of Alo-3: a new knottin-type antifungal peptide from the insect Acrocinus longimanus. Biochemistry 42 (49):14434–14442. doi:10.1021/bi035400o

Benincasa M, Scocchi M, Pacor S, Tossi A, Nobili D, Basaglia G et al (2006) Fungicidal activity of five cathelicidin peptides against clinically isolated yeasts. J Antimicrob Chemother 58(5):950–959. doi:10.1093/jac/dkl382

Blondelle SE, Lohner K (2010) Optimization and high-throughput screening of antimicrobial peptides. Curr Pharm Des 16(28):3204–3211

Bowman SM, Free SJ (2006) The structure and synthesis of the fungal cell wall. BioEssays: News Rev Mol Cell Dev Biol 28(8):799–808. doi:10.1002/bies.20441

Brogden KA (2005) Antimicrobial peptides: pore formers or metabolic inhibitors in bacteria? Nat Rev Microbiol 3(3):238–250. doi:10.1038/nrmicro1098

Carotenuto A, Malfi S, Saviello MR, Campiglia P, Gomez-Monterrey I, Mangoni ML et al (2008) A different molecular mechanism underlying antimicrobial and hemolytic actions of temporins A and L. J Med Chem 51(8):2354–2362. doi:10.1021/jm701604t

Chae JH, Kurokawa K, So YI, Hwang HO, Kim MS, Park JW et al (2012) Purification and characterization of tenecin 4, a new anti-Gram-negative bacterial peptide, from the beetle Tenebrio molitor. Dev Comp Immunol 36(3):540–546. doi:10.1016/j.dci.2011.09.010

Charlton ST, Whetstone J, Fayinka ST, Read KD, Illum L, Davis SS (2008) Evaluation of direct transport pathways of glycine receptor antagonists and an angiotensin antagonist from the nasal cavity to the central nervous system in the rat model. Pharm Res 25(7):1531–1543. doi:10.1007/s11095-008-9550-2

Chen C, Wu S, Li X, Zhang X, Yan M (2011) [Structure, function and molecular design strategies of antibacterial peptide SMAP-29: a review]. Sheng Wu Gong Cheng Xue Bao Chin J Biotechnol 27 (6):846–859

Conlon JM, Sonnevend A, Patel M, Davidson C, Nielsen PF, Pal T et al (2003) Isolation of peptides of the brevinin-1 family with potent candidacidal activity from the skin secretions of the frog Rana boylii. J Pept Res 62(5):207–213

Da Silva P, Jouvensal L, Lamberty M, Bulet P, Caille A, Vovelle F (2003) Solution structure of termicin, an antimicrobial peptide from the termite Pseudacanthotermes spiniger. Protein Sci: Publ Protein Soc 12(3):438–446. doi:10.1110/ps.0228303

Dawson RM, Liu CQ (2010) Disulphide bonds of the peptide protegrin-1 are not essential for antimicrobial activity and haemolytic activity. Int J Antimicrob Agents 36(6):579–580. doi:10.1016/j.ijantimicag.2010.08.011

Dawson RM, Liu CQ (2011) Analogues of peptide SMAP-29 with comparable antimicrobial potency and reduced cytotoxicity. Int J Antimicrob Agents 37 (5):432–437. doi:10.1016/j.ijantimicag.2011.01.007

Day CL, Anderson BF, Tweedie JW, Baker EN (1993) Structure of the recombinant N-terminal lobe of human lactoferrin at 2.0 A resolution. J Mol Biol 232 (4):1084–1100. doi:10.1006/jmbi.1993.1462

De Lucca AJ, Bland JM, Jacks TJ, Grimm C, Cleveland TE, Walsh TJ (1997) Fungicidal activity of cecropin A. Antimicrob Agents Chemother 41(2):481–483

De Lucca AJ, Bland JM, Vigo CB, Jacks TJ, Peter J, Walsh TJ (2000) D-cecropin B: proteolytic resistance, lethality for pathogenic fungi and binding properties. Med Mycol 38(4):301–308

De Smet K, Contreras R (2005) Human antimicrobial peptides: defensins, cathelicidins and histatins. Biotechnol Lett 27(18):1337–1347. doi:10.1007/s10529-005-0936-5

Del Sal G, Storici P, Schneider C, Romeo D, Zanetti M (1992) cDNA cloning of the neutrophil bactericidal peptide indolicidin. Biochem Biophys Res Commun 187(1):467–472

den Hertog AL, van Marle J, van Veen HA, Van't Hof W, Bolscher JG, Veerman EC et al (2005) Candidacidal effects of two antimicrobial peptides: histatin 5 causes small membrane defects, but LL-37 causes massive disruption of the cell membrane. Biochem J 388 (Pt 2):689–695. doi:10.1042/BJ20042099

den Hertog AL, van Marle J, Veerman EC, Valentijn-Benz M, Nazmi K, Kalay H et al (2006) The human cathelicidin peptide LL-37 and truncated variants induce segregation of lipids and proteins in the plasma membrane of Candida albicans. Biol Chem 387 (10–11):1495–1502. doi:10.1515/BC.2006.187

Dhople V, Krukemeyer A, Ramamoorthy A (2006) The human beta-defensin-3, an antibacterial peptide with multiple biological functions. Biochim Biophys Acta 1758(9):1499–1512. doi:10.1016/j.bbamem.2006.07.007

Driscoll J, Zuo Y, Xu T, Choi JR, Troxler RF, Oppenheim FG (1995) Functional comparison of native and recombinant human salivary histatin 1. J Dent Res 74(12):1837–1844

Durr UH, Sudheendra US, Ramamoorthy A (2006) LL-37, the only human member of the cathelicidin family of antimicrobial peptides. Biochim Biophys Acta 1758(9):1408–1425. doi:10.1016/j.bbamem.2006.03.030

Eckert R (2011) Road to clinical efficacy: challenges and novel strategies for antimicrobial peptide development. Future Microbiol 6(6):635–651. doi:10.2217/fmb.11.27

Edgerton M, Koshlukova SE, Lo TE, Chrzan BG, Straubinger RM, Raj PA (1998) Candidacidal activity of salivary histatins. Identification of a histatin 5-binding protein on Candida albicans. J Biol Chem 273(32):20438–20447

Fahrner RL, Dieckmann T, Harwig SS, Lehrer RI, Eisenberg D, Feigon J (1996) Solution structure of protegrin-1, a broad-spectrum antimicrobial peptide from porcine leukocytes. Chem Biol 3 (7):543–550

Fanos V, Cataldi L (2001) Renal transport of antibiotics and nephrotoxicity: a review. J Chemother 13 (5):461–472. doi:10.1179/joc.2001.13.5.461

Fisher MC, Henk DA, Briggs CJ, Brownstein JS, Madoff LC, McCraw SL et al (2012) Emerging fungal threats to animal, plant and ecosystem health. Nature 484 (7393):186–194. doi:10.1038/nature10947

Fitzgerald DH, Coleman DC, O'Connell BC (2003) Binding, internalisation and degradation of histatin 3 in histatin-resistant derivatives of Candida albicans. FEMS Microbiol Lett 220(2):247–253

Fjell CD, Hiss JA, Hancock RE, Schneider G (2012) Designing antimicrobial peptides: form follows function. Nat Rev Drug Discov 11(1):37–51. doi:10.1038/nrd3591

Fontaine T, Simenel C, Dubreucq G, Adam O, Delepierre M, Lemoine J et al (2000) Molecular organization of the alkali-insoluble fraction of Aspergillus fumigatus cell wall. J Biol Chem 275 (36):27594–27607. doi:10.1074/jbc.M909975199

Fosgerau K, Hoffmann T (2015) Peptide therapeutics: current status and future directions. Drug Discov Today 20(1):122–128. doi:10.1016/j.drudis.2014.10.003

Fox JL (2013) Antimicrobial peptides stage a comeback. Nat Biotechnol 31(5):379–382. doi:10.1038/nbt.2572

Friedrich CL, Rozek A, Patrzykat A, Hancock RE (2001) Structure and mechanism of action of an indolicidin peptide derivative with improved activity against gram-positive bacteria. J Biol Chem 276 (26):24015–24022. doi:10.1074/jbc.M009691200

Fukuoka T, Johnston DA, Winslow CA, de Groot MJ, Burt C, Hitchcock CA et al (2003) Genetic basis for differential activities of fluconazole and voriconazole against Candida krusei. Antimicrob Agents Chemother 47(4):1213–1219

Gabrielska J, Gagos M, Gubernator J, Gruszecki WI (2006) Binding of antibiotic amphotericin B to lipid membranes: a 1H NMR study. FEBS Lett 580 (11):2677–2685. doi:10.1016/j.febslet.2006.04.021

Ganz T (2003) Defensins: antimicrobial peptides of innate immunity. Nat Rev Immunol 3(9):710–720. doi:10.1038/nri1180

Ganz T (2005) Defensins and other antimicrobial peptides: a historical perspective and an update. Comb Chem High Throughput Screen 8(3):209–217

Ganz T, Lehrer RI (1995) Defensins. Pharmacol Ther 66 (2):191–205

Garibotto FM, Garro AD, Masman MF, Rodriguez AM, Luiten PG, Raimondi M et al (2010) New small-size peptides possessing antifungal activity. Bioorg Med Chem 18(1):158–167. doi:10.1016/j.bmc.2009.11.009

Gennaro R, Zanetti M (2000) Structural features and biological activities of the cathelicidin-derived antimicrobial peptides. Biopolymers 55(1):31–49. doi:10.1002/1097-0282(2000)55:1<31::AID-BIP40>3.0.CO;2-9

Goraya J, Wang Y, Li Z, O'Flaherty M, Knoop FC, Platz JE et al (2000) Peptides with antimicrobial activity from four different families isolated from the skins of the North American frogs Rana luteiventris, Rana berlandieri and Rana pipiens. Eur J Biochem/FEBS 267(3):894–900

Gupta V, Hwang BH, Lee J, Anselmo AC, Doshi N, Mitragotri S (2013) Mucoadhesive intestinal devices for oral delivery of salmon calcitonin. J Control Release: Off J Control Release Soc 172(3):753–762. doi:10.1016/j.jconrel.2013.09.004

Gyurko C, Lendenmann U, Troxler RF, Oppenheim FG (2000) Candida albicans mutants deficient in respiration are resistant to the small cationic salivary antimicrobial peptide histatin 5. Antimicrob Agents Chemother 44(2):348–354

Hale JD, Hancock RE (2007) Alternative mechanisms of action of cationic antimicrobial peptides on bacteria. Expert Rev Anti Infect Ther 5(6):951–959. doi:10.1586/14787210.5.6.951

Hancock RE, Patrzykat A (2002) Clinical development of cationic antimicrobial peptides: from natural to novel antibiotics. Curr Drug Targets Infect Disord 2 (1):79–83

Hancock RE, Rozek A (2002) Role of membranes in the activities of antimicrobial cationic peptides. FEMS Microbiol Lett 206(2):143–149

Hancock RE, Nijnik A, Philpott DJ (2012) Modulating immunity as a therapy for bacterial infections. Nat Rev Microbiol 10(4):243–254. doi:10.1038/nrmicro2745

Hay DI (1975) Fractionation of human parotid salivary proteins and the isolation of an histidine-rich acidic peptide which shows high affinity for hydroxyapatite surfaces. Arch Oral Biol 20(9):553–558

Helmerhorst EJ, Van't Hof W, Veerman EC, Simoons-Smit I, Nieuw Amerongen AV (1997) Synthetic histatin analogues with broad-spectrum antimicrobial activity. Biochem J 326(Pt 1):39–45

Helmerhorst EJ, Reijnders IM, van't Hof W, Simoons-Smit I, Veerman EC, Amerongen AV (1999a) Amphotericin B- and fluconazole-resistant Candida spp., Aspergillus fumigatus, and other newly emerging pathogenic fungi are susceptible to basic antifungal peptides. Antimicrob Agents Chemother 43(3):702–704

Helmerhorst EJ, Breeuwer P, van't Hof W, Walgreen-Weterings E, Oomen LC, Veerman EC et al (1999b) The cellular target of histatin 5 on Candida albicans is the energized mitochondrion. J Biol Chem 274 (11):7286–7291

Helmerhorst EJ, Troxler RF, Oppenheim FG (2001) The human salivary peptide histatin 5 exerts its antifungal activity through the formation of reactive oxygen species. Proc Natl Acad Sci U S A 98(25):14637–14642. doi:10.1073/pnas.141366998

Helmerhorst EJ, Murphy MP, Troxler RF, Oppenheim FG (2002) Characterization of the mitochondrial respiratory pathways in Candida albicans. Biochim Biophys Acta 1556(1):73–80

Holt A, Killian JA (2010) Orientation and dynamics of transmembrane peptides: the power of simple models. Eur Biophys J: EBJ 39(4):609–621. doi:10.1007/s00249-009-0567-1

Hsu CH, Chen C, Jou ML, Lee AY, Lin YC, Yu YP et al (2005) Structural and DNA-binding studies on the bovine antimicrobial peptide, indolicidin: evidence for multiple conformations involved in binding to membranes and DNA. Nucleic Acids Res 33 (13):4053–4064. doi:10.1093/nar/gki725

Hujakka H, Ratilainen J, Korjamo T, Lankinen H, Kuusela P, Santa H et al (2001) Synthesis and antimicrobial activity of the symmetric dimeric form of Temporin A based on 3-N, N-di(3-aminopropyl) amino propanoic acid as the branching unit. Bioorg Med Chem 9(6):1601–1607

Jenssen H, Hamill P, Hancock RE (2006) Peptide antimicrobial agents. Clin Microbiol Rev 19(3):491–511. doi:10.1128/CMR.00056-05

Joly S, Maze C, McCray PB Jr, Guthmiller JM (2004) Human beta-defensins 2 and 3 demonstrate strain-selective activity against oral microorganisms. J Clin Microbiol 42(3):1024–1029

Kahn A, Carey EJ, Blair JE (2014) Universal fungal prophylaxis and risk of coccidioidomycosis in liver transplant recipients living in an endemic area. Liver Transplant: Off Publ Am Assoc Study Liver Dis Int Liver Transplant Soc. doi:10.1002/lt.24055

Kaminski DM (2014) Recent progress in the study of the interactions of amphotericin B with cholesterol and ergosterol in lipid environments. Eur Biophys J: EBJ 43(10–11):453–467. doi:10.1007/s00249-014-0983-8

Kaspar AA, Reichert JM (2013) Future directions for peptide therapeutics development. Drug Discov Today 18(17–18):807–817. doi:10.1016/j.drudis.2013.05.011

Kavanagh K, Dowd S (2004) Histatins: antimicrobial peptides with therapeutic potential. J Pharm Pharmacol 56(3):285–289. doi:10.1211/0022357022971

Kim DH, Lee YT, Lee YJ, Chung JH, Lee BL, Choi BS et al (1998) Bacterial expression of tenecin 3, an insect antifungal protein isolated from Tenebrio molitor, and its efficient purification. Mol Cells 8(6):786–789

Kim DH, Lee DG, Kim KL, Lee Y (2001) Internalization of tenecin 3 by a fungal cellular process is essential for its fungicidal effect on Candida albicans. Eur J Biochem/FEBS 268(16):4449–4458

Knappe D, Henklein P, Hoffmann R, Hilpert K (2010) Easy strategy to protect antimicrobial peptides from fast degradation in serum. Antimicrob Agents Chemother 54(9):4003–4005. doi:10.1128/AAC.00300-10

Kosciuczuk EM, Lisowski P, Jarczak J, Strzalkowska N, Jozwik A, Horbanczuk J et al (2012) Cathelicidins: family of antimicrobial peptides. A review. Mol Biol Rep 39(12):10957–10970. doi:10.1007/s11033-012-1997-x

Koshlukova SE, Lloyd TL, Araujo MW, Edgerton M (1999) Salivary histatin 5 induces non-lytic release of ATP from Candida albicans leading to cell death. J Biol Chem 274(27):18872–18879

Koshlukova SE, Araujo MW, Baev D, Edgerton M (2000) Released ATP is an extracellular cytotoxic mediator in salivary histatin 5-induced killing of Candida albicans. Infect Immun 68(12):6848–6856

Kumar R, Chadha S, Saraswat D, Bajwa JS, Li RA, Conti HR et al (2011) Histatin 5 uptake by Candida albicans utilizes polyamine transporters Dur3 and Dur31 proteins. J Biol Chem 286(51):43748–43758. doi:10.1074/jbc.M111.311175

Kwon MY, Hong SY, Lee KH (1998) Structure-activity analysis of brevinin 1E amide, an antimicrobial peptide from Rana esculenta. Biochim Biophys Acta 1387 (1–2):239–248

Ladokhin AS, White SH (2001) 'Detergent-like' permeabilization of anionic lipid vesicles by melittin. Biochim Biophys Acta 1514(2):253–260

Lakshminarayanan R, Liu S, Li J, Nandhakumar M, Aung TT, Goh E et al (2014) Synthetic multivalent antifungal peptides effective against fungi. PLoS One 9(2): e87730. doi:10.1371/journal.pone.0087730

Lamberty M, Zachary D, Lanot R, Bordereau C, Robert A, Hoffmann JA et al (2001) Insect immunity. Constitutive expression of a cysteine-rich antifungal and a linear antibacterial peptide in a termite insect. J Biol Chem 276(6):4085–4092. doi:10.1074/jbc.M002998200

Lee SY, Moon HJ, Kurata S, Natori S, Lee BL (1995) Purification and cDNA cloning of an antifungal protein from the hemolymph of Holotrichia diomphalia larvae. Biol Pharm Bull 18(8):1049–1052

Lee DG, Kim DH, Park Y, Kim HK, Kim HN, Shin YK et al (2001) Fungicidal effect of antimicrobial peptide, PMAP-23, isolated from porcine myeloid against Candida albicans. Biochem Biophys Res Commun 282 (2):570–574. doi:10.1006/bbrc.2001.4602

Lee DG, Kim PI, Park Y, Park SC, Woo ER, Hahm KS (2002a) Antifungal mechanism of SMAP-29 (1–18) isolated from sheep myeloid mRNA against Trichosporon beigelii. Biochem Biophys Res Commun 295(3):591–596

Lee DG, Kim PI, Park Y, Woo ER, Choi JS, Choi CH et al (2002b) Design of novel peptide analogs with potent fungicidal activity, based on PMAP-23 antimicrobial peptide isolated from porcine myeloid. Biochem Biophys Res Commun 293(1):231–238. doi:10.1016/S0006-291X(02)00222-X

Lee DG, Kim HK, Kim SA, Park Y, Park SC, Jang SH et al (2003) Fungicidal effect of indolicidin and its interaction with phospholipid membranes. Biochem Biophys Res Commun 305(2):305–310

Legrand D, Pierce A, Elass E, Carpentier M, Mariller C, Mazurier J (2008) Lactoferrin structure and functions. Adv Exp Med Biol 606:163–194. doi:10.1007/978-0-387-74087-4_6

Lehrer RI, Ganz T, Szklarek D, Selsted ME (1988) Modulation of the in vitro candidacidal activity of human neutrophil defensins by target cell metabolism and divalent cations. J Clin Invest 81(6):1829–1835. doi:10.1172/JCI113527

Levay PF, Viljoen M (1995) Lactoferrin: a general review. Haematologica 80(3):252–267

Li XS, Sun JN, Okamoto-Shibayama K, Edgerton M (2006) Candida albicans cell wall ssa proteins bind and facilitate import of salivary histatin 5 required for toxicity. J Biol Chem 281(32):22453–22463. doi:10.1074/jbc.M604064200

Lim K, Chua RR, Ho B, Tambyah PA, Hadinoto K, Leong SS (2015) Development of a catheter functionalized by a polydopamine peptide coating with antimicrobial and antibiofilm properties. Acta Biomater 15:127–138. doi:10.1016/j.actbio.2014.12.015

Liu AY, Destoumieux D, Wong AV, Park CH, Valore EV, Liu L et al (2002) Human beta-defensin-2 production in keratinocytes is regulated by interleukin-1, bacteria, and the state of differentiation. J Invest Dermatol 118(2):275–281. doi:10.1046/j.0022-202x.2001.01651.x

Lopez-Garcia B, Lee PH, Yamasaki K, Gallo RL (2005) Anti-fungal activity of cathelicidins and their potential role in Candida albicans skin infection. J Invest Dermatol 125(1):108–115. doi:10.1111/j.0022-202X.2005.23713.x

Lupetti A, Paulusma-Annema A, Welling MM, Senesi S, van Dissel JT, Nibbering PH (2000) Candidacidal activities of human lactoferrin peptides derived from the N terminus. Antimicrob Agents Chemother 44 (12):3257–3263

Mangoni ML, Rinaldi AC, Di Giulio A, Mignogna G, Bozzi A, Barra D et al (2000) Structure-function relationships of temporins, small antimicrobial peptides from amphibian skin. Eur J Biochem/FEBS 267(5):1447–1454

Marenah L, Flatt PR, Orr DF, McClean S, Shaw C, Abdel-Wahab YH (2004) Brevinin-1 and multiple insulin-releasing peptides in the skin of the frog Rana palustris. J Endocrinol 181(2):347–354

Martinez J, Temesgen Z (2006) Opportunistic infections in patients with HIV and AIDS. Fungal and parasitic infections. J Med Liban Liban Med J 54(2):84–90

Marukutira T, Huprikar S, Azie N, Quan SP, Meier-Kriesche HU, Horn DL (2014) Clinical characteristics and outcomes in 303 HIV-infected patients with invasive fungal infections: data from the Prospective Antifungal Therapy Alliance registry, a multicenter, observational study. Hiv/Aids 6:39–47. doi:10.2147/HIV.S53910

Masman MF, Rodriguez AM, Raimondi M, Zacchino SA, Luiten PG, Somlai C et al (2009) Penetratin and derivatives acting as antifungal agents. Eur J Med Chem 44(1):212–228. doi:10.1016/j.ejmech.2008.02.019

Matsuzaki K (2009) Control of cell selectivity of antimicrobial peptides. Biochim Biophys Acta 1788 (8):1687–1692. doi:10.1016/j.bbamem.2008.09.013

Maurya IK, Pathak S, Sharma M, Sanwal H, Chaudhary P, Tupe S et al (2011) Antifungal activity of novel synthetic peptides by accumulation of reactive oxygen species (ROS) and disruption of cell wall against Candida albicans. Peptides 32(8):1732–1740. doi:10.1016/j.peptides.2011.06.003

McNeil MM, Nash SL, Hajjeh RA, Phelan MA, Conn LA, Plikaytis BD et al (2001) Trends in mortality due to invasive mycotic diseases in the United States, 1980–1997. Clin Infect Dis: Off Publ Infect Dis Soc Am 33(5):641–647. doi:10.1086/322606

Milletti F (2012) Cell-penetrating peptides: classes, origin, and current landscape. Drug Discov Today 17 (15–16):850–860, doi:http://dx.doi.org/10.1016/j.drudis.2012.03.002

Moore A (2003) The big and small of drug discovery. EMBO Rep 4(2):114–117. doi:10.1038/sj.embor.embor748

Morikawa N, Hagiwara K, Nakajima T (1992) Brevinin-1 and -2, unique antimicrobial peptides from the skin of the frog, Rana brevipoda porsa. Biochem Biophys Res Commun 189(1):184–190

Murakami Y, Nagata H, Shizukuishi S, Nakashima K, Okawa T, Takigawa M et al (1994) Histatin as a synergistic stimulator with epidermal growth factor of rabbit chondrocyte proliferation. Biochem Biophys Res Commun 198(1):274–280. doi:10.1006/bbrc.1994.1038

Neville F, Ivankin A, Konovalov O, Gidalevitz D (2010) A comparative study on the interactions of SMAP-29 with lipid monolayers. Biochim Biophys Acta 1798 (5):851–860. doi:10.1016/j.bbamem.2009.09.017

Nguyen LT, Schibli DJ, Vogel HJ (2005) Structural studies and model membrane interactions of two peptides derived from bovine lactoferricin. J Pept Sci: Off Publ Eur Pept Soc 11(7):379–389. doi:10.1002/psc.629

Nguyen LT, Chau JK, Perry NA, de Boer L, Zaat SA, Vogel HJ (2010) Serum stabilities of short tryptophan- and arginine-rich antimicrobial peptide analogs. PLoS One 5(9):e12684. doi:10.1371/journal.pone.0012684

Nguyen LT, Haney EF, Vogel HJ (2011) The expanding scope of antimicrobial peptide structures and their modes of action. Trends Biotechnol 29(9):464–472, doi:http://dx.doi.org/10.1016/j.tibtech.2011.05.001

Novkovic M, Simunic J, Bojovic V, Tossi A, Juretic D (2012) DADP: the database of anuran defense peptides. Bioinformatics 28(10):1406–1407. doi:10.1093/bioinformatics/bts141

Oppenheim FG, Yang YC, Diamond RD, Hyslop D, Offner GD, Troxler RF (1986) The primary structure and functional characterization of the neutral histidine-rich polypeptide from human parotid secretion. J Biol Chem 261(3):1177–1182

Oppenheim JJ, Biragyn A, Kwak LW, Yang D (2003) Roles of antimicrobial peptides such as defensins in innate and adaptive immunity. Ann Rheum Dis 62 (Suppl 2):ii17–ii21

Ostrosky-Zeichner L, Casadevall A, Galgiani JN, Odds FC, Rex JH (2010) An insight into the antifungal pipeline: selected new molecules and beyond. Nat Rev Drug Discov 9(9):719–727. doi:10.1038/nrd3074

Otvos L Jr, Wade JD (2014) Current challenges in peptide-based drug discovery. Front Chem 2:62. doi:10.3389/fchem.2014.00062

Oudhoff MJ, Blaauboer ME, Nazmi K, Scheres N, Bolscher JG, Veerman EC (2010) The role of salivary histatin and the human cathelicidin LL-37 in wound healing and innate immunity. Biol Chem 391 (5):541–548. doi:10.1515/BC.2010.057

Pal T, Abraham B, Sonnevend A, Jumaa P, Conlon JM (2006) Brevinin-1BYa: a naturally occurring peptide from frog skin with broad-spectrum antibacterial and antifungal properties. Int J Antimicrob Agents 27 (6):525–529. doi:10.1016/j.ijantimicag.2006.01.010

Park SH, Kim HE, Kim CM, Yun HJ, Choi EC, Lee BJ (2002a) Role of proline, cysteine and a disulphide bridge in the structure and activity of the antimicrobial peptide gaegurin 5. Biochem J 368 (Pt 1):171–182. doi:10.1042/BJ20020385

Park K, Oh D, Shin SY, Hahm KS, Kim Y (2002b) Structural studies of porcine myeloid antibacterial peptide PMAP-23 and its analogues in DPC micelles by NMR spectroscopy. Biochem Biophys Res Commun 290(1):204–212. doi:10.1006/bbrc.2001.6173

Pena OM, Afacan N, Pistolic J, Chen C, Madera L, Falsafi R et al (2013) Synthetic cationic peptide IDR-1018 modulates human macrophage differentiation. PLoS One 8(1):e52449. doi:10.1371/journal.pone.0052449

Peschel A, Sahl HG (2006) The co-evolution of host cationic antimicrobial peptides and microbial resistance. Nat Rev Microbiol 4(7):529–536. doi:10.1038/nrmicro1441

Raj PA, Dentino AR (2002) Current status of defensins and their role in innate and adaptive immunity. FEMS Microbiol Lett 206(1):9–18

Raj PA, Marcus E, Sukumaran DK (1998) Structure of human salivary histatin 5 in aqueous and nonaqueous solutions. Biopolymers 45(1):51–67. doi:10.1002/(sici)1097-0282(199801)45:1<51::aid-bip5>3.0.co;2-y

Riciluca KC, Sayegh RS, Melo RL, Silva PI Jr (2012) Rondonin an antifungal peptide from spider (Acanthoscurria rondoniae) haemolymph. Results Immunol 2:66–71. doi:10.1016/j.rinim.2012.03.001

Risso A, Braidot E, Sordano MC, Vianello A, Macri F, Skerlavaj B et al (2002) BMAP-28, an antibiotic peptide of innate immunity, induces cell death through opening of the mitochondrial permeability transition pore. Mol Cell Biol 22(6):1926–1935

Rodriguez-Cerdeira C, Arenas R, Moreno-Coutino G, Vasquez E, Fernandez R, Chang P (2014) Systemic fungal infections in patients with human immunodeficiency virus. Actas Dermosifiliogr 105 (1):5–17. doi:10.1016/j.ad.2012.06.017

Rollins-Smith LA, Carey C, Conlon JM, Reinert LK, Doersam JK, Bergman T et al (2003) Activities of temporin family peptides against the chytrid fungus (Batrachochytrium dendrobatidis) associated with global amphibian declines. Antimicrob Agents Chemother 47(3):1157–1160

Rothstein DM, Spacciapoli P, Tran LT, Xu T, Roberts FD, Dalla Serra M et al (2001) Anticandida activity is retained in P-113, a 12-amino-acid fragment of histatin 5. Antimicrob Agents Chemother 45 (5):1367–1373. doi:10.1128/aac.45.5.1367-1373.2001

Roversi D, Luca V, Aureli S, Park Y, Mangoni ML, Stella L (2014) How many antimicrobial peptide molecules kill a bacterium? The case of PMAP-23. ACS Chem Biol 9(9):2003–2007. doi:10.1021/cb500426r

Ryu JS, Cho AY, Seo SW, Min H (2014) Engineering bioactive peptide-based therapeutic molecules. Methods Mol Biol 1088:35–50. doi:10.1007/978-1-62703-673-3_3

Sabatini LM, Azen EA (1989) Histatins, a family of salivary histidine-rich proteins, are encoded by at least two loci (HIS1 and HIS2). Biochem Biophys Res Commun 160(2):495–502

Sachdeva V, Zhou Y, Banga AK (2013) In vivo transdermal delivery of leuprolide using microneedles and iontophoresis. Curr Pharm Biotechnol 14 (2):180–193

Saravanan R, Joshi M, Mohanram H, Bhunia A, Mangoni ML, Bhattacharjya S (2013) NMR structure of temporin-1 ta in lipopolysaccharide micelles: mechanistic insight into inactivation by outer membrane. PLoS One 8(9):e72718. doi:10.1371/journal.pone.0072718

Savelyeva A, Ghavami S, Davoodpour P, Asoodeh A, Los MJ (2014) An overview of Brevinin superfamily: structure, function and clinical perspectives. Adv Exp Med Biol 818:197–212. doi:10.1007/978-1-4471-6458-6_10

Schoffelmeer EA, Klis FM, Sietsma JH, Cornelissen BJ (1999) The cell wall of Fusarium oxysporum. Fungal Genet Biol: FG & B 27(2–3):275–282. doi:10.1006/fgbi.1999.1153

Selsted ME, Ouellette AJ (2005) Mammalian defensins in the antimicrobial immune response. Nat Immunol 6 (6):551–557. doi:10.1038/ni1206

Selsted ME, Novotny MJ, Morris WL, Tang YQ, Smith W, Cullor JS (1992) Indolicidin, a novel bactericidal tridecapeptide amide from neutrophils. J Biol Chem 267(7):4292–4295

Shai Y (1999) Mechanism of the binding, insertion and destabilization of phospholipid bilayer membranes by alpha-helical antimicrobial and cell non-selective membrane-lytic peptides. Biochim Biophys Acta 1462(1–2):55–70

Shimamura M, Yamamoto Y, Ashino H, Oikawa T, Hazato T, Tsuda H et al (2004) Bovine lactoferrin inhibits tumor-induced angiogenesis. Int J Cancer J Int Cancer 111(1):111–116. doi:10.1002/ijc.20187

Shin SY, Park EJ, Yang ST, Jung HJ, Eom SH, Song WK et al (2001) Structure-activity analysis of SMAP-29, a sheep leukocytes-derived antimicrobial peptide. Biochem Biophys Res Commun 285(4):1046–1051. doi:10.1006/bbrc.2001.5280

Simmaco M, Mignogna G, Canofeni S, Miele R, Mangoni ML, Barra D (1996) Temporins, antimicrobial peptides from the European red frog Rana temporaria. Eur J Biochem/FEBS 242(3):788–792

Situ H, Balasubramanian SV, Bobek LA (2000) Role of alpha-helical conformation of histatin-5 in candidacidal activity examined by proline variants. Biochim Biophys Acta 1475(3):377–382

Skerlavaj B, Benincasa M, Risso A, Zanetti M, Gennaro R (1999) SMAP-29: a potent antibacterial and antifungal peptide from sheep leukocytes. FEBS Lett 463 (1–2):58–62

Steckbeck JD, Deslouches B, Montelaro RC (2014) Antimicrobial peptides: new drugs for bad bugs? Expert Opin Biol Ther 14(1):11–14. doi:10.1517/14712598.2013.844227

Steiner H, Andreu D, Merrifield RB (1988) Binding and action of cecropin and cecropin analogues: antibacterial peptides from insects. Biochim Biophys Acta 939(2):260–266

Steinstraesser L, Kraneburg U, Jacobsen F, Al-Benna S (2011) Host defense peptides and their antimicrobial-immunomodulatory duality. Immunobiology 216 (3):322–333, doi:http://dx.doi.org/10.1016/j.imbio.2010.07.003

Steinstraesser L, Hirsch T, Schulte M, Kueckelhaus M, Jacobsen F, Mersch EA et al (2012) Innate defense regulator peptide 1018 in wound healing and wound infection. PLoS One 7(8):e39373. doi:10.1371/journal.pone.0039373

Suarez-Carmona M, Hubert P, Delvenne P, Herfs M (2014) Defensins: "Simple" antimicrobial peptides or broad-spectrum molecules? Cytokine Growth Factor Rev. doi:10.1016/j.cytogfr.2014.12.005

Sugiyama K, Suzuki Y, Furuta H (1985) Isolation and characterization of histamine-releasing peptides from human parotid saliva. Life Sci 37(5):475–480

Tack BF, Sawai MV, Kearney WR, Robertson AD, Sherman MA, Wang W et al (2002) SMAP-29 has two LPS-binding sites and a central hinge. Eur J Biochem/FEBS 269(4):1181–1189

Takahashi D, Shukla SK, Prakash O, Zhang G (2010) Structural determinants of host defense peptides for antimicrobial activity and target cell selectivity. Biochimie 92(9):1236–1241, doi:http://dx.doi.org/10.1016/j.biochi.2010.02.023

Tanida T, Okamoto T, Ueta E, Yamamoto T, Osaki T (2006) Antimicrobial peptides enhance the candidacidal activity of antifungal drugs by promoting the efflux of ATP from Candida cells. J Antimicrob Chemother 57(1):94–103. doi:10.1093/jac/dki402

Theis T, Stahl U (2004) Antifungal proteins: targets, mechanisms and prospective applications. Cell Mol Life Sci: CMLS 61(4):437–455. doi:10.1007/s00018-003-3231-4

Tielens AG, Rotte C, van Hellemond JJ, Martin W (2002) Mitochondria as we don't know them. Trends Biochem Sci 27(11):564–572

Tsai H, Bobek LA (1997) Human salivary histatin-5 exerts potent fungicidal activity against Cryptococcus neoformans. Biochim Biophys Acta (BBA) Gen Subj 1336(3):367–369, doi:http://dx.doi.org/10.1016/S0304-4165(97)00076-7

van der Weerden NL, Bleackley MR, Anderson MA (2013) Properties and mechanisms of action of naturally occurring antifungal peptides. Cell Mol Life Sci: CMLS 70(19):3545–3570. doi:10.1007/s00018-013-1260-1

Viejo-Diaz M, Andres MT, Perez-Gil J, Sanchez M, Fierro JF (2003) Potassium efflux induced by a new lactoferrin-derived peptide mimicking the effect of native human lactoferrin on the bacterial cytoplasmic membrane. Biochem Biokhim 68(2):217–227

Viejo-Diaz M, Andres MT, Fierro JF (2004) Effects of human lactoferrin on the cytoplasmic membrane of Candida albicans cells related with its candidacidal activity. FEMS Immunol Med Microbiol 42 (2):181–185. doi:10.1016/j.femsim.2004.04.005

Wade D, Silberring J, Soliymani R, Heikkinen S, Kilpelainen I, Lankinen H et al (2000) Antibacterial activities of temporin A analogs. FEBS Lett 479 (1–2):6–9

Wang G (2008) Structures of human host defense cathelicidin LL-37 and its smallest antimicrobial peptide KR-12 in lipid micelles. J Biol Chem 283 (47):32637–32643. doi:10.1074/jbc.M805533200

Wang P, Nan YH, Yang ST, Kang SW, Kim Y, Park IS et al (2010) Cell selectivity and anti-inflammatory activity of a Leu/Lys-rich alpha-helical model antimicrobial peptide and its diastereomeric peptides. Peptides 31(7):1251–1261. doi:10.1016/j.peptides.2010.03.032

Ward PP, Paz E, Conneely OM (2005) Multifunctional roles of lactoferrin: a critical overview. Cell Mol Life Sci: CMLS 62(22):2540–2548. doi:10.1007/s00018-005-5369-8

Wisplinghoff H, Bischoff T, Tallent SM, Seifert H, Wenzel RP, Edmond MB (2004) Nosocomial bloodstream infections in US hospitals: analysis of 24,179 cases from a prospective nationwide surveillance study. Clin Infect Dis: Off Publ Infect Dis Soc Am 39(3):309–317. doi:10.1086/421946

Wong JH, Ye XJ, Ng TB (2013) Cathelicidins: peptides with antimicrobial, immunomodulatory, anti-inflammatory, angiogenic, anticancer and procancer activities. Curr Protein Pept Sci 14(6):504–514

Wong-Beringer A, Jacobs RA, Guglielmo BJ (1998) Lipid formulations of amphotericin B: clinical efficacy and toxicities. Clin Infect Dis: Off Publ Infect Dis Soc Am 27(3):603–618

Wunder D, Dong J, Baev D, Edgerton M (2004) Human salivary histatin 5 fungicidal action does not induce programmed cell death pathways in Candida albicans. Antimicrob Agents Chemother 48(1):110–115

Xu X, Lai R (2015) The chemistry and biological activities of peptides from amphibian skin secretions. Chem Rev 115(4):1760–1846. doi:10.1021/cr4006704

Xu Y, Ambudkar I, Yamagishi H, Swaim W, Walsh TJ, O'Connell BC (1999) Histatin 3-mediated killing of Candida albicans: effect of extracellular salt concentration on binding and internalization. Antimicrob Agents Chemother 43(9):2256–2262

Yang S, Jung, H., Kim J (2009) Solution structure of BMAP-27. doi:10.2210/pdb2ket/pdb

Yang L, Harroun TA, Weiss TM, Ding L, Huang HW (2001) Barrel-stave model or toroidal model? A case study on melittin pores. Biophys J 81(3):1475–1485. doi:10.1016/S0006-3495(01)75802-X

Yeaman MR, Yount NY (2003) Mechanisms of antimicrobial peptide action and resistance. Pharmacol Rev 55(1):27–55. doi:10.1124/pr.55.1.2

Yount NY, Bayer AS, Xiong YQ, Yeaman MR (2006) Advances in antimicrobial peptide immunobiology. Biopolymers 84(5):435–458. doi:10.1002/bip.20543

Yu S, Chai X, Wang Y, Cao Y, Zhang J, Wu Q et al (2014) Triazole derivatives with improved in vitro antifungal activity over azole drugs. Drug Des Devel Ther 8:383–390. doi:10.2147/DDDT.S58680

Zanetti M (2005) The role of cathelicidins in the innate host defenses of mammals. Curr Issues Mol Biol 7 (2):179–196

Zanetti M, Gennaro R, Romeo D (1995) Cathelicidins: a novel protein family with a common proregion and a variable C-terminal antimicrobial domain. FEBS Lett 374(1):1–5

Zasloff M (2002) Antimicrobial peptides of multicellular organisms. Nature 415(6870):389–395. doi:10.1038/415389a

Zhai B, Zhou H, Yang L, Zhang J, Jung K, Giam CZ et al (2010) Polymyxin B, in combination with fluconazole, exerts a potent fungicidal effect. J Antimicrob Chemother 65(5):931–938. doi:10.1093/jac/dkq046

Zhang J, Movahedi A, Xu J, Wang M, Wu X, Xu C et al (2015) In vitro production and antifungal activity of peptide ABP-dHC-cecropin A. J Biotechnol 199C:47–54. doi:10.1016/j.jbiotec.2015.02.018

Zhao L, Tolbert WD, Ericksen B, Zhan C, Wu X, Yuan W et al (2013) Single, double and quadruple alanine substitutions at oligomeric interfaces identify hydrophobicity as the key determinant of human neutrophil alpha defensin HNP1 function. PLoS One 8 (11):e78937. doi:10.1371/journal.pone.0078937

Lipopeptides: Status and Strategies to Control Fungal Infection

4

Piyush Baindara and Suresh Korpole

Abstract

Global food security is become central focus, specifically because of threatening plant diseases caused by fungal phyto-pathogens and massive economic losses thereof. In the context of bio-control of fungal phyto-pathogens, lipopeptides produced by *Bacillus* sp. have been studied well. The three families of *Bacillus* lipopeptides are surfactin, iturins and fengycins, confirmed for their antagonistic activities against various fungal phyto-pathogens. In recent past lipopeptides produced by *Pseudomonas* sp. has also proven effective bio-control agents, specifically against fungal phyto-pathogens. On other hand echinocandins are novel class of anti-fungal lipopeptides produced by various *Aspergillus* sp., used successfully in treatment of serious fungal infections and currently in clinical trials. Here we summarized all available information and data of lipopeptides in focus of their use as bio-control agent for plant protection as well as in treatment of fungal diseases in human caused by different pathogenic fungi.

4.1 Introduction

Host defense peptides or antimicrobial peptides (AMPs) play an essential protective role in the innate immune system of all organisms. These antimicrobial peptides are multifunctional compounds with multiple utilities (Franco 2011). The increasing resistance of bacteria and fungi to multiple antibiotics is a major public health concern worldwide that leads to enormous efforts for developing new antibiotics with new modes of action. Two promising families of drugs that meet these criteria are antimicrobial peptides (AMPs) and lipopeptides (Boman 1998; Zasloff 2002). Lipopeptides differ from AMPs as the earlier are produced only in bacteria and fungi but later is produced by all forms of life. Lipopeptides are small molecules produced during cultivation on various carbon sources. They are formed as short linear peptides linked with a lipid tail or other lipophilic molecules to form cyclic structures (Arnusch et al. 2012; Raaijmakers et al. 2010; Mandal et al. 2013).

P. Baindara • S. Korpole (✉)
MTCC and Gene Bank, CSIR-Institute of Microbial Technology, Sector-39A, Chandigarh 160036, India
e-mail: korpole@gmail.com; suresh@imtech.res.in

A. Basak et al. (eds.), *Recent Trends in Antifungal Agents and Antifungal Therapy*,
DOI 10.1007/978-81-322-2782-3_4

Polymyxin A was the first antimicrobial lipopeptide described in detail. It was isolated in the year 1949 from the bacterium *Bacillus polymyxa* (Mandal et al. 2013) which is reclassified as *Paenibacillus polymyxa*. However, the effective lipopeptide production as a bio-surfactant was first reported from *Bacillus subtilis* IAM1213 (Jones 1949). Since then, various types of antimicrobial lipopeptides have been isolated from other *Bacillus* strains and their biosynthesis and antimicrobial activity studied in detail (Arima et al. 1968; Peypoux et al. 1984; Grangemard et al. 1999; Rivardo et al. 2009; Tareq et al. 2014). Other genera like *Pseudomonas*, *Streptomyces*, and *Aspergillus* representing Gram-negative bacteria, actinobacteria, and fungi are also reported for lipopeptide production (Cochrane and Vederas 2014; Morikawa et al. 2000; Higginbotham and Murphy 2010; Rao et al. 2013; Nielsen and Sørensen 2003). Recently, lipopeptides have received great attention for their cytotoxic, antitumor, immunosuppressant, and surfactant properties apart from antimicrobial property (Sørensen et al. 2001; Cameotra and Makkar 2004; Donadio et al. 2007; Song et al. 2011). Some of them like daptomycin (Song et al. 2014), caspofungin (Robbel and Marahiel 2010), micafungin (Ngai et al. 2011), and anidulafungin (Emiroglu 2011) have even reached the market with antibiotic status. Lipopeptides showed remarkable membrane-active and surface-interface properties resulting in a number of excellent biological activities, which are of great relevance in health care and biotechnology-based processes. These properties make lipopeptides as potent candidate drugs for the resolution of a number of global issues in medicine industry and environmental protection (George and Reboli 2012).

4.1.1 Origin and Structural Diversity of Lipopeptides

4.1.1.1 Lipopeptides Produced by Gram-Positive Bacteria (*Bacillus* sp.)

Broadly there are three types of lipopeptides, namely, surfactin, iturin, and fengycin, that are produced by various *Bacillus* species.

4.1.1.1.1 Iturin

Iturins usually have a molecular mass of ~1.1 kDa. They consist of two major parts: a peptide part composed of short amino acid sequence and carbon hydrophobic tail. This composition clearly indicates an amphiphilic character of these compounds and suggests cell membrane-based mechanism of action (Banat et al. 2010). The iturins are cyclic peptides containing seven amino acids (heptapeptide) linked to a fatty acid (β-amino) chain that varies between C14 and C17 carbon molecules. Bacillomycin D and bacillomycin L are important members of the iturin family and peptide-lipid antibiotics. Structural determination indicated that bacillomycins D and L consist of a heptapeptide chain linked to a liposoluble β-amino acid. Further aspartyl, glutamyl, asparaginyl, and glutarninyl amino acids were found as peptidic moieties (Arima et al. 1968; Stachelhaus et al. 1995; Moyne et al. 2001). Bacillomycin Lc, a new antifungal antibiotic of the iturin class, was isolated from a strain of *Bacillus subtilis*. The structure determined by chemical and spectrometric analyses showed differences from bacillomycin L. It also exhibited sequence changes from aspartate-1 to asparagine-1 and from glutamine-5 to glutamate-5 (Tenoux et al. 1991). Bacillomycin F, another variant, also displayed differences with other members of the iturin family. In fact, it is a mixture of homologous peptide-lipids, essentially containing $C_{51}H_{80}N_{12}0_{14}$ and $C_{52}N_{82}N_{12}0_{14}$. The lipid moiety consists of minor *iso*C_{15}, *anteiso*C_{15} β-amino acids, and major *iso*C_{16}, *iso*C_{17} and *anteiso*C_{17} p-amino acids (Eshita et al. 1995). Mycosubtilin, a peptide-lipid antibiotic, was isolated from *Bacillus subtilis*. It was identified as a cyclopeptide composed of seven α-amino acids in an LDDLLDL sequence closed by a β-amino acid linkage similar to that found in other antibiotics of the iturin group (Peypoux et al. 1985). The unique biological and physicochemical properties make these molecules interesting for biocontrol, food, and pharmaceutical applications (Fig. 4.1).

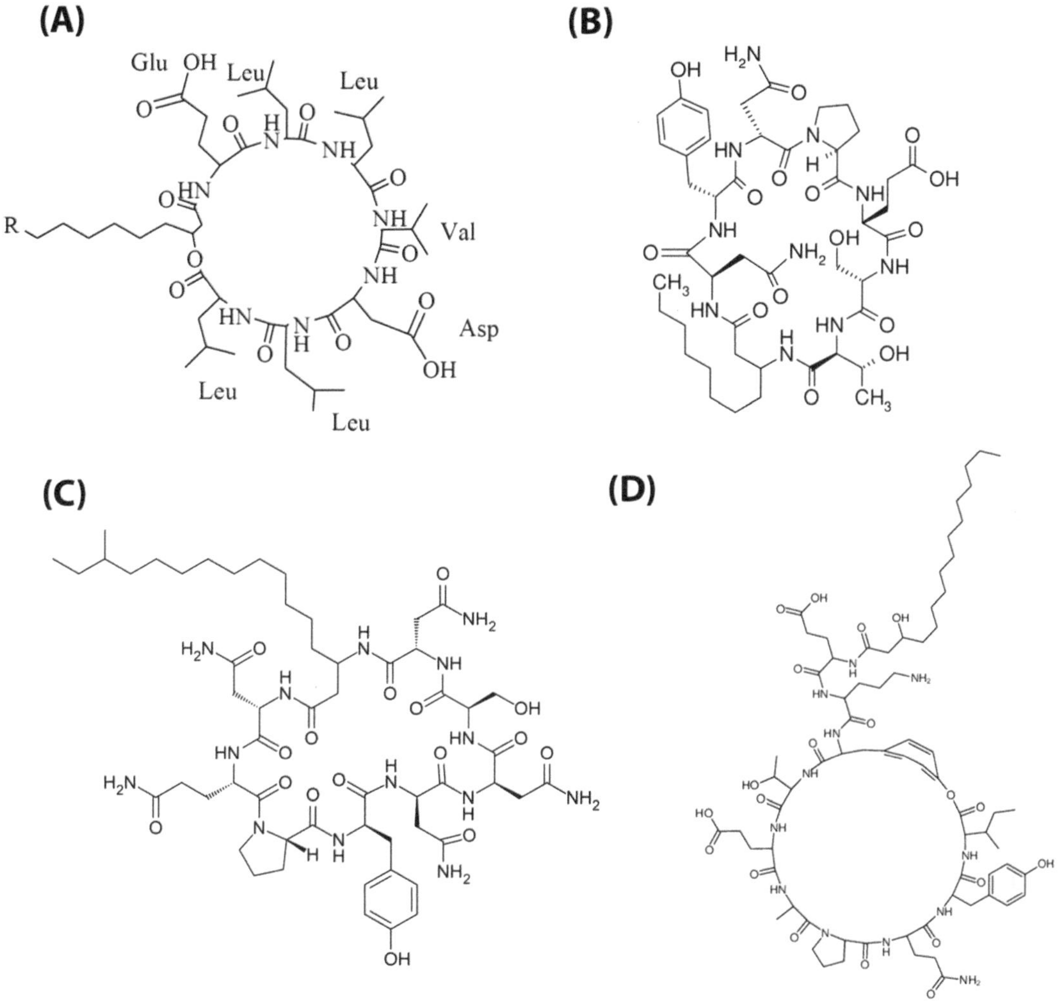

Fig. 4.1 Structure of different lipopeptides produced by *Bacillus* sp. (**a**) Surfactin (Peypoux et al. 1986). (**b**) Bacillomycin D (Liu et al. 2008). (**c**) Mycosubtilin (Peypoux et al. 1985). (**d**) Fengycin/plipastatin A (Peypoux et al. 1981)

4.1.1.1.2 Surfactin

Surfactin (~1.36 kDa) is an amphipathic cyclic heptapeptide composed of Glu-Leu-Leu-Val-Asp-Leu-Leu (ELLVDLL) with the chiral sequence LLDLLDL interlinked with β-hydroxy fatty acid with the chain length of 12–16 carbon atoms to form a cyclic lactone ring structure (Jacques 2011). The order of amino acids and the size of lipid moiety differ in a variety of surfactins (Tsan et al. 2007). In surfactin molecules while the hydrophobic amino acids are located at positions 2, 3, 4, 6, and 7, the Glu and Asp residues are located at positions 1 and 5, respectively. Usually, surfactin isoforms coexist in the cell as a mixture of several peptidic variants with different aliphatic chain lengths and differ in their biological properties accordingly (Roongsawang et al. 2002). However, the culture conditions decide the pattern of amino acids and β-hydroxy fatty acids in the surfactin molecules (Raaijmakers et al. 2010). The β-turn may be formed by an intramolecular hydrogen bond; thus, the β-sheet may depend on an intermolecular hydrogen bond (Fzb et al. 2004). The lipopeptide lichenysin produced by *B. licheniformis* structurally resembles surfactin of *B. subtilis*, but the glutamic acid at position 1 in surfactins is replaced by glutaminyl residue in lichenysins. This local variation caused significant changes in various properties of the molecule compared to surfactin. Lichenysin exhibits superior surfactant power, low critical micellar concentration, and higher hemolytic activity. Lichenysin is also a better chelating agent because of increased association constants with Ca^{2+} and Mg^{2+} by a factor of 4 and 16, respectively (Peypoux et al. 1984; Romero et al. 2007a). Pumilacidins A, B, C, D, E, F, and G are cyclic acyl heptapeptide composed of a β-hydroxy fatty acid, 2 L-leucine, 2 D-leucine, L-glutamic acid, L-aspartic acid, and L-isoleucine (or L-valine) (Grangemard et al. 2001). They were first isolated from the culture broth of a strain of *B. pumilus*.

A surfactant named BL-86 is produced by *B. licheniformis* strain 86. It is a mixture of lipopeptides with the major component size ranging between 979 and 1,091 Da with variation in increments of 14 Da. The variation in molecular weight represents changes in the number of methylene groups of the lipid and/or peptide portion of the surfactant. It showed an ester carbonyl structure, which could be a part of a lactone ring connecting the β-position of the lipid to one of the carbonyl groups in the peptide (Naruse et al. 1990).

4.1.1.1.3 Fengycin

Fengycin represents the third family of lipopeptides after surfactin and iturin and is also called plipastatin (Akpa et al. 2001). These are bioactive lipopeptides produced by several strains of *B. subtilis* F-29-3. Fengycins effectively inhibit the growth of filamentous fungi but are ineffective against yeast and bacteria. The inhibition is antagonized by sterols, phospholipids, and oleic acid; however, addition of two other unsaturated fatty acids is found to increase the antifungal effect. Fengycins are broadly classified into two groups that differed by one amino acid. Fengycin A is composed of 1 D-Ala, 1 L-Ile, 1 L-Pro, 1 D-allo-Thr, 3 L-Glu, 1 D-Tyr, 1 L-Tyr, and 1 D-Orn, whereas fengycin B D-Val was replaced by D-Ala (Horowitz and Griffin 1991; Akpa et al. 2001). These bioactive molecules are lipodecapeptides containing a lactone ring with β-hydroxy fatty acid chain (either saturated or unsaturated). As mentioned the structure of fengycin contains a peptide chain of ten amino acids, cyclic decapeptides formed by lactonization (Il et al. 2010). The fatty acid moieties vary from C14 to C17 carbon atoms. The presence of different fatty acids yields different homologous compounds and isomers. Members of fengycin family also exhibit heterogeneity at the sixth position in peptide moiety. Based on variations in single amino acid (at the sixth position), fengycins are classified into two classes, namely, fengycin A and fengycin B. While fengycin A contains Ala at sixth position, it is replaced by valine in the case of fengycin B (Vanittanakom and Loeffler 1986; Bonmatin et al. 2003).

4.1.1.2 Lipopeptides Produced by Gram-Negative Bacteria (*Pseudomonas* sp.)

4.1.1.2.1 Viscosinamide

Viscosinamide produced by *Pseudomonas fluorescens* DR54 showed antagonistic properties against plant pathogenic fungi like *Pythium ultimum* and *Rhizoctonia solani*. However, bio-surfactant properties of viscosinamide differed from the known bio-surfactant viscosin that contained glutamine rather than glutamate at the second amino acid position. Isolation and determination of structure of this new compound with antibiotic and antifungal properties reveals promising perspectives for the application of *P. fluorescens* DR54 as a biological control agent. Tests performed in plants also showed that purified viscosinamide reduced the aerial mycelium development of both *P. ultimum* and *R. solani* (Lee et al. 2010; De Faria et al. 2011; Nielsen et al. 1999).

Since *P. ultimum* is a root pathogen, its growth inhibition using the viscosinamide has been studied on sugar beet seeds. Seeds infected with *P. ultimum* and *P. fluorescens* DR54 have improved plant emergence after 7 days compared to a control without the biocontrol strain. The

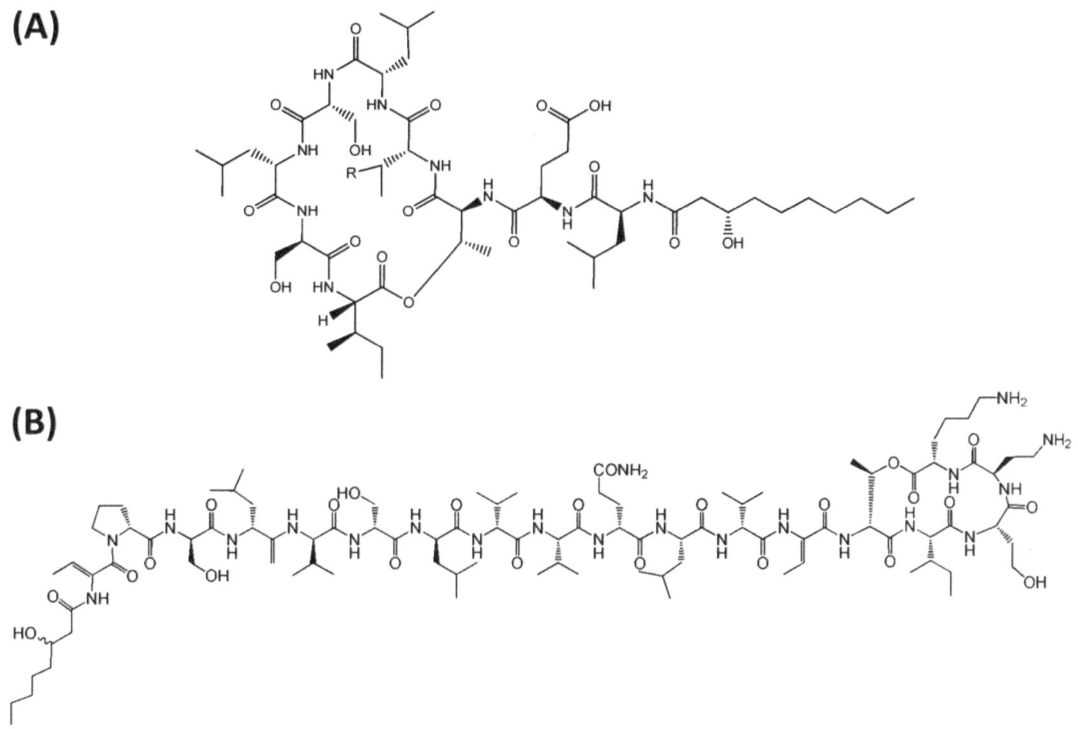

Fig. 4.2 Structure of lipopeptides produced by *Pseudomonas* sp. (**a**) Viscosin (R = H) or viscosinamide and massetolide (R = CH3) (Thrane et al. 2000). (**b**) Tolaasin (Hansen et al. 2000)

impact of *P. fluorescens* DR54 on the growth and activity of *P. ultimum* was also studied by direct microscopy after staining with the vital fluorescent dye like Calcofluor White and fluorescein diacetate. *P. fluorescens* DR54 caused reduction in *P. ultimum* mycelial density, oospore formation, and intracellular activity. Further, *P. ultimum* oospore formation was completely absent in the presence of *P. fluorescens* DR54. In vitro studies confirmed that purified viscosinamide induced encystment of *Pythium* zoospores (De Faria et al. 2011; Nielsen et al. 1999; Thrane et al. 1999) (Fig. 4.2).

4.1.1.2.2 Tolaasin

Tolaasin (tolaasin I, tolaasin II, and tolaasins A, B, C, D, and E are the available analogs) and few other metabolites are produced by *P. tolaasii*. All these analogs showed differences in their amino acid composition. All observed lipodepsipeptides of bacterial origin are maintained by the β-hydroxy octanoyl phi chain at the N-terminus, except tolaasin A, in which the acyl moiety is a gamma-carboxy butanoyl phi moiety (Mazzola et al. 2009; Andolfi et al. 2008). *P. tolaasii* is a causal organism of brown blotch disease to *Agaricus bisporus* and causes yellowing of *Pleurotus ostreatus* (Molinaro et al. 2003; Nutkins et al. 1991; Moquet et al. 1996; Cho et al. 2007).

Tolaasin is a lipodepsipeptide, found to be produced by a pathogen of mushroom species. During development of lysis bioassay for tolaasin, it was also found to cause hemolysis. Hemolysis by tolaasin involves a dose-dependent manner that is maximal between pH 6.0 and 7.0 and increased with increase in temperature. Tolaasin can also be produced in a growth-dependent manner, and its production always initiates during exponential growth of *P. tolaasii* and continued till the stationary phase. Ultrastructural studies performed to identify the mechanism of action showed disruption of the *A. bisporus* plasma membrane and vacuole membranes by tolaasin. Multilamellar liposomes inoculated with tolaasin demonstrated loss of lytic activity and prevented tolaasin-induced hemolysis, suggesting tolaasin induces partitions and forms pores in erythrocyte membranes that cause lysis by a colloid osmotic mechanism. Tolaasin is phytotoxic when infiltrated into leaves of *Nicotiana tabacum* and was shown to be active against a range of basidiomycetes and Gram-positive bacteria (Soler-Rivas and Arpin 1999; Bassarello et al. 2004; Coraiola et al. 2006).

4.1.1.2.3 Tensin

Tensin is an antifungal cyclic lipopeptide produced by *P. fluorescens* strain 96.578. The molecular formula is $C_{67}H_{116}N_{12}O_{20}$, representing an exact molecular weight of 1408.84 Da. The NMR data obtained for tensin predicted it as a cyclic lipopeptide composed of a 3-hydroxydecanoyl residue in combination with 11 amino acid residues that include 5 leucine (Leu), 1 isoleucine (Ile), 1 aspartic acid (Asp), 1 glutamine (Gln), 1 glutamic acid (Glu), 1 threonine (Thr), and 1 serine (Ser) amino acids (Rainey et al. 1991). *P. fluorescens* is already used as a biocontrol agent against fungi such as *R. solani*. Various studies suggested that tensin is effective against radial growth of *R. solani* (Cho et al. 2010).

4.1.1.2.4 Syringotoxin and Syringopeptin

Syringotoxin is produced by *P. syringae* pv. *syringae* which is a known pathogen of various species of citrus trees. Peptide moiety of bioactive molecule contains amino acids in order of Ser-Dab-Gly-Hse-Orn-aThr-Dhb-(3-OH)Asp-(4-Cl)Thr with the terminal carboxy group closing a macrocyclic ring on the OH group of the N-terminal Ser (Nielsen et al. 2000). Syringopeptin, another major phytotoxic antibiotic produced by *P. syringae* pv. *syringae*, is also a lipodepsipeptide. The amino acid sequence is Ser-Ser-Dab-Dab-Arg-Phe-Dhb-4(Cl)Thr-3(OH)Asp with the beta-carboxy group of the C-terminal residue closing a macrocyclic ring on the OH group of the N-terminal Ser (Couillerot et al. 2009; Ballio et al. 1990) (Fig. 4.3).

4.1.1.2.5 Pseudophomins

The pseudophomins are also cyclic lipodepsipeptides grouped into pseudophomins A and B. They are isolated from *P. fluorescens* strain BRG100, a bacterium with potential application as a biocontrol agent for plant pathogens and agricultural weeds. Pseudophomin B showed higher antifungal activity against the phytopathogens such as *Phoma lingam*/*Leptosphaeria maculans* and *Sclerotinia sclerotiorum*. In contrast, pseudophomin A showed stronger inhibition toward green foxtail (*Setaria viridis*) and induced root germination than pseudophomin B (Segre et al. 1989; Galonić et al. 2007).

4.1.1.2.6 Pseudomycins

Pseudomycins are peptide antimycotics that are isolated from *P. syringae*, another plant-associated bacterium. Pseudomycins contain three analogs classified as A–C that contain amino acids hydroxyl aspartic acid, aspartic acid, serine, arginine, lysine, and diaminobutyric acid. The molecular mass of pseudomycins A–C is 1,224, 1,208, and 1,252 Da, respectively, as determined by plasma desorption mass spectrometry. Pseudomycin D, on the other hand, has a molecular mass of 2,401 Da and is more complex than pseudomycins A–C (Wilson Quail et al. 2002; Pedras et al. 2003). Pseudomycin A is the predominant peptide of the pseudomycin family that possesses selective phytotoxicity and is effective against human pathogen *Candida albicans* (Harrison et al. 1991).

4.1.1.2.7 Massetolide A

Massetolide A is isolated from *P. fluorescens* SS101, identified as a biocontrol agent. *P. fluorescens* SS101 was effective in preventing infection of tomato (*Lycopersicon esculentum*) leaves by *P. infestans* and significantly reduced the expansion of existing late blight lesions. Purified massetolide A displayed significant control of *P. infestans* both locally and systemically via induced resistance. This study suggests that

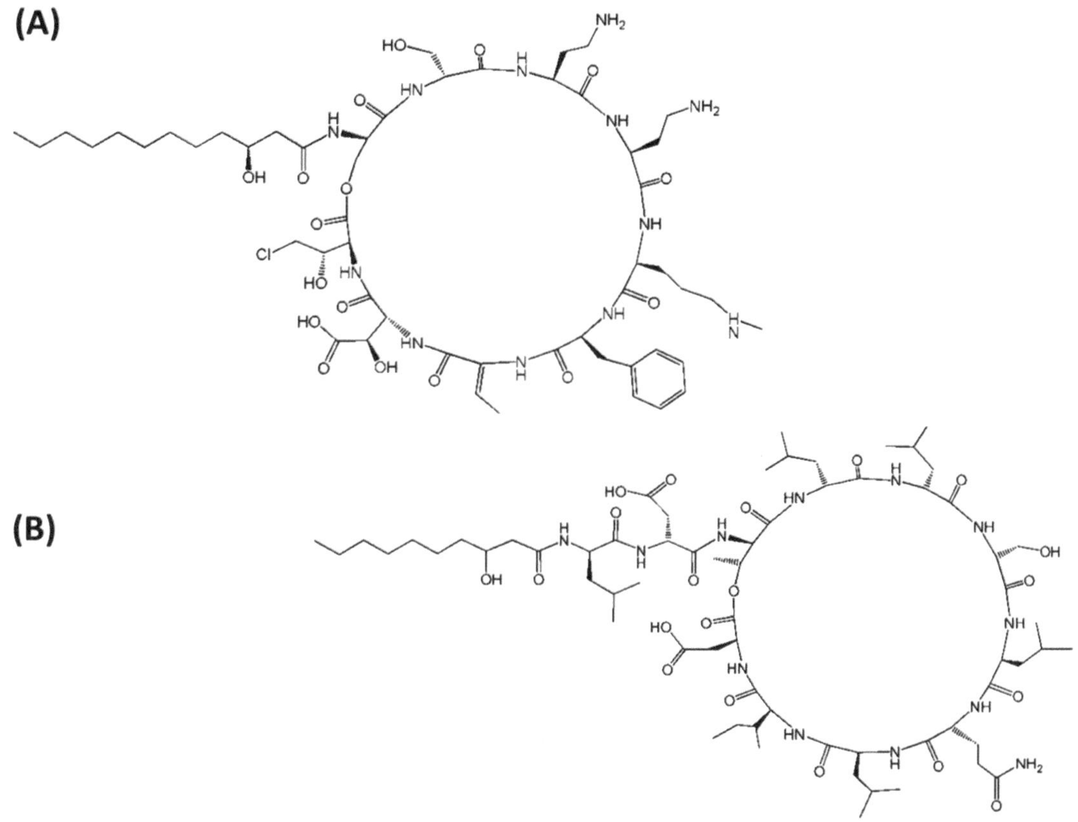

Fig. 4.3 Structure of lipopeptides produced by *Pseudomonas* sp. (**a**) Syringomycin (Mátyus et al. 2008). (**b**) Amphisin (Nielsen and Sørensen 2003)

the cyclic lipopeptide massetolide A is a metabolite with versatile functions in the ecology of *P. fluorescens* SS101 and in interactions with tomato plants and the late blight pathogen *P. infestans* (Ballio et al. 1994; Sun et al. 2001). Thus, the number of *Pseudomonas* strains has shown promising results in biological control of late blight caused by *P. infestans*.

4.1.1.3 Fungal Lipopeptides

4.1.1.3.1 Echinocandins

The echinocandins produced by fungi comprise a new class of antifungal lipopeptide agents. Echinocandins isolated for the first time from *Aspergillus rugulosus* and *A. nidulans* exhibited antifungal and anti-yeast activities (Tran et al. 2007; De Bruijn et al. 2008). Compounds under this group are amphiphilic, cyclic hexapeptides with an N-linked acyl lipid side chain. The molecular weight of these compounds is around 1,200 Da (Tóth et al. 2012). So far, three echinocandin compounds have been determined for structural elucidation. While the echinocandin B (ECB) structure was determined by chemical degradation studies in combination with X-ray crystallographic analysis, structures of echinocandins C and D were elucidated by their conversion into a common intermediate isoform derived from the echinocandin B (Benz et al. 1974; Denning 2003). They are all acylated by a linoleic acid molecule at the N-terminus. The synthesis of echinocandins C and D was carried out using the stereocontrolled synthesis of all constituents such as (2S,3S,4S)-3-hydroxy-4-methylproline, (2S,3R)-3-hydroxyhomotyrosine, and γ,δ-dihydroxyornithine (Benz et al. 1974; Messik and Oberthür 2013). The common feature to these lipopeptides is a cyclic peptide and an N-terminal fatty acid group; however, the fatty acid group is usually either branched or unbranched with a chain length of 14–18 carbon atoms. Both the cyclic structure (Shamala et al. 1976; Kurokawa and Ohfune 1993) and an acyl side chain are essentially required for the biological activity of the compound. This emphasizes the importance of the acyl group for the mechanism of antimicrobial action (Journet et al. 1999; Rodriguez et al. 1999; Debono et al. 1989).

Deacylation of ECB has been accomplished using an *Actinoplanes utahensis* culture to design new derivatives with enhanced efficiency (Debono et al. 1989). The biological activity was restored with the reintroduction of the N-acyl group that might play certain structural requirements. Indeed, small groups such as acetyl, benzoyl, or cyclohexanoyl were effective in restoring activity, and also increase in the chain length from C12 up to C17–C18 was proportional to the increase in activity (Debono et al. 1995).

The echinocandin drugs available in the market consist of three agents, caspofungin (Cancidas; Merck and Co., Whitehouse Station, New Jersey, USA), micafungin (Mycamine; Astellas Pharmaceuticals, Grand Island, New York, USA), and anidulafungin (Eraxis; Pfizer Pharmaceuticals, New York, USA). Echinocandins exhibit antimicrobial activity by blocking the synthesis of 1-3-β-D-glucan, a critical component of the fungal cell wall, through noncompetitive inhibition of the enzyme 1-3-β-D-glucan synthase. Osmotic disruption after loss of cell wall integrity is mainly responsible for fungicidal activity (Boeck et al. 1989; Hobbs et al. 1988). The echinocandin antifungal spectrum is restricted with few exceptions to *Candida* spp. and *Aspergillus* spp. Echinocandin antifungal agents have been studied in various clinical settings ranging from invasive, oral, and esophageal candidiasis and for treatment of invasive aspergillosis. At present echinocandin drugs are available only as intravenous injections in the market (Fig. 4.4).

(a) *Caspofungin*

Caspofungin (caspofungin acetate) is a semisynthetic water-soluble lipopeptide produced as a fermentation product of the fungus *Glarea lozoyensis*. This also belongs to the echinocandin family and is a derivative of pneumocandin B_0. Studies demonstrated that caspofungin has antifungal activity against yeasts of the genus *Candida* (including isolates resistant to azoles and amphotericin B). This also includes species of the genus *Candida* that are resistant (*Candida krusei*) or less susceptible to azoles (*Candida dubliniensis*, *Candida glabrata*) or resistant to amphotericin B (Diekema et al. 2005; Douglas 2006; Barchiesi et al. 1999). Several species of *Aspergillus* and certain dimorphic fungi, such as *Histoplasma*, *Blastomyces*, and *Coccidioides*, were also inhibited by caspofungin (Vazquez et al. 1997; Bachmann et al. 2002; Kurtz et al. 1994). In vitro and in vivo experiments confirmed the additive or synergic activity of caspofungin with amphotericin B and triazoles against different fungi. It also possesses activity against *Pneumocystis carinii* (Marco et al. 1998).

The β(1,3)-D-Glucan is found to be an essential component of the cell walls of numerous fungal species as the solid three-dimensional matrix formed by chains of β(1,3)-D-glucan gives shape and mechanical strength to the cell wall. Caspofungin blocks the synthesis of β(1,3)-D-glucan by noncompetitive inhibition of the enzyme β(1,3)-D-glucan synthase and showed potential antifungal effects (Pfaller et al. 1998; Zhang et al. 2006; Georgopapadakou 1997). Clinical trials proved caspofungin to be well tolerated and effective for invasive aspergillosis patients, oropharyngeal candidiasis, esophageal candidiasis, and invasive candidiasis. It showed an efficacy that is equivalent to amphotericin B (Robbel and Marahiel 2010; Douglas 2001; Georgopapadakou 2001; Groetzner et al. 2008; Groll and Walsh 2001; Herbrecht et al. 2010; Mora-Duarte et al. 2002). Experimental studies carried out in several models showed correlation with the in vitro susceptibility data of caspofungin.

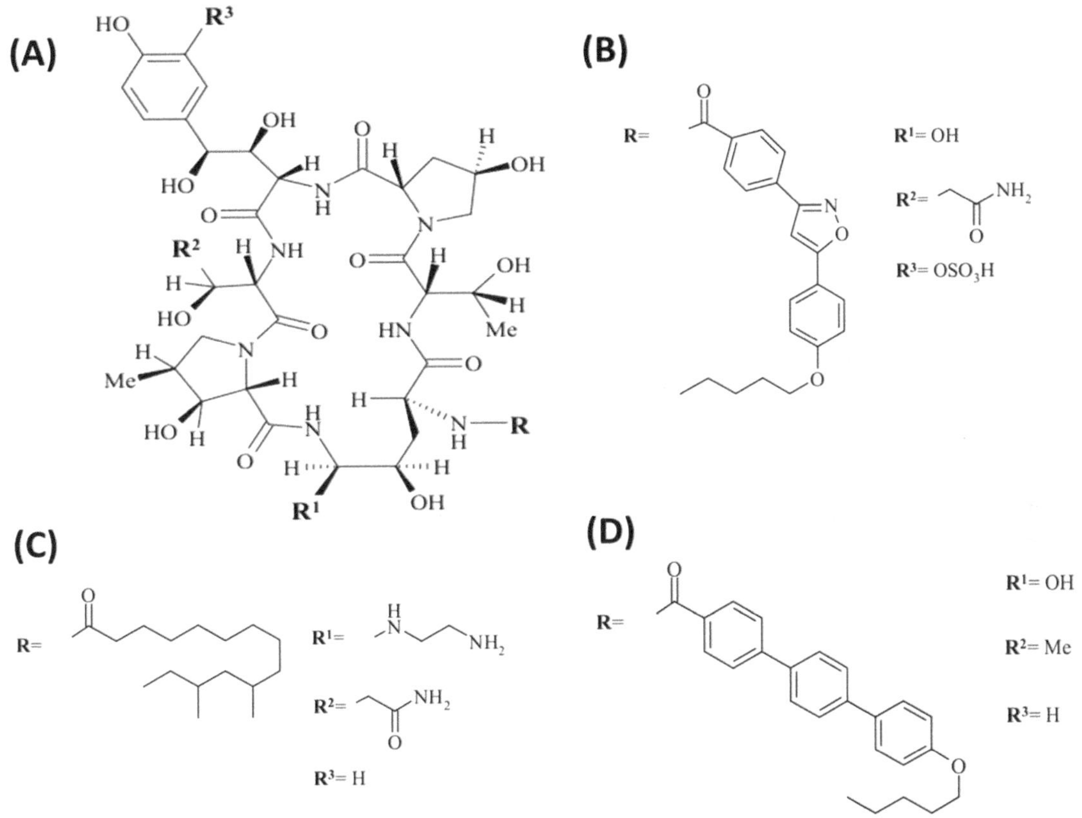

Fig. 4.4 Structure of some fungal lipopeptides (**a**) Echinocandin B. (**b**) Micafungin. (**c**) Caspofungin. (**d**) Anidulafungin

Several studies have also demonstrated that caspofungin is effective against disseminated candidiasis in immune-competent or immunosuppressed mice (Kartsonis et al. 2005a; Keating and Figgitt 2003; Abruzzo et al. 1997). Experimental data confirmed its efficacy in immunosuppressed mice infected with in vitro fluconazole-resistant or dose-dependent susceptible isolates of *C. krusei*, *C. glabrata*, and *C. albicans* (Graybill et al. 1997a, b). Caspofungin administered (0.05–5 mg/kg/day) by the intravenous or intraperitoneal route into the infected animals significantly prolonged the survival and reduced the renal fungal load.

Several experimental studies of invasive aspergillosis in immunocompromised rodents have demonstrated the efficacy of caspofungin (Abruzzo et al. 2000; Sionov et al. 2006; Chabrol et al. 2010; Kartsonis et al. 2005b). However, caspofungin did not protect mice against lethal infection with *C. neoformans* even at high doses of 40 mg/kg/day. These experimental data confirm the resistance of *C. neoformans* observed in vitro (Kartsonis et al. 2005a).

Caspofungin was compared with the reference treatment trimethoprim plus sulfamethoxazole in mice pretreated with steroids and infected with *P. carinii* (Maertens et al. 2004) by administering through subcutaneous or oral route. Results suggested 90 % reduction in number of pulmonary cysts in subcutaneous administration. Though similar results were observed for the reference treatment, the effect was more rapid with caspofungin (4 days) when compared to reference treatment (14 days). However, the oral route of administration was always found to be less effective for caspofungin. Moreover, a prophylactic effect was observed in the same models upon daily subcutaneous administration of caspofungin (Madureira et al. 2007; Powles et al. 1998).

(b) *Micafungin*

Micafungin is used worldwide in chemotherapy of life-threatening fungal infections. It is the second approved antifungal agent in the echinocandin series. FR901379, which is known as a seed compound of micafungin, was first discovered at Fujisawa Pharmaceutical Co., Ltd, Japan, in the year 1989 (Fortún et al. 2009). Micafungin is a water-soluble antifungal agent with molecular weight of 1292.26 Da that was derived from *Coleophoma empetri* (Spreghini et al. 2012; Hashimoto 2009; Mikamo et al. 2000). The water solubility of micafungin is attributed to a sulfate moiety in the molecule. Micafungin is a potent inhibitor of 1,3-β-D-glucan synthase, an enzyme involved in cell wall biosynthesis in several pathogenic fungal species. Micafungin acts in a concentration-dependent manner as a noncompetitive inhibitor in the formation of the enzyme 1,3-β-D-glucan synthase. This enzyme is necessary for synthesis of 1,3-*β*-D-glucan, a glucose polymer crucial to the structure and integrity of the cell wall of several fungal pathogens (Tóth et al. 2012; Barrett 2002; Vicente et al. 2003; Onishi et al. 2000; Tawara et al. 2000). Fungal cells that are unable to synthesize this polysaccharide cannot maintain their shape and lack adequate rigidity to resist osmotic pressure, which results in fungal cell lysis. This mechanism is found to be unique to the echinocandin class of antifungal agents. It contains the potential in additive or synergistic activity with other antifungals like polyenes and azoles. Apart from cell wall structure, glucan is also involved in cell growth and division (Nishiyama et al. 2002). Micafungin demonstrates a prolonged concentration-dependent post-antifungal effect. Fungal cell walls are usually made up of chitin as a major component along with mannoproteins. The selective antifungal effect of echinocandin activity is because of varied quantities of fungal cell wall components among different fungal species (Wiederhold et al. 2008; Groll et al. 2001). Of significance, the cell walls of zygomycetes and cryptococci lack 1,3-*β*-D-glucan, which explains the poor activity of echinocandins, including micafungin, against selective fungal pathogens (Vicente et al. 2003; Andes et al. 2008).

Micafungin is the second antimicrobial agent in the echinocandins class that is approved for use in clinical practice. Though it is a potent antifungal, it showed slow fungicidal activity against clinical isolates like *Candida albicans*, *C. dubliniensis*, *C. tropicalis*, *C. glabrata*, and *C. krusei*. The activity was somewhat higher (MIC90s) for *C. parapsilosis*, *C. lusitaniae*, and *C. guilliermondii* (Vicente et al. 2003; Mariné et al. 2007; Nakai et al. 2002; Laverdiere et al. 2002). Micafungin did not show activity against basidiomycetous yeasts, *Cryptococcus neoformans* or *Trichosporon* species in vitro, but it displayed inhibitory activity against *Aspergillus* species at lower concentrations than amphotericin B and itraconazole (Vicente et al. 2003; Andes et al. 2008; Laverdiere et al. 2002; Ostrosky-Zeichner et al. 2003).

Although clinical data are still lacking, the role of micafungin appears to be similar to that of caspofungin. The initial approval has been granted for treatment of esophageal candidiasis and prophylaxis in subjects with neutropenia. Pharmacokinetic and pharmacodynamic studies revealed no adverse effects, and safety is reported similar to those of other agents in the echinocandin class.

4.1.1.4 Clinical Efficacy

Micafungin is found to be clinically effective in treatment of *Candida* and *Aspergillus* infections as most trials that were conducted in Japan, South Africa, the United States, Latin America, and Germany produced positive results. The US FDA has approved micafungin as a therapeutic agent for the treatment of esophageal candidiasis. However, the FDA did not approve micafungin to use for children as safety and efficacy studies in this population have not been established.

Candida Infections The safety and efficacy of micafungin in the treatment of esophageal candidiasis in AIDS patients was evaluated (Uchida et al. 2000; Fujie et al. 2001), and doses of micafungin were determined as 12.5, 50, 75, and 100 mg/day through intravenous injection. Clinical improvement was noted for patients who received 75 or 100 mg/day showing favorable clinical and endoscopic outcomes with rapid response within a duration of 3–5 days (Fujie et al. 2001; Pettengell et al. 2004). In a multinational, double-blind, non-inferiority study, a single high dose of micafungin (150 mg/day) with fluconazole (200 mg/day) (Fujie et al. 2001) also showed similar results. In another open-label, non-comparative study of micafungin for the treatment of candidemia patients, the overall success was noted in 83.2 % of patients (De Wet et al. 2004). Similarly, results of several smaller trials also supported the efficacy of micafungin in the management of candidemia (Kohno et al. 2013; Denning et al. 2006). An interesting report noted that a successful outcome with topical application of micafungin is in the treatment of refractory yeast-related corneal ulcers (Hachem et al. 2008).

Mold Infections Although data is available on using micafungin for the treatment of refractory aspergillosis, it is from open trials and not examined in a randomized, controlled manner. In a phase 2 multinational study, efficacy of micafungin as primary or salvage therapy for invasive aspergillosis was assessed (Kohno et al. 2004). Another study involving stem cell transplant (SCT) recipients evaluated the safety and efficacy of micafungin in combination with other antifungal drugs for the treatment of refractory aspergillosis (De Wet et al. 2004; Matsumoto et al. 2005). Overall, significant *Aspergillus*-infected patients had a satisfactory response. In addition, several case reports have described success with micafungin in severely compromised hosts with refractory aspergillosis (Moretti et al. 2014; Kontoyiannis et al. 2009; Ota et al. 2004). Considering the poor prognosis in patients with refractory aspergillosis, the aforementioned data from these unpublished trials and case reports are encouraging. Most available data were in the setting of refractory aspergillosis with micafungin used as salvage therapy in combination with other drugs. Data for its use as monotherapy, particularly for the initial treatment of invasive aspergillosis, are lacking.

A phase 3, large, randomized, double-blind, multi-institutional comparative study was conducted in the United States and Canada that evaluated the efficacy of micafungin as prophylaxis during the pre-engraftment period of neutropenia (Yokote et al. 2004; Chandrasekar et al. 2004). However, whether micafungin should be preferred to fluconazole as prophylaxis during the pre-engraftment period in SCT recipients and whether the improved long-term survival seen with fluconazole will also be seen with micafungin are important questions that need to be addressed (Van Burik et al. 2004; Hiramatsu et al. 2009).

Urinary Sepsis *Candida glabrata* is frequently resistant to fluconazole, and in advanced renal failure, the safe use of this and other recommended drugs is limited. Patients suffering from renal and severe urinary sepsis by *C. glabrata* are successfully treated with micafungin. In fact, amphotericin B, fluconazole, and flucytosine are recommended for effective treatment of symptomatic candiduria (Marr et al. 2000; De Pauw and Donnelly 2007). However, in renal failure these agents are contradicted, and due to significant toxicity risk, their utilities are substantially limited. Fluconazole is extensively excreted by the kidneys reaching high urinary concentrations. Non-*C. albicans* species, notably *C. glabrata*, is nowadays frequently resistant to fluconazole therapy. Echinocandins are potent antifungal agents used to treat these strains as they exert activity by inhibiting β-D-glucan synthase (Franco 2011; Zasloff 2002), a major component of the fungal cell wall (Ullmann et al. 2012). Micafungin, like the entire class of echinocandin drugs, has a broad spectrum of activity on

Candida species, and it is now considered a first-line therapy in candidiasis. In general, micafungin is well tolerated and has a low potential of drug interaction, its clearance is independent from glomerular filtration, and there are no contradictions for dose adjustment even in severe renal failures. All echinocandins are highly protein bound (99 %) and share the major pharmacokinetic disadvantage of poor glomerular filtration and tubular secretion resulting in very low urinary concentrations (Jones 1949; Arima et al. 1968). For that reason, even if they have a very favorable profile in terms of efficacy and safety, their use seems precluded in fungal urinary tract infections. However, despite micafungin being minimally excreted in urine, its wide distribution in many organs and tissues, including the kidneys, liver, and spleen, has been shown in animal models (Nishiyama et al. 2002; Mermel et al. 2009; Sucher et al. 2009). The observed tissue concentrations were severalfold in excess of the MIC against clinical isolates of *Candida* spp. and *Aspergillus* spp. Few reports in the literature are available on the safe and effective use of echinocandins (i.e., caspofungin and micafungin) in candiduria (Petraitis et al. 2002; Niwa et al. 2004; Lagrotteria et al. 2007; Kauffman 2005). Of these, six were of candiduria successfully treated with caspofungin (collected by Merck Research Laboratories retrospectively from a phase 2–3 clinical study). To our knowledge, only three cases of candiduria treated with micafungin are reported in the literature. There are some features in treatment where firstly the patient was treated with a dose of 200 mg daily of micafungin (a dose higher than the standard recommended dose) without adverse effects. The choice of using a dosage greater than usual was empirical and was motivated by the possibility of achieving higher renal tissue concentration with such dosage. In conclusion the present case shows that micafungin can achieve clinically relevant fungal sterilization even in urine and there is room to consider micafungin in the armamentarium against *C. glabrata* urinary infections, notably in the context of renal failure where other antifungal effective drugs are unsafe or contradicted.

(c) *Anidulafungin*

Anidulafungin is a semisynthetic product of echinocandin B which is a fermentation product of the mold *A. nidulans*. It was developed by Eli Lily, underwent preclinical and clinical studies at Vicuron Pharmaceuticals, and was sold to Pfizer and marketed under the name Eraxis™. Anidulafungin (Eraxis; Pfizer) is the newest echinocandin antifungal approved by the US Food and Drug Administration recently and currently used for the treatment of esophageal candidiasis, candidemia, and deep-tissue candidiasis.

Anidulafungin is a novel echinocandin and has several advantages over existing antifungals. The unique features of anidulafungin are its slow degradation in humans where it undergoes biotransformation rather than being metabolized. It has potent in vitro activity against *Aspergillus* and *Candida* species, including the strains that resist fluconazole or amphotericin B. Results of various clinical trials indicate anidulafungin as effective in treatment of esophageal candidiasis, including azole-refractory disease. The results of a recent study comparing fluconazole versus anidulafungin demonstrated the superiority of anidulafungin in the treatment of candidemia and invasive candidiasis (IC). Studies evaluating the concomitant use of anidulafungin, amphotericin B, voriconazole, or cyclosporine did not show significant drug-drug interactions or any adverse effects. To date, anidulafungin appears to have an excellent safety profile. Based on the early clinical experience, it appears that anidulafungin will be a valuable asset in the management of serious and difficult-to-treat fungal infections.

Anidulafungin has potent in vitro fungicidal activity against a broad range of *Candida* species, including *C. albicans*, *C. glabrata*, *C. tropicalis*, *C. parapsilosis*, *C. famata*, *C. rugosa*, and *C. stellatoidea* (Table 4.1) (Cochrane and Vederas 2014; Morikawa et al. 2000; Higginbotham and Murphy 2010). Anidulafungin is also effective against species of *Candida* that are intrinsically resistant to azoles (*C. krusei*), amphotericin B (*C. lusitaniae*), or other echinocandins (*C. parapsilosis*) (Rao

Table 4.1 Selected antifungal lipopeptide antibiotics currently in preclinical and clinical development

Compound name (commercial name)	Therapeutic area	Method of manufacture	Lead compound and producing organism	Company	Phase
Caspofungin (Cancidas)	Antifungal	Semisynthetic	Echinocandins and *Glarea lozoyensis*	Merck and Co.	Market
Micafungin (Mycamine)	Antifungal	Semisynthetic	Echinocandins and *Coleophoma empetri*	Astellas Pharmaceuticals	Market
Anidulafungin (Eraxis)	Antifungal	Semisynthetic	Echinocandins and *Aspergillus nidulans*	Pfizer Pharmaceuticals	Market

Adopted from Giovanna Pirri et al. (2009)

Table 4.2 Lipopeptides as biocontrol of fungal phytopathogens

Plant disease	Phytopathogen	Lipopeptide-producing microorganism	Lipopeptide inhibiting the phytopathogen	Ref.
Damping-off bean	*Pythium ultimum*	*Bacillus subtilis* M4	Iturin/fengycin	Ongena et al. (2005)
Gray mold disease of apple	*Botrytis cinerea*	*Bacillus subtilis* M4	Fengycin	Ongena et al. (2005)
Powdery mildew of cucurbits	*Podosphaera fusca*	*Bacillus subtilis*	Iturin/fengycin	Romero et al. (2007b)
Fusarium head blight (FHB) in wheat and barley and ear rot in corn	*Gibberella zeae* (anamorph of *Fusarium graminearum*)	*Bacillus subtilis* JA; JA026	Fengycin	Liu et al. (2005)
Sugar beet seed infection	*Rhizoctonia solani*	*Pseudomonas fluorescens* strain 96.578	Tensin	Nielsen et al. (2000)
Sclerotinia stem rot disease	*Sclerotinia sclerotiorum*	*Bacillus amyloliquefaciens*	Surfactin/fengycin	Alvarez et al. (2012a)
Rice blast	*Magnaporthe grisea*	*Chromobacterium* sp. C61	Chromobactomycin	Kim et al. (2014)

et al. 2013; Nielsen and Sørensen 2003). Anidulafungin has also demonstrated excellent in vitro activity against several species of *Aspergillus* (Table 4.2) (Nielsen and Sørensen 2003; Sørensen et al. 2001; Cameotra and Makkar 2004). Anidulafungin also demonstrates additive effects in vitro in combination with amphotericin B against species of the genera *Aspergillus* and *Fusarium* by synergistic activity when combined with itraconazole or voriconazole against *Aspergillus* species (Donadio et al. 2007; Song et al. 2011).

Mechanism of Action Anidulafungin is a semisynthetic lipopeptide synthesized from fermentation products of *A. nidulans*. The compound is a noncompetitive inhibitor of 1,3-β-D-glucan synthase, which results in the selective inhibition of the synthesis of glucan, a major structural component of the cell wall of many pathogenic fungi that is absent in mammalian cells. A difference in glucan content determines the excellent activity of anidulafungin in fungi and the paucity of adverse effects in humans (Rivardo et al. 2009).

4.1.2 Lipopeptides Against Fungal Phytopathogens

Some bacterial species live in association with plant roots and other parts. The natural antagonistic property showed by these microorganisms (notably belonging to the *Bacillus* and closely related *Paenibacillus* genera and *Pseudomonas* sp.) has emerged as promising alternatives to reduce the use of chemical pesticides and other toxic substances in agriculture (Sobel et al. 2000; McSpadden Gardener 2004; Cawoy et al. 2011; Jacobsen et al. 2004). The disease protection properties of these microorganisms rely on three main traits.

The *first* and foremost among these is a high ecological fitness to efficiently colonize in the roots, which is a prerequisite to proficiently compete for space and nutrients in the microenvironment of the rhizosphere. *Second* is their capacity to secrete highly active antimicrobial substances with strong antagonistic activity toward various plant pathogens. The *third* is their ability to trigger plant immune response in tissues which imparts resistance state that makes the host less susceptible to subsequent infection (Pérez-García et al. 2011; Lugtenberg and Kamilova 2009; Chen et al. 2000). The efficient antimicrobial substance production which leads to direct antagonism of phytopathogens is a key biocontrol mechanism (Berendsen et al. 2012; Chen et al. 2009). Antimicrobial production efficiency and a high rhizosphere fitness possibly explain the strong biocontrol potential of *Bacillus* both in vitro and under field conditions for their successful marketing commercially (Rückert et al. 2011; Pal and Mc Spadden Gardener 2006; Janisiewicz and Korsten 2002; Larkin and Tavantzis 2013; Spadaro and Gullino 2005; Király et al. 2008). Among the *Bacillus* antimicrobial products, cyclic lipopeptides (LPs) such as surfactin, iturin, and fengycin families are of high interest not only due to their high production rate in bioreactors by *B. subtilis* or *B. amyloliquefaciens* but also because of the secretion of these compounds in relevant amounts under natural conditions in the rhizosphere environment (Yang et al. 2013; Kinsella et al. 2009; Nihorimbere et al. 2009, 2012; Dietel et al. 2013; Yaryura et al. 2008) (Fig. 4.5).

One of the *B. subtilis* strain designated as JA was found to antagonize the growth of *Gibberella zeae*. The electrospray ionization mass spectrometry (ESI/MS) analysis of the product revealed the antifungal lipopeptide production by this strain. A mutant of this strain exhibited favorable properties including the high yield of antifungal lipopeptide production with relatively faster growth over the parent strain, which suggested that this strain would be a promising biocontrol candidate in agriculture (Makovitzki et al. 2007).

The potential of *B. subtilis* strain M4 at protecting plants against fungal diseases was demonstrated in different pathosystems, and fengycin-like lipopeptide production was reported by this strain. The protective effect against damping-off of bean seedlings which is caused by *Pythium ultimum* (gray mold of apple and in other postharvest diseases) was demonstrated. The protection ability was also established by the strong biocontrol activity of lipopeptide-enriched extracts applied onto infected tissues of plants. Apart from this, root pre-inoculation with strain M4 enabled the host plant to respond more efficiently to subsequent pathogen infection on leaves and other parts of the plant. The mechanisms by which *Bacillus* spp. suppress disease in infected plants are yet to be discovered. These antifungal properties of stain M4 emphasize the interest of *B. subtilis* as a pathogen antagonist and plant defense-inducing agent. The production of cyclic fengycin-type lipopeptides may be closely related to the expression of these two biocontrol qualities (Liu et al. 2005).

Apart from members of *Bacillus* genus, a *Chromobacterium* sp. strain C61 that displayed antifungal activities was also used successfully as a biocontrol agent for various plant diseases under field conditions. The analysis of C61 culture filtrates exposed an antifungal cyclic lipopeptide, *chromobactomycin*, that contained a unique nonameric peptide ring. The chromobactomycin inhibited the growth of several phytopathogenic fungi in vitro as well as plant applications. It significantly reduced disease severity caused by several pathogens and inhibited the mycelial growth of *R. solani*, *Pyricularia grisea*, *B. cinerea*, *Alternaria longipes*, and *C. gloeosporioides* (Ongena et al. 2005) (Fig. 4.6).

Isolation and characterization of bacterial antagonist for use as biological control of phytopathogenic fungi like rice blast fungus has been studied. The antimicrobial substance was further

Fig. 4.5 Panel 1: Fungicidal activities of the lipopeptides (C14-KLLK and C16-KLLK) upon artificial infection of gray mold (*Botrytis cinerea*) on cucumber leaves for 4 days after infection and 3 days after treatment. Panel 2: Cucumber fruits infected with *B. cinerea* conidia 3 days after the last treatment with double-distilled water (*A*) or 30 mg/l C14-KLLK (*B*). Panel 3: Corn leaves infected with *Cochliobolus heterostrophus*, untreated (*A*) or treated with 30 mg/l C14-KLLK (Debois et al. 2014)

purified and characterized the antifungal molecule produced by the antagonist. The strain was isolated from soil and identified as *B. licheniformis* BC98 that showed high antagonist antifungal activity against the rice blast fungus *Magnaporthe grisea*. This bacterial strain also inhibited the growth of other phytopathogens such as *Curvularia lunata* and *Rhizoctonia bataticola*. Biochemical and mass studies of biologically active fractions revealed it as a lipopeptide with molecular mass of 1,035 Da. However, it was identified as surfactin after NMR analysis. Microscopic analysis of the effect of the antagonist on *M. grisea* revealed bulbous hyphae showing patchy and vacuolated cytoplasm under the electron microscope. The antagonist inhibited germination of *M. grisea*, and therefore it is considered as a potential candidate for control of rice blast disease (Romero et al. 2007b).

Podosphaera fusca is the main causal agent of cucurbit *powdery mildew*, and four *Bacillus subtilis* strains, UMAF6614, UMAF6619, UMAF6639, and UMAF8561, were found to inhibit the growth of *P. fusca*. Therefore, they were studied for their ability to suppress the

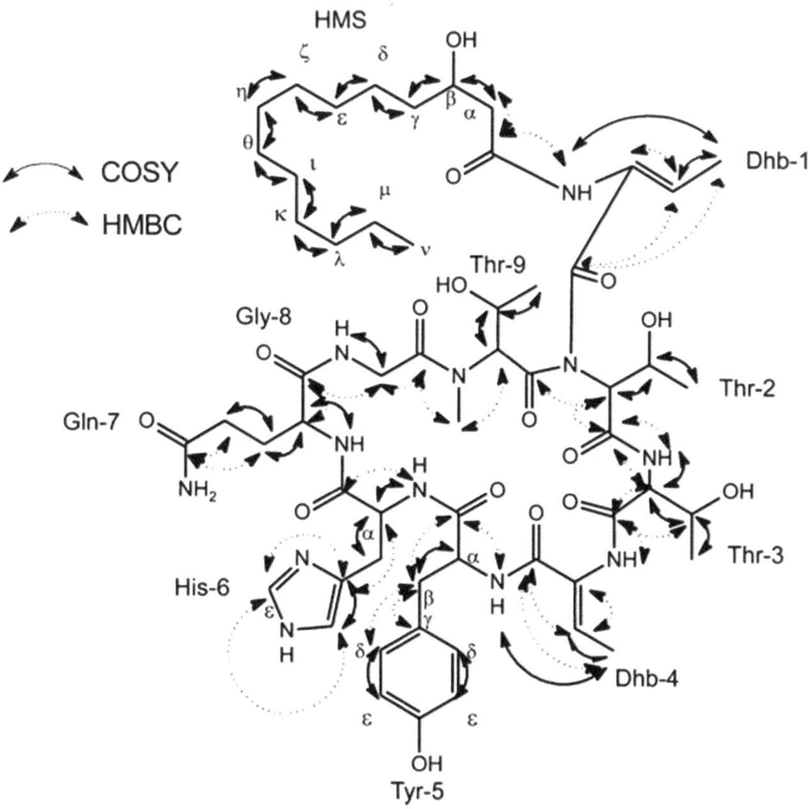

Fig. 4.6 Chemical structure of chromobactomycin. *Dhb* dihydroxybutyric acid, *Gln* glutamine, *Gly* glycine, *His* histidine, *HMS* *β*-hydroxymyristate, *Thr* threonine,*Tyr* tyrosine (Ongena et al. 2005)

disease on melon. Experiments were performed using detached leafs and seedling assays. They were also further subjected to elucidate the mode of action involved in their biocontrol performance. Three lipopeptide antibiotics, i.e., surfactin, fengycin, and iturin A or bacillomycin, were identified in butanolic extracts of *B. subtilis* culture filtrate. The purified lipopeptide fractions (bacillomycin, fengycin, and iturin A) have shown strong inhibitory effects on *P. fusca* conidia germination and provided interesting evidence of their presumed involvement in the antagonistic activity. This also suggested that the iturin and fengycin families of lipopeptides have a major role in the antagonism of *B. subtilis* toward *P. fusca* (Alvarez et al. 2012a) (Fig. 4.7).

The antifungal compounds identified as iturin, surfactin, and fengycin isoforms produced by two previously isolated *Bacillus* sp. strains, ARP23 and MEP218, were found to be effective against *Sclerotinia sclerotiorum*. These strains were further assessed for their ability to control *sclerotinia stem rot* in soybean. The field trials showed effective results. While the surfactin C15 and fengycin A (C16–C17) and B (C16) isoforms were produced by strain ARP23, the major lipopeptide produced by strain MEP218 was iturin A C15. Mycelial growth, morphology, and sclerotial germination were altered in the presence of lipopeptides. Foliar application of *Bacillus amyloliquefaciens* strains on soybean plants prior to *S. sclerotiorum* infection also revealed significant protection against sclerotinia stem rot. Strains ARP23 and MEP218 were renamed or identified as strains belonging to the species *B. amyloliquefaciens* and were concluded to produce antifungal compounds belonging to the cyclic lipopeptide family. As sclerotinia stem rot was considered as one of the most severe soybean diseases worldwide, these results proposed the potential of *B. amyloliquefaciens* strains ARP23 and MEP218 to control plant diseases caused by *S. sclerotiorum* as biocontrol agent (Kim et al. 2014) (Fig. 4.8).

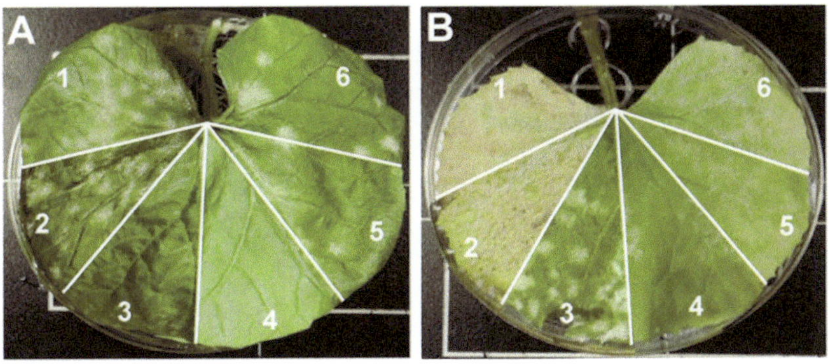

Fig. 4.7 Powdery mildew symptoms on melon leaves following treatments with butanolic extracts of cell-free filtrates of the antagonistic strain *Bacillus subtilis* UMAF6639. Treatments are *1* untreated control, *2* nutrient broth control, *3* washed cells from stationary phase, *4* cell-free filtrate nondiluted, and *5* and *6* cell-free filtrates 1:4- and 1:16-fold diluted, respectively. Panels are compositions of photographs of leaves inoculated with conidia of *Podosphaera fusca* and treated as described, taken 7 (**a**) and 16 (**b**) days after treatments (Fzb et al. 2004)

Fig. 4.8 Systemic protection against *Botrytis cinerea* B05 infection by synthetic ultrashort lipopeptides in cucumber and *Arabidopsis* seedlings. (**a**) The first leaf of cucumber plants (*white arrows*) was infiltrated with 100 μl of water (MOCK) or 100 μl of a 12.5 M solution of P1 24 h before inoculation of the second and third leaves with *B. cinerea* mycelium. Symptoms were assessed 3–4 days after infection. (**b**) *Arabidopsis rosette* leaves 3–4 days after infection with *B. cinerea* spores. Four leaves from each plant rosette were treated with 100 μl of water (MOCK) or 100 μl of 12.5 M peptides. Pathogen inoculation was done on non-treated leaves. Three to four days later, infected leaves were detached and analyzed (Tendulkar et al. 2007)

4.2 Conclusion

Lipopeptides are amphiphilic molecules with diverse physiochemical properties. Because of their broad-spectrum antimicrobial activities, they are used as various biocontrol agents. Moreover, presence-specific amino acid residues import certain functions to the lipopeptides that are important for biological activity. Though the molecular pattern is not conserved in all lipopeptides, structural homologues are interestingly more active in comparison to others. In fact, this is the reason why some *Bacillus* strains are more efficient than others against pathogenic strains. Therefore, the identification and understanding of putative receptors and their molecular mechanism is essential. However, efficiency of biological control properties of lipopeptides can be improved by using different combinations of production mechanism or by improving the production strains or in combination. Large-scale production of these lipopeptides is essential for biotechnological studies including therapeutic applications.

References

Abruzzo GK, Flattery AM, Gill CJ, Kong L, Smith JG, Pikounis VB, Balkovec JM, Bouffard AF, Dropinski JF, Rosen H, Kropp H, Bartizal K (1997) Evaluation of the echinocandin antifungal MK-0991 (L-743,872): efficacies in mouse models of disseminated aspergillosis, candidiasis, and cryptococcosis. Antimicrob Agents Chemother 41:2333–2338

Abruzzo GK, Gill CJ, Flattery AM, Kong L, Leighton C, Smith JG, Pikounis VB, Bartizal K, Rosen H (2000) Efficacy of the echinocandin caspofungin against disseminated aspergillosis and candidiasis in cyclophosphamide-induced immunosuppressed mice. Antimicrob Agents Chemother 44:2310–2318

Akpa E, Jacques P, Wathelet B, Paquot M, Fuchs R, Budzikiewicz H, Thonart P (2001) Influence of culture conditions on lipopeptide production by Bacillus subtilis. Appl Biochem Biotechnol 91–93:551–561

Alvarez F, Castro M, Príncipe A, Borioli G, Fischer S, Mori G, Jofré E (2012) The plant-associated Bacillus amyloliquefaciens strains MEP 218 and ARP 23 capable of producing the cyclic lipopeptides iturin or surfactin and fengycin are effective in biocontrol of sclerotinia stem rot disease. J Appl Microbiol 112:159–174

Andes DR, Diekema DJ, Pfaller MA, Marchillo K, Bohrmueller J (2008) In vivo pharmacodynamic target investigation for micafungin against Candida albicans and C. glabrata in a neutropenic murine candidiasis model. Antimicrob Agents Chemother 52:3497–3503

Andolfi A, Cimmino A, Lo Cantore P, Iacobellis NS, Evidente A (2008) Bioactive and structural metabolites of Pseudomonas and Burkholderia species causal agents of cultivated mushrooms diseases. Perspect Med Chem 2008:81–112

Arima K, Kakinuma A, Tamura G (1968) Surfactin, a crystalline peptidelipid surfactant produced by Bacillus subtilis: isolation, characterization and its inhibition of fibrin clot formation. Biochem Biophys Res Commun 31:488–494

Arnusch CJ, Ulm H, Josten M, Shadkchan Y, Osherov N, Sahl HG, Shai Y (2012) Ultrashort peptide bioconjugates are exclusively antifungal agents and synergize with cyclodextrin and amphotericin B. Antimicrob Agents Chemother 56:1–9

Bachmann SP, Patterson TF, López-Ribot JL (2002) In vitro activity of caspofungin (MK-0991) against Candida albicans clinical isolates displaying different mechanisms of azole resistance. J Clin Microbiol 40:2228–2230

Ballio A, Bossa F, Collina A, Gallo M, Iacobellis NS, Paci M, Pucci P, Scaloni A, Segre A, Simmaco M (1990) Structure of syringotoxin, a bioactive metabolite of Pseudomonas syringae pv. syringae. FEBS Lett 269:377–380

Ballio A, Bossa F, Di Giorgio D, Ferranti P, Paci M, Pucci P, Scaloni A, Segre A, Strobel GA (1994) Novel bioactive lipodepsipeptides from Pseudomonas syringae: the pseudomycins. FEBS Lett 355:96–100

Banat IM, Franzetti A, Gandolfi I, Bestetti G, Martinotti MG, Fracchia L, Smyth TJ, Marchant R (2010) Microbial biosurfactants production, applications and future potential. Appl Microbiol Biotechnol 87:427–444

Barchiesi F, Schimizzi AM, Fothergill AW, Scalise G, Rinaldi MG (1999) In vitro activity of the new echinocandin antifungal, MK-0991, against common and uncommon clinical isolates of Candida species. Eur J Clin Microbiol Infect Dis 18:302–304

Barrett D (2002) From natural products to clinically useful antifungals. Biochim Biophys Acta – Mol Basis Dis 1587:224–233

Bassarello C, Lazzaroni S, Bifulco G, Lo Cantore P, Iacobellis NS, Riccio R, Gomez-Paloma L, Evidente A, Tolaasins A-E (2004) Five new lipodepsipeptides produced by Pseudomonas tolaasii. J Nat Prod 67:811–816

Benz VF, Knüsel F, Nüesch J, Treichler H, Voser W, Nyfeler R, Keller-Schierlein W (1974) Echinocandin B, ein neuartiges polypeptid-antibioticum aus *Aspergillus nidulans* var. *echinulatus*: Isolierung und bausteune. Helv Chim Acta 57:2459–2477

Berendsen RL, Pieterse CMJ, Bakker PAHM (2012) The rhizosphere microbiome and plant health. Trends Plant Sci 17:478–486

Boeck LD, Fukuda DS, Abbott BJ, Debono M (1989) Deacylation of echinocandin B by Actinoplanes utahensis. J Antibiot (Tokyo) 42:382–388

Boman HG (1998) Gene-encoded peptide antibiotics and the concept of innate immunity: an update review. Scand J Immunol 48:15–25

Bonmatin J-M, Laprévote O, Peypoux F (2003) Diversity among microbial cyclic lipopeptides: iturins and surfactins. Activity-structure relationships to design new bioactive agents. Comb Chem High Throughput Screen 6:541–556

Cameotra SS, Makkar RS (2004) Recent applications of biosurfactants as biological and immunological molecules. Curr Opin Microbiol 7:262–266

Cawoy H, Bettiol W, Fickers P, Ongena M (2011) Bacillus-based biological control of plant diseases. In: Pesticides in the modern world – pesticides use and management. InTech, Croatia, pp 273–303

Chabrol A, Cuzin L, Huguet F, Alvarez M, Verdeil X, Linas MD, Cassaing S, Giron J, Tetu L, Attal M, Récher C (2010) Prophylaxis of invasive aspergillosis with voriconazole or caspofungin during building work in patients with acute leukemia. Haematologica 95:996–1003

Chandrasekar PH, Cutright JL, Manavathu EK (2004) Efficacy of voriconazole plus amphotericin B or micafungin in a guinea-pig model of invasive pulmonary aspergillosis. Clin Microbiol Infect 10:925–928

Chen C, Belanger RR, Benhamou N, Paulitz TC (2000) Defense enzymes induced in cucumber roots by treatment with plant growth-promoting rhizobacteria (PGPR) and Pythium aphanidermatum. Physiol Mol Plant Pathol 56:13–23

Chen XH, Koumoutsi A, Scholz R, Schneider K, Vater J, Süssmuth R, Piel J, Borriss R (2009) Genome analysis of Bacillus amyloliquefaciens FZB42 reveals its potential for biocontrol of plant pathogens. J Biotechnol 140:27–37

Cho K-H, Kim S-T, Kim Y-K (2007) Purification of a pore-forming peptide toxin, tolaasin, produced by Pseudomonas tolaasii 6264. J Biochem Mol Biol 40:113–118

Cho KH, Wang HS, Kim YK (2010) Temperature-dependent hemolytic activity of membrane pore-forming peptide toxin, tolaasin. J Pept Sci 16:85–90

Cochrane SA, Vederas JC (2014) Lipopeptides from Bacillus and Paenibacillus spp.: a gold mine of antibiotic candidates. Med Res Rev 36:4–31

Coraiola M, Lo Cantore P, Lazzaroni S, Evidente A, Iacobellis NS, Dalla Serra M (2006) WLIP and tolaasin I, lipodepsipeptides from Pseudomonas reactans and Pseudomonas tolaasii, permeabilise model membranes. Biochim Biophys Acta – Biomembr 1758:1713–1722

Couillerot O, Prigent-Combaret C, Caballero-Mellado J, Moënne-Loccoz Y (2009) Pseudomonas fluorescens and closely-related fluorescent pseudomonads as biocontrol agents of soil-borne phytopathogens. Lett Appl Microbiol 48:505–512

De Bruijn I, De Kock MJD, De Waard P, Van Beek TA, Raaijmakers JM (2008) Massetolide A biosynthesis in Pseudomonas fluorescens. J Bacteriol 190:2777–2789

De Faria AF, Stéfani D, Vaz BG, Silva ÍS, Garcia JS, Eberlin MN, Grossman MJ, Alves OL, Durrant LR (2011) Purification and structural characterization of fengycin homologues produced by Bacillus subtilis LSFM-05 grown on raw glycerol. J Ind Microbiol Biotechnol 38:863–871

De Pauw BE, Donnelly JP (2007) Prophylaxis and aspergillosis – has the principle been proven? N Engl J Med 356:409–411

De Wet N, Llanos-Cuentas A, Suleiman J, Baraldi E, Krantz EF, Della Negra M, Diekmann-Berndt H (2004) A randomized, double-blind, parallel-group, dose-response study of micafungin compared with fluconazole for the treatment of esophageal candidiasis in HIV-positive patients. Clin Infect Dis 39:842–849

Debois D, Jourdan E, Smargiasso N, Thonart P, De Pauw E, Ongena M (2014) Spatiotemporal monitoring of the antibiome secreted by Bacillus biofilms on plant roots using MALDI mass spectrometry imaging. Anal Chem 86:4431–4438

Debono M, Abbott BJ, Fukuda DS, Barnhart M, Willard KE, Molloy RM, Michel KH, Turner JR, Butler TF, Hunt AH (1989) Synthesis of new analogs of echinocandin B by enzymatic deacylation and chemical reacylation of the echinocandin B peptide: synthesis of the antifungal agent cilofungin (LY121019). J Antibiot (Tokyo) 42:389–397

Debono M, Turner WW, LaGrandeur L, Burkhardt FJ, Nissen JS, Nichols KK, Rodriguez MJ, Zweifel MJ, Zeckner DJ, Gordee RS, Tang J, Parr TR (1995) Semisynthetic chemical modification of the antifungal lipopeptide echinocandin B (ECB): structure-activity studies of the lipophilic and geometric parameters of polyarylated acyl analogs of ECB. J Med Chem 38:3271–3281

Denning DW (2003) Echinocandin antifungal drugs. Lancet 362:1142–1151

Denning DW, Marr KA, Lau WM, Facklam DP, Ratanatharathorn V, Becker C, Ullmann AJ, Seibel NL, Flynn PM, van Burik JAH, Buell DN, Patterson TF (2006) Micafungin (FK463), alone or in combination with other systemic antifungal agents, for the treatment of acute invasive aspergillosis. J Infect 53:337–349

Diekema DJ, Petroelje B, Messer SA, Hollis RJ, Pfaller MA (2005) Activities of available and investigational antifungal agents against Rhodotorula species. J Clin Microbiol 43:476–478

Dietel K, Beator B, Budiharjo A, Fan B, Borriss R (2013) Bacterial traits involved in colonization of Arabidopsis thaliana roots by Bacillus amyloliquefaciens FZB42. Plant Pathol J 29:59–66

Donadio S, Monciardini P, Sosio M (2007) Polyketide synthases and nonribosomal peptide synthetases: the emerging view from bacterial genomics. Nat Prod Rep 24:1073–1109

Douglas CM (2001) Fungal beta(1,3)-D-glucan synthesis. Med Mycol 39(Suppl 1):55–66

Douglas CM (2006) Understanding the microbiology of the Aspergillus cell wall and the efficacy of caspofungin. Med Mycol 44:95–99

Emiroglu M (2011) Micafungin use in children. Expert Rev Anti Infect Ther 9:821–834

Eshita SM, Roberto NH, Beale JM, Mamiya BM, Workman RF (1995) Bacillomycin Lc, a new antibiotic of the iturin group: isolations, structures, and antifungal activities of the congeners. J Antibiot (Tokyo) 48:1240–1247

Fortún J, Martín-Dávila P, Montejo M, Muñoz P, Cisneros JM, Ramos A, Aragón C, Blanes M, San Juan R, Gavaldá J, Llinares P (2009) Prophylaxis with caspofungin for invasive fungal infections in high-risk liver transplant recipients. Transplantation 87:424–435

Franco OL (2011) Peptide promiscuity: an evolutionary concept for plant defense. FEBS Lett 585:995–1000

Fujie A, Iwamoto T, Sato B, Muramatsu H, Kasahara C, Furuta T, Hori Y, Hino M, Hashimoto S (2001) FR131535, a novel water-soluble echinocandin-like lipopeptide: synthesis and biological properties. Bioorg Med Chem Lett 11:399–402

Fzb S, Koumoutsi A, Chen X, Henne A, Liesegang H, Hitzeroth G, Franke P, Vater J, Borriss R (2004) Structural and functional characterization of gene clusters directing nonribosomal synthesis of bioactive cyclic lipopeptides in Bacillus. J Bacteriol 186:1084–1096

Galonić DP, Barr EW, Walsh CT, Bollinger JM, Krebs C (2007) Two interconverting Fe(IV) intermediates in aliphatic chlorination by the halogenase CytC3. Nat Chem Biol 3:113–116

George J, Reboli AC (2012) Anidulafungin: when and how? The clinician's view. Mycoses 55:36–44

Georgopapadakou NH (1997) Antifungals targeted to the cell wall. Expert Opin Investig Drugs 6:147–150

Georgopapadakou NH (2001) Update on antifungals targeted to the cell wall: focus on beta-1,3-glucan synthase inhibitors. Expert Opin Investig Drugs 10:269–280

Grangemard I, Bonmatin JM, Bernillon J, Das BC, Peypoux F (1999) Lichenysins G, a novel family of lipopeptide biosurfactants from Bacillus licheniformis IM 1307: production, isolation and structural evaluation by NMR and mass spectrometry. J Antibiot (Tokyo) 52:363–373

Grangemard I, Wallach J, Maget-Dana R, Peypoux F (2001) Lichenysin: a more efficient cation chelator than surfactin. Appl Biochem Biotechnol 90:199–210

Graybill JR, Najvar LK, Luther MF, Fothergill AW (1997a) Treatment of murine disseminated candidiasis with L-743,872. Antimicrob Agents Chemother 41:1775–1777

Graybill JR, Bocanegra R, Luther M, Fothergill A, Rinaldi MJ (1997b) Treatment of murine Candida krusei or Candida glabrata infection with L- 743,872. Antimicrob Agents Chemother 41:1937–1939

Groetzner J, Kaczmarek I, Wittwer T, Strauch J, Meiser B, Wahlers T, Daebritz S, Reichart B (2008) Caspofungin as first-line therapy for the treatment of invasive aspergillosis after thoracic organ transplantation. J Heart Lung Transplant 27:1–6

Groll AH, Walsh TJ (2001) Caspofungin: pharmacology, safety and therapeutic potential in superficial and invasive fungal infections. Expert Opin Investig Drugs 10:1545–1558

Groll AH, Mickiene D, Petraitis V, Petraitiene R, Ibrahim KH, Piscitelli SC, Bekersky I, Walsh TJ (2001) Compartmental pharmacokinetics and tissue distribution of the antifungal echinocandin lipopeptide micafungin (FK463) in rabbits. Antimicrob Agents Chemother 45:3322–3327

Hachem R, Hanna H, Kontoyiannis D, Jiang Y, Raad I (2008) The changing epidemiology of invasive candidiasis: Candida glabrata and candida krusei as the leading causes of candidemia in hematologic malignancy. Cancer 112:2493–2499

Hansen M, Thrane C, Olsson S, Sørensen J, Srensen J (2000) Confocal imaging of living fungal hyphae challenged with the fungal antagonist viscosinamide. Mycologia 92:216–221

Harrison L, Teplow DB, Rinaldi M, Strobel G (1991) Pseudomycins, a family of novel peptides from Pseudomonas syringae possessing broad-spectrum antifungal activity. J Gen Microbiol 137:2857–2865

Hashimoto S (2009) Micafungin: a sulfated echinocandin. J Antibiot (Tokyo) 62:27–35

Herbrecht R, Maertens J, Baila L, Aoun M, Heinz W, Martino R, Schwartz S, Ullmann AJ, Meert L, Paesmans M, Marchetti O, Akan H, Ameye L, Shivaprakash M, Viscoli C (2010) Caspofungin first-line therapy for invasive aspergillosis in allogeneic hematopoietic stem cell transplant patients: an European Organisation for Research and Treatment of Cancer study. Bone Marrow Transplant 45:1227–1233

Higginbotham SJ, Murphy CD (2010) Identification and characterisation of a Streptomyces sp. isolate exhibiting activity against methicillin-resistant Staphylococcus aureus. Microbiol Res 165:82–86

Hiramatsu Y, Maeda Y, Fujii N, Saito T, Nawa Y, Hara M, Yano T, Asakura S, Sunami K, Tabayashi T, Miyata A, Matsuoka KI, Shinagawa K, Ikeda K, Matsuo K, Tanimoto M (2009) Use of micafungin versus fluconazole for antifungal prophylaxis in neutropenic patients receiving hematopoietic stem cell transplantation. Int J Hematol 88:588–595

Hobbs M, Perfect J, Durack D (1988) Evaluation of in vitro antifungal activity of LY121019. Eur J Clin Microbiol Infect Dis 7:77–80

Horowitz S, Griffin WM (1991) Structural analysis of Bacillus licheniformis 86 surfactant. J Ind Microbiol 7:45–52

Jacobsen BJ, Zidack NK, Larson BJ (2004) The role of bacillus-based biological control agents in integrated pest management systems: plant diseases. Phytopathology 94:1272–1275

Jacques P (2011) Surfactin and other lipopeptides from *Bacillus spp.* In: Biosurfactants. Springer, Berlin, pp 57–92

Janisiewicz WJ, Korsten L (2002) Biological control of postharvest diseases of fruits. Annu Rev Phytopathol 40:411–441

Jones TSG (1949) Chemical evidence for the multiplicity of the antibiotics produced by Bacillus polymyxa. Ann N Y Acad Sci 51:909–916

Journet M, Cai D, Dimichele LM, Hughes DL, Larsen RD, Verhoeven TR, Reider PJ (1999) Semisynthesis of an antifungal lipopeptide echinocandin. J Org Chem 64:2411–2417

Kartsonis N, Killar J, Mixson L, Hoe CM, Sable C, Bartizal K, Motyl M (2005a) Caspofungin susceptibility testing of isolates from patients with esophageal candidiasis or invasive candidiasis: relationship of MIC to treatment outcome. Antimicrob Agents Chemother 49:3616–3623

Kartsonis NA, Saah AJ, Lipka CJ, Taylor AF, Sable CA (2005b) Salvage therapy with caspofungin for invasive aspergillosis: results from the caspofungin compassionate use study. J Infect 50:196–205

Kauffman CA (2005) Candiduria. Clin Infect Dis 41 (Suppl 6):S371–S376

Keating GM, Figgitt DP (2003) Caspofungin: a review of its use in oesophageal candidiasis, invasive candidiasis and invasive aspergillosis. Drugs 63:2235–2263

Kim PI, Ryu J, Kim YH, Chi YT (2010) Production of biosurfactant lipopeptides iturin A, fengycin, and surfactin A from Bacillus subtilis CMB32 for control of Colletotrichum gloeosporioides. J Microbiol Biotechnol 20:138–145

Kim HJ, Choi HS, Yang SY, Kim IS, Yamaguchi T, Sohng JK, Park SK, Kim JC, Lee CH, Gardener BM, Kim YC (2014) Both extracellular chitinase and a new cyclic lipopeptide, chromobactomycin, contribute to the biocontrol activity of Chromobacterium sp. C61. Mol. Plant Pathol 15:122–132

Kinsella K, Schulthess CP, Morris TF, Stuart JD (2009) Rapid quantification of Bacillus subtilis antibiotics in the rhizosphere. Soil Biol Biochem 41:374–379

Király L, Hafez YM, Fodor J, Király Z (2008) Suppression of tobacco mosaic virus-induced hypersensitive-type necrotization in tobacco at high temperature is associated with downregulation of NADPH oxidase and superoxide and stimulation of dehydroascorbate reductase. J Gen Virol 89:799–808

Kohno S, Masaoka T, Yamaguchi H, Mori T, Urabe A, Ito A, Niki Y, Ikemoto H (2004) A multicenter, open-label clinical study of micafungin (FK463) in the treatment of deep-seated mycosis in Japan. Scand J Infect Dis 36:372–379

Kohno S, Izumikawa K, Yoshida M, Takesue Y, Oka S, Kamei K, Miyazaki Y, Yoshinari T, Kartsonis NA, Niki Y (2013) A double-blind comparative study of the safety and efficacy of caspofungin versus micafungin in the treatment of candidiasis and aspergillosis. Eur J Clin Microbiol Infect Dis 32:387–397

Kontoyiannis DP, Ratanatharathorn V, Young JA, Raymond J, Laverdière M, Denning DW, Patterson TF, Facklam D, Kovanda L, Arnold L, Lau W, Buell D, Marr KA (2009) Micafungin alone or in combination with other systemic antifungal therapies in hematopoietic stem cell transplant recipients with invasive aspergillosis: short communication. Transpl Infect Dis 11:89–93

Kurokawa N, Ohfune Y (1993) Synthetic studies on antifungal cyclic-peptides, echinocandins – stereoselective total synthesis of echinocandin-d via a novel peptide coupling. Tetrahedron 49:6195–6222

Kurtz MB, Heath IB, Marrinan J, Dreikorn S, Onishi J, Douglas C (1994) Morphological effects of lipopeptides against Aspergillus fumigatus correlate with activities against (1,3)-??-D-glucan synthase. Antimicrob Agents Chemother 38:1480–1489

Lagrotteria D, Rotstein C, Lee CH (2007) Treatment of candiduria with micafungin: a case series. Can J Infect Dis Med Microbiol 18:149–150

Larkin RP, Tavantzis S (2013) Use of biocontrol organisms and compost amendments for improved control of soilborne diseases and increased potato production. Am J Potato Res 90:261–270

Laverdiere M, Hoban D, Restieri C, Habel F (2002) In vitro activity of three new triazoles and one echinocandin against Candida bloodstream isolates from cancer patients. J Antimicrob Chemother 50:119–123

Lee SC, Kim SH, Park IH, Chung SY, Subhosh Chandra M, Choi YL (2010) Isolation, purification, and characterization of novel fengycin S from bacillus amyloliquefaciens LSC04 degrading-crude oil. Biotechnol Bioproc Eng 15:246–253

Liu J, Liu M, Wang J, Yao JM, Pan RR, Yu ZL (2005) Enhancement of the Gibberella zeae growth inhibitory lipopeptides from a Bacillus subtilis mutant by ion beam implantation. Appl Microbiol Biotechnol 69:223–228

Liu XY, Yang SZ, Mu BZ (2008) Isolation and characterization of a C12-lipopeptide produced by *Bacillus subtilis* HSO 121. J Pept Sci 14:864–875

Lugtenberg B, Kamilova F (2009) Plant-growth-promoting rhizobacteria. Annu Rev Microbiol 63:541–556

Madureira A, Bergeron A, Lacroix C, Robin M, Rocha V, de Latour RP, Ferry C, Devergie A, Lapalu J, Gluckman E, Socié G, Ghannoum M, Ribaud P (2007) Breakthrough invasive aspergillosis in allogeneic haematopoietic stem cell transplant recipients treated with caspofungin. Int J Antimicrob Agents 30:551–554

Maertens J, Raad I, Petrikkos G, Boogaerts M, Selleslag D, Petersen FB, Sable CA, Kartsonis NA, Ngai A, Taylor A, Patterson TF, Denning DW, Walsh TJ (2004) Efficacy and safety of caspofungin for treatment of invasive aspergillosis in patients refractory to or intolerant of conventional antifungal therapy. Clin Infect Dis: Off Publ Infect Dis Soc Am 39:1563–1571

Makovitzki A, Viterbo A, Brotman Y, Chet I, Shai Y (2007) Inhibition of fungal and bacterial plant pathogens in vitro and in planta with ultrashort cationic lipopeptides. Appl Environ Microbiol 73:6629–6636

Mandal SM, Barbosa AEAD, Franco OL (2013) Lipopeptides in microbial infection control: scope and reality for industry. Biotechnol Adv 31:338–345

Marco F, Pfaller MA, Messer SA, Jones RN (1998) Activity of MK-0991 (l-743,872), a new echinocandin, compared with those of LY303366 and four other antifungal agents tested against blood stream isolates of Candida spp. Diagn Microbiol Infect Dis 32:33–37

Mariné M, Serena C, Pastor J, Quindós G, Carrillo AJ, Guarro J (2007) In vitro activity of micafungin combined with itraconazole against Candida spp. Int J Antimicrob Agents 30:463–465

Marr KA, Seidel K, Slavin MA, Bowden RA, Schoch HG, Flowers ME, Corey L, Boeckh M (2000) Prolonged fluconazole prophylaxis is associated with persistent protection against candidiasis-related death in allogeneic marrow transplant recipients: long-term follow-up of a randomized, placebo-controlled trial. Blood 96:2055–2061

Matsumoto Y, Dogru M, Goto E, Fujishima H, Tsubota K (2005) Successful topical application of a new antifungal agent, micafungin, in the treatment of refractory fungal corneal ulcers: report of three cases and literature review. Cornea 24:748–753

Mátyus E, Blaskó K, Fidy J, Tieleman DP (2008) Structure and dynamics of the antifungal molecules Syringotoxin-B and Syringopeptin-25A from molecular dynamics simulation. Eur Biophys J 37:495–502

Mazzola M, De Bruijn I, Cohen MF, Raaijmakers JM (2009) Protozoan-induced regulation of cyclic lipopeptide biosynthesis is an effective predation defense mechanism for Pseudomonas fluorescens. Appl Environ Microbiol 75:6804–6811

McSpadden Gardener BB (2004) Ecology of Bacillus and Paenibacillus spp. in agricultural systems. Phytopathology 94:1252–1258

Mermel LA, Allon M, Bouza E, Craven DE, Flynn P, O'Grady NP, Raad II, Rijnders BJA, Sherertz RJ, Warren DK (2009) Clinical practice guidelines for the diagnosis and management of intravascular catheter-related infection: 2009 update by the Infectious Diseases Society of America. Clin Infect Dis 49:1–45

Messik F, Oberthür M (2013) Total synthesis of the antifungal agent echinocandin c. Angew Chem Int Ed Engl 52:5871–5875

Mikamo H, Sato Y, Tamaya T (2000) In vitro antifungal activity of FK463, a new water-soluble echinocandin-like lipopeptide. J Antimicrob Chemother 46:485–487

Molinaro A, Bedini E, Ferrara R, Lanzetta R, Parrilli M, Evidente A, Lo Cantore P, Iacobellis NS (2003) Structural determination of the O-specific chain of the lipopolysaccharide from the mushrooms pathogenic bacterium Pseudomonas tolaasii. Carbohydr Res 338:1251–1257

Moquet F, Mamoun M, Olivier JM (1996) Pseudomonas tolaasii and tolaasin: comparison of symptom induction on a wide range of Agaricus bisporus strains. FEMS Microbiol Lett 142:99–103

Mora-Duarte J, Betts R, Rotstein C, Colombo AL, Thompson-Moya L, Smietana J, Lupinacci R, Sable C, Kartsonis N, Perfect J (2002) Comparison of caspofungin and amphotericin B for invasive candidiasis. N Engl J Med 347:2020–2029

Moretti S, Bozza S, Massi-Benedetti C, Prezioso L, Rossetti E, Romani L, Aversa F, Pitzurra L (2014) An immunomodulatory activity of micafungin in preclinical aspergillosis. J Antimicrob Chemother 69:1065–1074

Morikawa M, Hirata Y, Imanaka T (2000) A study on the structure-function relationship of lipopeptide biosurfactants. Biochim Biophys Acta – Mol Cell Biol Lipids 1488:211–218

Moyne AL, Shelby R, Cleveland TE, Tuzun S (2001) Bacillomycin D: an iturin with antifungal activity against Aspergillus flavus. J Appl Microbiol 90:622–629

Nakai T, Uno J, Otomo K, Ikeda F, Tawara S, Goto T, Nishimura K, Miyaji M (2002) In vitro activity of FK463, a novel lipopeptide antifungal agent, against a variety of clinically important molds. Chemotherapy 48:78–81

Naruse N, Tenmyo O, Kobaru S, Kamei H, Miyaki T, Konishi M, Oki T (1990) Pumilacidin, a complex of new antiviral antibiotics. Production, isolation, chemical properties, structure and biological activity. J Antibiot (Tokyo) 43:267–280

Ngai AL, Bourque MR, Lupinacci RJ, Strohmaier KM, Kartsonis NA (2011) Overview of safety experience with caspofungin in clinical trials conducted over the first 15 years: a brief report. Int J Antimicrob Agents 38:540–544

Nielsen TH, Sørensen J (2003) Production of cyclic lipopeptides by Pseudomonas fluorescens strains in bulk soil and in the sugar beet rhizosphere. Appl Environ Microbiol 69:861–868

Nielsen TH, Christophersen C, Anthoni U, Sørensen J (1999) Viscosinamide, a new cyclic depsipeptide with surfactant and antifungal properties produced by Pseudomonas fluorescens DR54. J Appl Microbiol 87:80–90

Nielsen TH, Thrane C, Christophersen C, Anthoni U, Sørensen J (2000) Structure, production characteristics and fun gel antagonism of tensin – a new antifungal cyclic lipopeptide from Pseudomonas fluorescens strain 96.578. J Appl Microbiol 89:992–1001

Nihorimbere V, Fickers P, Thonart P, Ongena M (2009) Ecological fitness of Bacillus subtilis BGS3 regarding production of the surfactin lipopeptide in the rhizosphere. Environ Microbiol Rep 1:124–130

Nihorimbere V, Cawoy H, Seyer A, Brunelle A, Thonart P, Ongena M (2012) Impact of rhizosphere factors on cyclic lipopeptide signature from the plant beneficial strain Bacillus amyloliquefaciens S499. FEMS Microbiol Ecol 79:176–191

Nishiyama Y, Uchida K, Yamaguchi H (2002) Morphological changes of Candida albicans induced by micafungin (FK463), a water-soluble echinocandin-like lipopeptide. J Electron Microsc (Tokyo) 51:247–255

Niwa T, Yokota Y, Tokunaga A, Yamato Y, Kagayama A, Fujiwara T, Hatakeyama J,

Anezaki M, Ohtsuka Y, Takagi A (2004) Tissue distribution after intravenous dosing of micafungin, an antifungal drug, to rats. Biol Pharm Bull 27:1154–1156

Nutkins JC, Mortishire-Smith RJ, Packman LC, Brodey CL, Rainey PB, Johnstone K, Williams DH (1991) Structure determination of tolaasin, an extracellular lipodepsipeptide produced by the mushroom pathogen, Pseudomonas tolaasii Paine. J Am Chem Soc 113:2621–2627

Ongena M, Jacques P, Touré Y, Destain J, Jabrane A, Thonart P (2005) Involvement of fengycin-type lipopeptides in the multifaceted biocontrol potential of Bacillus subtilis. Appl Microbiol Biotechnol 69:29–38

Onishi J, Meinz M, Thompson J, Curotto J, Dreikorn S, Rosenbach M, Douglas C, Abruzzo G, Flattery A, Kong L, Cabello A, Vicente F, Pelaez F, Diez MT, Martin I, Bills G, Giacobbe R, Dombrowski A, Schwartz R, Morris S, Harris G, Tsipouras A, Wilson K, Kurtz MB (2000) Discovery of novel antifungal (1,3)-beta-D-glucan synthase inhibitors. Antimicrob Agents Chemother 44:368–377

Ostrosky-Zeichner L, Rex JH, Pappas PG, Hamill RJ, Larsen RA, Horowitz HW, Powderly WG, Hyslop N, Kauffman CA, Cleary J, Mangino JE, Lee J (2003) Antifungal susceptibility survey of 2,000 bloodstream Candida isolates in the United States. Antimicrob Agents Chemother 47:3149–3154

Ota S, Tanaka J, Kahata K, Toubai T, Kondo K, Mori A, Toyoshima N, Musashi M, Asaka M, Imamura M (2004) Successful micafungin (FK463) treatment of invasive pulmonary aspergillosis in a patient with acute lymphoblastic leukemia in a phase II study. Int J Hematol 79:390–393

Pal KK, McSpadden Gardener B (2006) Biological control of plant pathogens. Plant Health Instr 2:1117–1142

Pedras MSC, Ismaila N, Quail JW, Boyetchko SM (2003) Structure, chemistry, and biological activity of pseudophomins A and B, new cyclic lipodepsipeptides isolated from the biocontrol bacterium Pseudomonas fluorescens. Phytochemistry 62:1105–1114

Pérez-García A, Romero D, de Vicente A (2011) Plant protection and growth stimulation by microorganisms: biotechnological applications of Bacilli in agriculture. Curr Opin Biotechnol 22:187–193

Petraitis V, Petraitiene R, Groll AH, Roussillon K, Hemmings M, Lyman CA, Sein T, Bacher J, Bekersky I, Walsh TJ (2002) Comparative antifungal activities and plasma pharmacokinetics of micafungin (FK463) against disseminated candidiasis and invasive pulmonary aspergillosis in persistently neutropenic rabbits. Antimicrob Agents Chemother 46:1857–1869

Pettengell K, Mynhardt J, Kluyts T, Lau W, Facklam D, Buell D, Baraldi E, Botes ME, Kluyts T, Malan DM, Elizabeth P, Mynhardt J, Pettengell K, Ross D, Smego RA, Soni P, Van Der Westhuizen IP, Marais C, Webber C, Lau W, Facklam D, Buell D (2004) Successful treatment of oesophageal candidiasis by micafungin: a novel systemic antifungal agent. Aliment Pharmacol Ther 20:475–481

Peypoux F, Besson F, Michel G, Delcambe L (1981) Structure of bacillomycin D, a new antibiotic of the iturin group. Eur J Biochem 118:323–327

Peypoux F, Pommier MT, Das BC, Besson F, Delcambe L, Michel G (1984) Structures of bacillomycin D and bacillomycin L peptidolipid antibiotics from Bacillus subtilis. J Antibiot (Tokyo) 37:1600–1604

Peypoux F, Marion D, Maget-Dana R, Ptak M, Das BC, Michel G (1985) Structure of bacillomycin F, a new peptidolipid antibiotic of the iturin group. Eur J Biochem 153:335–340

Peypoux F, Pommier MT, Marion D, Ptak M, Das BC, Michel G (1986) Revised structure of mycosubtilin, a peptidolipid antibiotic from Bacillus subtilis. J Antibiot (Tokyo) 39:636–641

Pfaller MA, Marco F, Messer SA, Jones RN (1998) In vitro activity of two echinocandin derivatives, LY303366 and MK-0991 (L-743,792), against clinical isolates of Aspergillus, Fusarium, Rhizopus, and other filamentous fungi. Diagn Microbiol Infect Dis 30:251–255

Powles MA, Liberator P, Anderson J, Karkhanis Y, Dropinski JF, Bouffard FA, Balkovec JM, Fujioka H, Aikawa M, Mcfadden D, Schmatz D (1998) Efficacy of MK-991 (L-743,872), a semisynthetic pneumocandin, in murine models of Pneumocystis carinii. Antimicrob Agents Chemother 42:1985–1989

Raaijmakers JM, de Bruijn I, Nybroe O, Ongena M (2010) Natural functions of lipopeptides from Bacillus and Pseudomonas: more than surfactants and antibiotics. FEMS Microbiol Rev 34:1037–1062

Rainey P, Brodey C, Johnstone K (1991) Biological properties and spectrum of activity of tolaasin, a lipodepsipeptide toxin produced by the mushroom pathogen Pseudomonas tolaasii. Physiol Mol Plant 39:57–70

Rao M, Wei W, Ge M, Chen D, Sheng X (2013) A new antibacterial lipopeptide found by UPLC-MS from an actinomycete Streptomyces sp. HCCB10043. Nat Prod Res 27:2190–2195

Rivardo F, Turner RJ, Allegrone G, Ceri H, Martinotti MG (2009) Anti-adhesion activity of two biosurfactants produced by Bacillus spp. prevents biofilm formation of human bacterial pathogens. Appl Microbiol Biotechnol 83:541–553

Robbel L, Marahiel MA (2010) Daptomycin, a bacterial lipopeptide synthesized by a nonribosomal machinery. J Biol Chem 285:27501–27508

Rodriguez MJ, Vasudevan V, Jamison JA, Borromeo PS, Turner WW (1999) The synthesis of water soluble prodrugs analogs of echinocandin B. Bioorg Med Chem Lett 9:1863–1868

Romero D, De Vicente A, Olmos JL, Dávila JC, Pérez-García A (2007a) Effect of lipopeptides of antagonistic strains of Bacillus subtilis on the morphology and

ultrastructure of the cucurbit fungal pathogen Podosphaera fusca. J Appl Microbiol 103:969–976
Romero D, de Vicente A, Rakotoaly RH, Dufour SE, Veening J-W, Arrebola E, Cazorla FM, Kuipers OP, Paquot M, Pérez-García A (2007b) The iturin and fengycin families of lipopeptides are key factors in antagonism of Bacillus subtilis toward Podosphaera fusca. Mol Plant Microbe Interact 20:430–440
Roongsawang N, Thaniyavarn J, Thaniyavarn S, Kameyama T, Haruki M, Imanaka T, Morikawa M, Kanaya S (2002) Isolation and characterization of a halotolerant Bacillus subtilis BBK-1 which produces three kinds of lipopeptides: Bacillomycin L, plipastatin, and surfactin. Extremophiles 6:499–506
Rückert C, Blom J, Chen X, Reva O, Borriss R (2011) Genome sequence of B. amyloliquefaciens type strain DSM7T reveals differences to plant-associated B. amyloliquefaciens FZB42. J Biotechnol 155:78–85
Segre A, Bachmann RC, Ballio A, Bossa F, Grgurina I, Iacobellis NS, Marino G, Pucci P, Simmaco M, Takemoto JY (1989) The structure of syringomycins A1, E and G. FEBS Lett 255:27–31
Shamala N, Row TNG, Venkatesan K (1976) Crystal and molecular structure of allo-4-hydroxy-L-proline dihydrate. Acta Crystallogr Sect B Struct Crystallogr Cryst Chem B32:3267–3270
Sionov E, Mendlovic S, Segal E (2006) Efficacy of amphotericin B or amphotericin B-intralipid in combination with caspofungin against experimental aspergillosis. J Infect 53:131–139
Sobel JD, Kauffman CA, McKinsey D, Zervos M, Vazquez JA, Karchmer AW, Lee J, Thomas C, Panzer H, Dismukes WE (2000) Candiduria: a randomized, double-blind study of treatment with fluconazole and placebo. The National Institute of Allergy and Infectious Diseases (NIAID) Mycoses Study Group. Clin Infect Dis 30:19–24
Soler-Rivas C, Arpin N (1999) The effects of tolaasin, the toxin produced by Pseudomonas tolaasii on tyrosinase activities and the induction of browning in Agaricus bisporus fruiting. Mol Plant 55:21–28
Song Y-C, Chou A-H, Homhuan A, Huang M-H, Chiang S-K, Shen K-Y, Chuang P-W, Leng C-H, Tao M-H, Chong P, Liu S-J (2011) Presentation of lipopeptide by dendritic cells induces anti-tumor responses via an endocytosis-independent pathway in vivo. J Leukoc Biol 90:323–332
Song YC, Liu HH, Chen IH, Chen HW, Chong P, Leng CH, Liu SJ (2014) A purified recombinant lipopeptide as adjuvant for cancer immunotherapy. Biomed Res Int 2014:349783
Sørensen D, Nielsen TH, Christophersen C, Sørensen J, Gajhede M (2001) Cyclic lipoundecapeptide amphisin from Pseudomonas sp. strain DSS73. Acta Crystallogr C Cryst Struct Commun 57:1123–1124
Spadaro D, Gullino ML (2005) Improving the efficacy of biocontrol agents against soilborne pathogens. Crop Prot 24:601–613
Spreghini E, Orlando F, Sanguinetti M, Posteraro B, Giannini D, Manso E, Barchiesia F (2012) Comparative effects of micafungin, caspofungin, and anidulafungin against a difficult-to-treat fungal opportunistic pathogen, Candida glabrata. Antimicrob Agents Chemother 56:1215–1222
Stachelhaus T, Schneider A, Marahiel MA (1995) Rational design of peptide antibiotics by targeted replacement of bacterial and fungal domains. Science 269:69–72
Sucher AJ, Chahine EB, Balcer HE (2009) Echinocandins: the newest class of antifungals. Ann Pharmacother 43:1647–1657
Sun X, Zeckner DJ, Current WL, Boyer R, McMillian C, Yumibe N, Chen SH (2001) N-acyloxymethyl carbamate linked prodrugs of pseudomycins are novel antifungal agents. Bioorg Med Chem Lett 11:1875–1879
Tareq FS, Lee MA, Lee HS, Lee YJ, Lee JS, Hasan CM, Islam MT, Shin HJ (2014) Gageotetrins A-C, noncytotoxic antimicrobial linear lipopeptides from a marine bacterium bacillus subtilis. Org Lett 16:928–931
Tawara S, Ikeda F, Maki K, Morishita Y, Otomo K, Teratani N, Goto T, Tomishima M, Ohki H, Yamada A, Kawabata K, Takasugi H, Sakane K, Tanaka H, Matsumoto F, Kuwahara S (2000) In vitro activities of a new lipopeptide antifungal agent, FK463, against a variety of clinically important fungi. Antimicrob Agents Chemother 44:57–62
Tendulkar SR, Saikumari YK, Patel V, Raghotama S, Munshi TK, Balaram P, Chattoo BB (2007) Isolation, purification and characterization of an antifungal molecule produced by Bacillus licheniformis BC98, and its effect on phytopathogen Magnaporthe grisea. J Appl Microbiol 103:2331–2339
Tenoux I, Besson F, Michel G (1991) Studies on the antifungal antibiotics: bacillomycin D and bacillomycin D methylester. Microbios 67:187–193
Thrane C, Olsson S, Harder Nielsen T, Sørensen J (1999) Vital fluorescent stains for detection of stress in Pythium ultimum and Rhizoctonia solani challenged with viscosinamide from Pseudomonas fluorescens DR54. FEMS Microbiol Ecol 30:11–23
Thrane C, Harder Nielsen T, Neiendam Nielsen M, Sørensen J, Olsson S (2000) Viscosinamide-producing Pseudomonas fluorescens DR54 exerts a biocontrol effect on Pythium ultimum in sugar beet rhizosphere. FEMS Microbiol Ecol 33:139–146
Tóth V, Nagy CT, Pócsi I, Emri T (2012) The echinocandin B producer fungus Aspergillus nidulans var. roseus ATCC 58397 does not possess innate resistance against its lipopeptide antimycotic. Appl Microbiol Biotechnol 95:113–122
Tran H, Ficke A, Asiimwe T, Höfte M, Raaijmakers JM (2007) Role of the cyclic lipopeptide massetolide a in biological control of Phytophthora infestans and in colonization of tomato plants by Pseudomonas fluorescens. New Phytol 175:731–742

Tsan P, Volpon L, Besson F, Lancelin JM (2007) Structure and dynamics of surfactin studied by NMR in micellar media. J Am Chem Soc 129:1968–1977

Uchida K, Nishiyama Y, Yokota N, Yamaguchi H (2000) In vitro antifungal activity of a novel lipopeptide antifungal agent, FK463, against various fungal pathogens. J Antibiot (Tokyo) 53:1175–1181

Ullmann AJ, Cornely OA, Donnelly JP, Akova M, Arendrup MC, Arikan-Akdagli S, Bassetti M, Bille J, Calandra T, Castagnola E, Garbino J, Groll AH, Herbrecht R, Hope WW, Jensen HE, Kullberg BJ, Lass-Flörl C, Lortholary O, Meersseman W, Petrikkos G, Richardson MD, Roilides E, Verweij PE, Viscoli C, Cuenca-Estrella M (2012) ESCMID guideline for the diagnosis and management of Candida diseases 2012: developing European guidelines in clinical microbiology and infectious diseases. Clin Microbiol Infect 18:1–8

Van Burik J-AH, Ratanatharathorn V, Stepan DE, Miller CB, Lipton JH, Vesole DH, Bunin N, Wall DA, Hiemenz JW, Satoi Y, Lee JM, Walsh TJ (2004) Micafungin versus fluconazole for prophylaxis against invasive fungal infections during neutropenia in patients undergoing hematopoietic stem cell transplantation. Clin Infect Dis 39:1407–1416

Vanittanakom N, Loeffler W (1986) Fengycin – a novel antifungal lipopeptide antibiotic produced by Bacillus subtilis F-29-3. J Antibiot (Tokyo) XXXIX:888–901

Vazquez JA, Lynch M, Boikov D, Sobel JD (1997) In vitro activity of a new pneumocandin antifungal, L-743,872, against azole-susceptible and -resistant Candida species. Antimicrob Agents Chemother 41:1612–1614

Vicente MF, Basilio A, Cabello A, Peláez F (2003) Microbial natural products as a source of antifungals. Clin Microbiol Infect 9:15–32

Wiederhold NP, Cota JM, Frei CR (2008) Micafungin in the treatment of invasive candidiasis and invasive aspergillosis. Infect Drug Resist 1:63–77

Wilson Quail J, Ismail N, Soledade M, Pedras C, Boyetchko SM (2002) Pseudophomins A and B, a class of cyclic lipodepsipeptides isolated from a Pseudomonas species. Acta Crystallogr C Cryst Struct Commun 58:o268–o271

Yang P, Sun Z x, Liu S y, Lu H x, Zhou Y, Sun M (2013) Combining antagonistic endophytic bacteria in different growth stages of cotton for control of Verticillium wilt. Crop Prot 47:17–23

Yaryura PM, León M, Correa OS, Kerber NL, Pucheu NL, García AF (2008) Assessment of the role of chemotaxis and biofilm formation as requirements for colonization of roots and seeds of soybean plants by Bacillus amyloliquefaciens BNM339. Curr Microbiol 56:625–632

Yokote T, Akioka T, Oka S, Fujisaka T, Yamano T, Hara S, Tsuji M, Hanafusa T (2004) Successful treatment with micafungin of invasive pulmonary aspergillosis in acute myeloid leukemia, with renal failure due to amphotericin B therapy. Ann Hematol 83:64–66

Zasloff M (2002) Antimicrobial peptides of multicellular organisms. Nature 415:389–395

Zhang JC, Dai JY, Fan J, Wu XP (2006) The treatment of pneumocystis Carinii pneumonia with caspofungin in elderly patients: a case report and literature review. Zhonghua Jie He He Hu Xi Za Zhi 29:463–465

Plant-Derived Antifungal Agents: Past and Recent Developments

5

G.M. Vidyasagar

Abstract

Fungal pathogens are increasing their vicinity and able to cause serious systemic and local infection in human beings. Although there are a few antifungal drugs available in the market, most of them are not completely safe and effective. On the other hand, plants have played an important role in the history of healthcare system. Several commonly available plant species belonging to algae, bryophytes, pteridophytes, and angiosperms are used in the treatment of fungal diseases by indigenous people and managed well the outburst of fungal diseases. Plant-based antimicrobials represent a vast untapped source of medicine, and they have enormous therapeutic potential as they can serve the purpose without any side effects that are often associated with synthetic antimicrobials. The present chapter deals with the historical background and the present status of medicinal plants used in the treatment of fungal diseases.

5.1 Introduction

5.1.1 Fungal Infections

Fungi are a large group with about 250,000 species available worldwide. Of which, more than 300 species have been reported as human pathogens (Guarro et al. 1997). The incidence of fungal infection has increased dramatically over the past few decades due to increase in the members of population susceptible to such infections.

The common systemic infections are aspergillosis, caused by species of *Aspergillus*, blastomycosis (*Blastomyces dermatitidis*); coccidioidomycosis, also called valley fever (*Coccidioides immitis*); cryptococcosis (*Cryptococcus neoformans*); histoplasmosis (*Histoplasma capsulatum*), candidiasis (*Candida* species); and *Pneumocystis* pneumonia (*Pneumocystis jirovecii* formerly known as *Pneumocystis carinii*). Superficial fungal infections include tinea capitis (*Microsporum canis*, *Trichophyton tonsurans*), tinea barbae (*T. rubrum*, *T. mentagrophytes*), tinea corporis (*M. canis*, *T. mentagrophytes*),

G.M. Vidyasagar (✉)
Department of Post Graduate Studies and Research in Botany, Gulbarga University, Kalaburagi 585106, Karnataka, India
e-mail: gmvidyasagar@gmail.com

A. Basak et al. (eds.), *Recent Trends in Antifungal Agents and Antifungal Therapy*,
DOI 10.1007/978-81-322-2782-3_5

tinea cruris (*T. rubrum*, *T. mentagrophytes*, *Epidermophyton floccosum*), tinea pedis (*T. rubrum*, *T. mentagrophytes*, *E. floccosum*), and tinea unguium (*T. rubrum*, *T. mentagrophytes*, *E. floccosum*). Approximately, 90 % of human fungal infections are caused by the species of *Aspergillus*, *Candida*, *Cladosporium*, *Epidermophyton*, *Microsporum*, and *Trichophyton* spp. Among these, aspergillosis and candidiasis have drawn more attention of scientists, policymakers, and medical epidemiologists.

Candida species are normal inhabitants of the skin, the gastrointestinal tract, the female genital tract, and urine. In contrast, molds produce conidia (spores) that are easily carried by wind and water, inhaled the floating spores through air, and cause sinus and lung infections. Sometimes the infection spreads to the brain and bones. Infections usually occur as a result of a decrease in natural human defenses or exposure to heavy fungal spores. Many infections are confined between the toes and a few spread over the skin. The internal infections become severe, if not taken care of in the initial stage and may cause permanent damage or even death. Therefore, the fungal infections are gaining prime importance in the healthcare system (Beck-Sague and Jarvis 1993). The increase in microbial infection in human beings and emergence of resistance in pathogen are the serious problems. The use of antimicrobial drugs for prophylactic or therapeutic purposes or for agricultural purposes has provided selective pressure favoring the survival and spread of resistant organisms. To overcome this, scientists are engaged in the development of new strategies to inhibit the fungal growth. Although a number of antifungal drugs are available in the market, many are proved to have serious side effects and hence, there is a great need of continuous search for newer, safer and cheaper antifungal agents from natural sources.

5.1.2 History of Plant Based Drug

There are about 250,000–500,000 species of plants on Earth (Borris 1996), and relatively a small group (1–10 %) of these is used as food by both humans and other animal species. Plants, in particular, have formed the basis of sophisticated traditional medicine systems with the earliest records dating from around 2600 BCE in Mesopotamia. About 1,000 plant-derived substances including oils of *Cedrus* species (cedar) and *Cupressus sempervirens* (cypress), *Glycyrrhiza glabra* (licorice), *Commiphora* species (myrrh), and *Papaver somniferum* (poppy juice) are documented and found useful even today. The best known record of Egyptian medicine is the "Ebers Papyrus" (1500 BCE), documented over 700 drugs of plant origin (Borchardt 2002). The Chinese materia medica reveals hundreds of important drugs starting from the first record dating from about 1100 BC to 659 AD (Huang 1999). Similarly, the Indian Ayurvedic system dates from before 1000 BC. Charaka and Sushruta Samhita reveal about 341 and 516 drugs, respectively (Dev 1999). Hippocrates in the late fifth century BC mentioned about 300–400 medicinal plants (Schuster 1966). Dioscorides, a Greek physician (100 CE), recorded the methods of collection, storage, and use of medicinal herbs during his travels with Roman armies. Galen (130–200 CE.), a practitioner and teacher of pharmacy and medicine in Rome, is well known for his complex herbal drug prescriptions. The Arabs, however, preserved much of the Greco-Roman expertise of the fifth to twelfth centuries and expanded it to include the use of their own resources along with Chinese and Indian herbs unknown to them.

Herbal drug systems are developed regionally in different parts of the world (Behl and Srivastava 2002), largely used in India and China (Xu 2004). In Europe and USA, the use of herbs declined due to the availability of purified and synthetic chemical drugs. In recent years, there has been a resurgence of the use of herbal drug as the side effects of chemical drugs became apparent and a call for green revolution and return to organic products.

5.1.3 Ethnomedicine

Ethnomedicine is practiced by various ethnic groups, especially by indigenous peoples. The word ethnomedicine is sometimes used as a synonym for traditional medicine. Ethnomedicine is rather old and dates back to the time when the early man became conscious of his environment. Ethnomedicine helped in the survival of ancestors and has given knowledge about medicinal plants and its uses. Cultural man struggled for his existence as a hunter-gatherer and, with his trial-and-error experience, must have learned to distinguish useful and harmful plants, particularly the plants with healing properties (Weideman 2005). The earliest record of human civilizations reveals that the elders and wise men of those times used herbal medicines to treat various diseases. Information regarding these medicinal herbs is available in the old literature, folklore, mythological stories, epic poems, medical treatises, and old manuscripts, on palm leaves and copper plates and other records preserved even today (Weideman 2005).

5.1.4 Potential of Herbal Remedies as a Source of New Drugs

Plants have played an important role in the healthcare system across the world. Their role in the development of new drugs could be by serving either as a natural blueprint for the development of new drugs or as a phytomedicine to be used for the treatment of disease. It is estimated that plant materials have provided the models for 50 % of the modern drugs. Many commercially available modern drugs were initially used in crude form in traditional or folk healing practice (Dagne 1996). Since the beginning of the nineteenth century, a large number of biologically active secondary metabolites of plant origin have been found to have commercial application as drugs. Recently, there has been an upsurge of interest in the use of plants with folkloric reputations as sources of potentially useful compounds (Mourice et al. 1999). Analysis of the number and sources of anticancer and anti-infective agents, reported from 1984 to 1995, indicates that over 60 % of the approved drugs and pre-NDA (New Drug Application) candidates are of natural origin. A recent review reported that at least 119 compounds of plant origin are used currently, and 77 % of these are being derived from traditional medicinal plants (Douglas 1987). Several new small molecules of natural product-derived drugs have been introduced into therapy in various countries in recent years, including acarbose, artemether, capsaicin, docetaxel, dronabinol, galantamine, irinotecan, paclitaxel, tacrolimus, and topotecan. This trend is likely to continue in the future, at least for the treatment of disease states such as cancer and infectious diseases. In a recent statistical survey, it was pointed out that the origin of 30,000 bioactive natural products could be divided between animals (13 %), bacteria (33 %), fungi (26 %), and higher plants (27 %) (Ernst 1999). Despite the rapid development in the field of chemistry of medicinal research, many plant-derived drugs such as, atropine, resperine, morphine, cocaine, ergotamine, and digitalis still cannot be synthetically produced (Ernavitha 2008). Thus, the isolation of drugs from plants still holds good.

Plant-based antimicrobials represent a vast untapped source of medicine, and they have enormous therapeutic potential as they can serve the purpose without any side effects that are often associated with synthetic antimicrobials. First scientific studies on the antimicrobial properties of plant drugs were recorded in the late nineteenth century (Zaika 1975). Since then, several plants have been identified with potential antifungal properties and used in the treatment of fungal diseases.

5.1.5 Algae as an Antifungal Agents

Algae are a large and diverse group of plants which ranges from unicellular to multicellular forms. Blue-green algae, green algae, red algae, brown algae, and marine algae are known for their food and medicinal properties. Algae have also been used in the preparation of biodiesel, bioethanol, biobutanol, and hydrogen gases (Raja et al. 2013) and could be used as antioxidants, antibiotics, and virostatic agents. The most powerful water-soluble antioxidants such as polyphenols, phycobiliproteins, and vitamins are found in algae (Plaza et al. 2008). These antioxidants are able to prevent the occurrence of cancer cell formation (Richardson 1993). Marine algae have been used as food and medicine for centuries. They produce a wide spectrum of biological activities such as antimicrobial (Bouhlal et al. 2011), antiviral (Kim and Karadeniz 2011), antifungal (De Felício et al. 2010), anti-allergic (Na et al. 2005), anticoagulant (Dayong et al. 2008), anticancer (Kim and Karadeniz 2011), antifouling, and antioxidant activities (Devi et al. 2011). The marine species such as *Rhodomela confervoides*, *Ulva lactuca*, *Cystoseira tamariscifolia*, and *Padina pavonica* from Bejaia coast are reported to have antifungal activity against *Aspergillus niger* (939 N), *Candida albicans* (ATCC 1024), and *Mucor ramaniannus* (NRRL1829). The extracts of *P. pavonica*, *R. confervoides*, and *U. lactuca* were very efficient against *M. ramaniannus* and *C. albicans* (Saidani et al. 2012). In another study (Tuney et al. 2006), the ethanolic extract of *P. pavonica* was active against *C. albicans*. In *Acanthaphora spicifera*, the methanolic extract showed higher antibacterial and antifungal activity against *E. coli*, *B. subtilis*, *Bacillus palmitus*, *P. aeruginosa*, and *C. albicans*, *M. gypseum*, and *A. niger* (Pandian et al. 2011). The purified carotenoid and chlorophyll pigments from *Chlorococcum humicola* (green alga) were effective against the harmful pathogens such as *E. coli*, *P. aeruginosa*, *Salmonella typhimurium*, *Klebsiella pneumoniae*, *V. cholera*, *S. aureus*, *B. subtilis*, *C. albicans*, *A. niger*, and *A. flavus* with a maximum of 80 % inhibition in benzene and ethyl acetate extracts (Bhagavathy et al. 2011). A report from Brazil reveals the chloroform and hexane fractions of *H. musciformis* inhibitory against *T. rubrum*, *T. tonsurans*, *T. mentagrophytes*, *M. canis*, *M. gypseum*, *C. albicans*, *C. krusei*, *C. guilliermondii*, and *C. parapsilosis*. The apolar extracts obtained from algae showed the best activity against pathogenic fungi (Guedes et al. 2012). *Asparagopsis taxiformis* (red alga) from Italy was found antifungal against *A. fumigatus*, *A. terreus*, and *A. flavus* with minimum inhibitory concentrations between <0.15 and >5 mg ml^{-1} against *Aspergillus* spp. Agar diffusion assays confirmed antifungal activity of *A. taxiformis* extracts against *Aspergillus* species (Genovese et al. 2013). Another red alga *Gracilaria arcuata* from Chabahar coasts showed very strong antibacterial and antifungal properties (Javad et al. 2014). *G. changii* is predominant in the mangrove areas and widely used in the traditional medicine in Malaysia. People in Malaysia use agar extracted from *Gracilaria* internally for coughs (Burkill 1935). Besides, *Gracilaria* sp. is boiled in vinegar and used to treat swollen knees and unhealthy sores (Burkill 1935). The methanol extract of *G. changii* exhibits inhibitory effect against candidiasis (Sasidharan et al. 2007). In Turkey, among the marine species such as *Cladophora glomerata*, *Enteromorpha linza*, *U. rigida*, *Cystoseira barbata*, *Padina pavonica*, *Corallina officinalis*, and *Ceramium ciliatum* studied from the coast of Vona, *Enteromorpha linza* and *P. pavonica* showed highest antifungal activity against *A. niger*, while *C. glomerata* showed highest antibacterial activity against *S. aureus* (Erturk and Tas 2011).

5.1.6 Bryophytes as Antifungal Agents

The bryophytes are a diverse group of land plants usually colonize in moist conditions. Alternation of generation with dominant free-living haploid gametophytic generation is an important feature of this group (Wyatt 1982). Bryophytes represent the simplest land plant group that are phylogenetically very old (Frahm 1994) and represent the level of evolution associated with transmigration to the land (Miller 1982). The secondary metabolites have provided numerous leads for the development of drugs, but the discovery of new drugs with novel structures has declined in the past few years. Bryophytes almost remain untouched in the drug discovery process. They are known to have numerous potential compounds, but much work remains to link medical effects with specific bryophyte species or compounds (Pant and Tewari 1990). Studies of their secondary metabolites are recent but reveal some original compounds, which have not seen even in higher plants. Mosses were traditionally used as antimicrobial agents in India (Frahm 2001) and as natural diapers (Ando and Mastuo 1984).

Bryophyte species actually produce broad-range antibiotics. Their usage in surgical dressings, diapers, and other human medicinal applications is well known not only in Asia (Frahm 2004) but also in Brazil (Pinheiro da Silva et al. 1989), England (Wren 1956), and Germany (Frahm 2004).

Chinese, Indians, and Native Americans have included bryophytes in their herbal medicine. Native Americans have used them for drugs, fibers, and clothing (University of Michigan, Dearborn 2003). Based on the concept of the doctrine of signatures, a variety of bryophytes, especially liverworts were used as herbal medicine (*Riccia* spp.), in the Himalayas to treat ringworm. Another study on *Riccia fluitans* from Florida indicated no ability to inhibit growth of *P. aeruginosa*, *S. aureus*, and *C. albicans* (Pates and Madsen 1955) including ringworm pathogen. The Gosiute native peoples in the USA use *Bryum*, *Mnium*, and *Philonotis* paste to reduce the pain of burns, bruises, and wounds (Flowers 1957). Similarly, the mixture of liverworts like *Conocephalum conicum* and *Marchantia polymorpha* thalli with vegetable oils is used in China for the treatment of bites, boils, burns, cuts, eczema, and wounds (Wu 1977; Ding 1982; Ando 1983). Indians in the Himalayan region use *M. polymorpha* or *M. palmata* to treat boils and abscesses as the young archegoniophore resembles a boil when it emerges from the thallus (Pant and Tewari 1989). The alcohol extract of *P. lyellii* was reported to be more active against *A. niger*, *A. fumigatus*, *F. oxysporum*, and *C. albicans*. *A. fumigates* was found to be more susceptible at 80 ng/disk concentration, and complete mycelial growth inhibition was seen at 1 mg/ml concentration which is better than ketoconazole (Subhisha and Subramoniam 2005). *Philonotis marchica*, *Grimmia pulvinata*, *Plagiomnium rugicum*, *Haplocladium* sp., *Bryum pallens*, *Drepanocladus aduncus*, *Pellia epiphylla*, and *Dumortiera hirsuta* were also found to have a broad spectrum of *antifungal* activity with maximum in ethanol extract (Shizadiann et al. 2009).

The antifungal compounds were also isolated from the liverwort *Plagiochila fasciculata* of New Zealand (Lorimer and Perry 1993, 1994). Scientists have found innumerable kinds of biological activity in compounds from bryophytes. Even in a single species, one might find multiple kinds of activity. For example, the liverworts *Plagiochasma japonica* and *Marchantia tosana* exhibit antitumor activity, antifungal and antimicrobial activity, inhibition of superoxide release, inhibition of thrombin activity, and muscle relaxation (Lahlou et al. 2000). One indication of the presence of unique and potential pharmaceutically important chemicals in bryophytes is the presence of unique odors especially in liverworts *Leptolejeunea* and *Moerckia* (Schuster 1966). *Lophozia bicrenata* has a pleasant odor, species of *Solenostoma* smell like carrots, *Geocalyx graveolens* has a turpentine-like odor, and *Conocephalum conicum* smells like mushrooms. The tropical *Plagiochila rutilans* smells like peppermint, due to the presence of several monoterpenoids (Heinrichs et al. 2001).

Although mosses are known to harbor fungi and will quickly become infected if kept moist in a plastic bag, some fungi are inhibited by many species of bryophytes, including many that cause skin infections. Jennings (1926) reported moss immunity to molds as early as 1926, but the possibility of using them as a source of antifungal activity seems to have been largely overlooked. Among these, *Hypnum cupressiforme* has remarkable antibacterial and antifungal effects. The absence of fungal diseases in liverworts led Pryce (1972) to suggest that lunularic acid, an aging hormone found in liverworts but not in mosses, might be responsible for liverwort antifungal activity. The degree of antibiotic activity in a given species may depend on the age of the gametophyte (Banerjee and Sen 1979), and this is well supported by the age-dependent antifungal activity in liverwort *Herbertus aduncus* against *Botrytis cinerea*, *Pythium debaryanum*, and *Rhizoctonia solani*. They subsequently isolated three aging substances, namely, alpha-herbertenol, beta-herbertenol, and alpha-formylherbertenol from it (Matsuo et al. 1982a, b, 1983). The paste of *Ceratodon purpureus* and *Bryum argenteum* to cure fungal infections of horses (Frahm 2004) was used by industrious horse owners and cured the fungal infections in horses within 24 h. The cinnamolide, the active compounds from *Porella* and *Makinoa*, showed antidermatophytic activity, and *Dumortiera hirsuta*, *Sphagnum portoricense*, and *Orthotrichum rupestre* showed antifungal activity against *C. albicans* (Subramoniam and Subhisha 2005a; Mc Cleary and Walkington 1966). A group of phenolic compounds isolated from liverwort have shown potent activity as an antifungal, antioxidant, and anticancer agent (Dai and Sun 2008), while plagiochin E from *Marchantia polymorpha* could reverse fungal resistance to fluconazole and induce apoptosis in *C. albicans*. Riccardin D from *Dumortiera hirsuta* inhibits hyphal formation and interferes with biofilm formation. Among the bisbibenzyls extracted from liverworts, isoriccardin C and BS-34 found better antifungal compounds in a dose-dependent manner (Sun 2009).

5.1.7 Pteridophytes as Antifungal Agents

The pteridophytes are an important group of plants that constitutes ferns and their allies. There are over 250 different genera and 12,000 species of ferns reported all over the world (Chang et al. 2011). Of these, between 2,200 and 2,600 species occur in China accounting about 22 % of the world total. Out of 1,250 species in India, 173 species are used as food, flavor, dye, medicine, bio-fertilizers, oil, fiber, and biogas producer (Manickam and Irudayaraj 1992). The secondary metabolites like alkaloids, glycosides, flavonoids, terpenoids, sterols, phenols, sesquiterpenes, etc., are important components in fern and are used in various industries (Kulandairaj and Britto 2000). Kirtikar et al. (1935) have described 27 species, Nayar (1959) 29 species, Chopra et al. (1956) 44 species, and Nadkarni (1954) 11 species of pteridophytes having medicinal importance. A detailed review on the medicinal uses of 105 ferns is published by May (1978). On the basis of phytochemical, pharmacological, and ethnobotanical studies, Singh (1999) reported 160 useful pteridophytes in India. *Selaginella bryopteris* (Linn.) Bak. is considered as a highly useful plant for unconsciousness condition, *Equisetum arvense* Linn. is used in nasal polyps and kidney infections and its ashes are useful in acidity, *E. debile* Roxb. is diuretic and given in gonorrhea, *Lycopodium clavatum* Linn. in the form of decoction is used in rheumatism and diseases of the lung and kidney, the leaf paste of *Ophioglossum reticulatum* Linn. is used in headache, *Botrychium virginianum* Sw. is used in dysentery, *Helminthostachys zeylanica* (Linn.) Hook. is used to revert impotency and also used as a brain tonic, and *Lygodium flexuosum* (Linn.) Sw. is an expectorant and also used in the treatment of ulcers, cut wounds, and sprains. The fronds of gleicheniaceous fern *Dicranopteris linearis* (Burm.) are used in the treatment of asthma and sterility in women, and *Osmunda regalis* Linn., the "royal fern," is used as a styptic and tonic, while the rhizome of *Angiopteris evecta* (Forst.) Hoffm. is used for scabies (Vasudev 1999). As per ancient literature, the pteridophytes have been known for about 2,000 years (Kirtikar and Basu 1975) and are used as medicine in Ayurvedic and Unani systems of medicine (Uddin et al. 1998). The antimicrobial potential of some ferns has been studied by Kumar and Kaushik (1999) and Parihar and Bohra (2002a, 2003). The paste of *Adiantum incisum* Forsk. and *A. venustum* is useful in the healing of wounds (Samant et al. 1998; Kholia and Punetha 2005).

The edible Malaysian fern species like *Stenochlaena palustris*, *Diplazium esculentum*, *Nephrolepis biserrata*, and *Acrostichum aureum* were screened against *A. niger*, *Rhizopus stolonifer*, and *C. albicans*, and a broad spectrum of antifungal activity was found in *Diplazium esculentum* leaves. The study shows that the fern extracts are favorable antifungal agents with potential applications in public heath against fungal diseases (Zakaria and Sanduran 2010). *Hemionitis arifolia* (Burm. f.) Moore., *Pteridium aquilinum* (Linn.) Kuhn., and *Christella parasitica* (Linn.) H. Lev. were evaluated against the fungal pathogens *Puccinia arachidis* and *Phaeoisariopsis personata* causing rust and early leaf spot diseases in groundnut, respectively, which are the pathogens sensitive to all three fern extracts. The chloroform extract of *H. arifolia* was found to have maximum antifungal activity against both fungi (Kitherian Sahayaraj et al. 2009). The aqueous and ethanolic frond extracts of *Cheilanthes anceps* were proved to be effective against *A. flavus* and *A. niger* (Mishra and Verma 2009). The Bheel tribe in Mt. Abu area use leaf juice of *Adiantum incisum* for the treatment of skin disease; however, in Goram Ghat area, the leaf powder mixed with butter is used for controlling internal burning in the body (Parihar and Parihar 2006). The phenolic compounds are abundantly present in lycopodiophyta (Pedersen and Ollgaard 1982), pteridophyta, angiosperm, and gymnosperm (Carnachan and Harris 2000). Ferns in particular serve as an important source of some phenolic compounds such as kaempferol, which is known to be extracted from *Phegopteris connectilis* (Adam 1999), and kaempferol-3-O-rutinoside, extracted from *Selliguea feei* (Baek et al. 1994). *Salvinia molesta*, a freshwater fern, is a good source of phenolics such as hypogallic acid, caffeic acid, paeoniflorin, and pikuroside (Choudhary et al. 2008).

Ethnomedicinally, *Adiantum* Linn. is an important plant popularly known as "Hansraj" in Ayurvedic System of Medicine and is used in the treatment of cold; tumors of the spleen, liver, and other viscera; skin diseases; bronchitis; and inflammatory diseases. Meenakshi Singh et al. (Jan 2008) have studies the antimicrobial property of four important species, i.e., *Adiantum capillus-veneris*, *Adiantum peruvianum*, *Adiantum venustum*, and *Adiantum caudatum*, against five gram-positive and six gram-negative bacteria (including multiresistant *S. aureus*) and eight fungal strains. The maximum activity was exhibited by the methanolic extract of *A. venustum* followed by *A. capillus-veneris*, *A. peruvianum*, and *A. caudatum*. *Pityrogramma* is a genus with about 17 species occurring mainly in tropical America (Smith et al. 2006). The ethanolic extract of *Pityrogramma calomelanos* (L.) Link showed good activity against *S. aureus* when associated with aminoglycosides and against species of *Candida* with benzoilmetronidazol. The results indicated that *P. calomelanos* should be studied as a possible source of natural products to combat bacteria and fungi either directly or by modulating the mechanisms of resistance of these microorganisms and enhancing the antimicrobial activity of these drugs (Souza et al. 2012). *Stenochlaena palustris* leaf extract tested for its fungicidal activity was found effective against food-borne pathogen *A. niger*. Morphological changes in *A. niger* treated with the fern leaf extract were observed through scanning electron microscope which reveal cell wall disruption with flattened appearance on hyphae (Sumathi and Parvathi 2010). Ethanol and methanol extracts of *Drynaria quercifolia* (L.) J. Smith rhizomes were also found to be inhibitory against *C. albicans* and *Cryptococcus neoformans*. Saponin, coumarin, and terpenoids were detected in ethanol extract and showed greater antibacterial and antifungal activity than methanol extract (Pargavi and Shivakumar 2014).

5.1.8 Higher Plants as Antifungal Agents

Herbal therapy for skin disorders has been used for thousands of years. Even our biologically close relatives, the great apes, use herbal self-medication (Huffman 2001). Specific herbs and their uses developed regionally, based on locally available plants and through trade in ethnobotanical remedies. The great importance of flowering plants in developing new therapeutic tools is evident. Medicinal plants and their derivatives are important for pharmacological research and drug development.

The natural products can be used directly as therapeutic agents, or as a source of raw materials for synthesis of new pharmacologically active models (Brazil 2006). In Chinese medicine, *Psoralea corylifolia* and *Eucalyptus globulus* have been used in the treatment of dermatomycosis caused by *T. mentagrophytes* and *T. rubrum* (Lau et al. 2010). The antifungal activity of *Eugenia umbelliflora* Berg. (Machado et al. 2009) against *E. floccosum*, *T. mentagrophytes*, *T. rubrum*, *M. canis*, and *M. gypseum* and *Wrightia tinctoria* against *T. rubrum*, *E. floccosum*, *A. niger*, and *Scopulariopsis brevicaulis* was reported (Ponnusamy et al. 2010). The major compound, identified as indirubin, is inhibitory against dermatophytes such as *E. floccosum*, *T. rubrum*, *T. tonsurans*, *T. mentagrophytes*, and *T. simii*. It was also active against *A. niger*, *C. albicans*, and *Cryptococcus* sp. (Newman and Cragg 2007). Lima et al. (2011) reported antifungal activity of the essential oils (EOs) of *Gymnophyton polycephalum*, *Satureja parvifolia*, and *Lippia integrifolia* against *M. gypseum*, *T. mentagrophytes*, and *T. rubrum*. The originality of many structures of natural products attracts attention to their use as a starting point for semi-synthesis and total synthesis (Butler 2005). The development of chemotherapeutic agents of synthetic origin and the discovery of powerful new antimicrobials isolated from natural sources account for invaluable contributions in the fight against fungal resistance (Silveira et al. 2006). The extracts obtained from plants such as *Euphorbia prostrata*, *Salvia texana*, *Colubrina greggii*, and *Clematis drummondii* have shown promising results for stimulating the search for new potential antifungal plant sources (Alanís-Garza et al. 2007).

Antifungal activity of *Indigofera suffruticosa* was effective against *T. rubrum* and *M. canis* (Sonia Periera et al. 2006). The oils of *Ocimum gratissimum* and *Trachyspermum ammi* exhibited strong antidermatophytic properties (Tiwari et al. 2003). Ranganathan (1996) reported the MIC of neem seed extract was lower than that of the neem leaf when tested against different species of *Trichophyton* and *Epidermophyton floccosum*. The MIC (31 μg/ml) of neem seed extract was also determined by Natarajan et al. (2003) against dermatophytes. The concentration at 15 μg/ml (below MIC) was observed to be sufficient for distorting the growth pattern of the organisms. In another study, the ethanolic extract at 100 μg/ml concentration was more active against dermatophytes (Pankajalakshmi 1994). The flower extract of *Tagetes erecta* showed maximum antimycotic activity against *F. oxysporum* and *T. mentagrophytes* followed by whole plant of *T. patula* and leaf extract of *T. erecta* (Rai and Acharya 1999). The *Capparis spinosa* and *Juglans regia* inhibited the growth of *M. canis*, *T. violaceum*, and *T. mentagrophytes* (Ali-Shtayeh and Abu Ghdeib 1999). *Nelumbo nucifera* rhizome extract was inhibitory against yeast (Mukherjee et al. 1995), and *Cinnamomum tamala* and while *Citrus maxima* oil exhibited complete inhibition of mycelial growth of *T. mentagrophytes* and *M. audouinii* (Dubey et al. 1998). *Wrightia tinctoria* leaves also possessed potent antimicrobial properties against dermatophytes. The methanol and ethanol extracts were found effective against bacteria and hexane extract against dermatophytes suggesting that the active principles may be useful in the topical treatment of superficial skin infections (Kannan et al. 2006).

Turmeric oil isolated from *Curcuma longa* L. by Apisariyakul et al. (1995) was inhibitory against 15 isolates of dermatophytes at 1:40–1:320 dilutions. A wide range of inhibition zone from 6.1 to 26.0 mm against 29 clinical strains of dermatophytes was reported by Mansuang et al. (2002) using crude ethanol extract of *Curcuma longa*. The antifungal activity of essential oils of higher plants against ringworm caused by *T. mentagrophytes* and *M. audouinii* was reported by Yadav and Dubey (1994). *Cinnamomum tamala*, *Citrus maxima*, *Cymbopogon citratus*, *Eucalyptus citriodora*, *Eupatorium cannabinum*, *Nepeta hindostana*, and *Ocimum canum* oils were absolutely toxic against both the test fungi. The essential oil from the fruits of *Luvunga scandens* was found very good against dermal infections (Garg and Jain 1999). In recent years, these reports have involved mainly the members of Lamiaceae and Asteraceae families. The antifungal effects of essential oils from several species of the Lamiaceae family, viz., *Satureja montana* L., *Lavandula angustifolia* Mill (D'Auria et al. 2005), *Lavandula hybrida* Reverchon, *Origanum vulgare* L., *Rosmarinus officinalis* L., and six chemotypes of *Thymus vulgaris* L., were studied against *C. albicans*. The greatest efficiency was obtained with the essential oil from *T. vulgaris*. An extensive study has been carried out on the antifungal activity of the essential oils from *Lavandula* and *Rosmarinus*. The essential oil from *Hyptis ovalifolia* leaves was found very effective against dermatophytes at a concentration of $<$500 μg/ml (eSouza et al. 2002).

In the Liliaceae family, reports on the antifungal activity concern mainly the *Allium* genus. By using an agar dilution assay, the antifungal activity of aqueous extracts prepared from *Allium cepa* L. and *Allium sativum* L. was evaluated against *Malassezia furfur*, *C. albicans*, as well as several strains of various dermatophyte species by Shams et al. (2006). The results indicate that onion and garlic might be promising sources of drugs for the treatment of fungal diseases caused by *Candida*, *Malassezia*, and the dermatophytes. Similar studies on the antifungal activity of onion and garlic were reported by Ghahfarokhia et al. (2004) against *T. rubrum* and *T. mentagrophytes*. According to Karunyal Samuel et al. (2001), 200 mg/ml concentration of *Allium sativum* aqueous bulb extract is effective against *T. rubrum*. Pyun. Shin (2006) studied the activity of essential oils from *Allium fistulosum* L., *A. sativum*, and *A. cepa* (Liliaceae) against three species of *Trichophyton* and found *A. sativum* oil as the strongest growth inhibitor against *T. rubrum*. There have been a large number of antifungal screening programs being carried out in plant drugs used in traditional medicine of Eastern Europe and Africa. Tadeg et al. (2005) screened the traditionally used Ethiopian medicinal plants, namely, *Acokanthera schimperi*, *Calpurnia aurea*, *Kalanchoe petitiana*, *Lippia adoensis*, *Malva parviflora*, *Olinia rochetiana*, *Phytolacca dodecandra*, and *Verbuscum sinaiticum*, against fungal pathogens and reported *L. adoensis* and *O. rochetiana* to be the effective antifungal plant drugs. Similarly, in the traditional practices in Southern Brazil, the aerial parts of *Pterocaulon alopecuroides* and *Pterocaulon polystachyum* are used to treat mycoses. In vitro studies indicated the crude methanol extract of *P. polystachyum* as the most active one for mycoses (Stein et al. 2005). In other studies, Lamidi et al. (2005) evaluated 77 crude extracts from leaves and stem barks of 15 Gabonese plants used in traditional medicine as antifungal agents and revealed ethanol extract of *Polyalthia suaveolens* as a good antifungal agent effective at 1 mg/ml concentration. From Balochistan, Pakistan, Zaidi and Crow (2005) reported very good antifungal activity in *Zygophyllum fabago* L. and *Vincetoxicum stocksii* against *C. albicans*.

Nigella sativa L. (Aljabre et al. 2005), *Melaleuca alternifolia* oil (Hammer et al. 2002), flowers of *Yucca gloriosa* L. (Favel et al. 2005), *Boerhavia diffusa* L. (Agrawal and Rangari 2003), *Hypericum* growing in Southern Brazil (Fenner et al. 2005), *Gentianella nitida* Griseb. (Rojas et al. 2004), *Croton urucurana* Baill. bark (Gurgel et al. 2005), *Inula viscosa* (Maoz and Neeman 1998), *Piper guineense* (Ngono Ngane et al. 2003), *Chimaphila umbellate* (Isabel et al. 2008), *Hypoestes serpens* (Rasoamiaranjanahary et al. 2003) and *Mitracarpus villosus* leaves (Irobi and Daramola 1993) were reported to have antifungal property. Similarly, the essential oils from *Cymbopogon citratus* (Lachoria et al. 2000); *Allium fistulosum* L., *A. sativum* L., and *A. cepa* L. (Pyun and Shin 2006); *Catharanthus roseus* (Singh and Singh,1997 and Rai and Upadhyay 1998); *Citrus sinensis* (Patra et al. 2003); *Syzygium aromaticum* L. (Merr. and Perry) (Taguchi et al. 2005); *Eucalyptus globulus* Labill., *Eucalyptus maculata* Hook., and *Eucalyptus viminalis* Labill. (Takahashi et al. 2004); *Eucalyptus rostrata* (Singh et al. 1988); *Satureja montana* L., *Lavandula angustifolia*, *L. hybrida* Reverchon, *Origanum vulgare* L., *Rosmarinus officinalis* L., and *Thymus vulgaris* L. (Tarfa et al. 2004); *Chrysactinia mexicana* Gray. (Cardenas et al. 2005); *Azadirachta indica* (Pant et al. 1986; Vaijayanthimala et al. 2004); *Cassia alata* (Ibrahim and Osman 1995); *C. fistula* (Lillykutty and Santhakumari 1969); mint (Kishore et al. 1993); *Artemisia herba-alba* and *A. judaica* (Charchari et al. 1996); *Acalypha indica* (Gopalakrishnan et al. 2000); *Acalypha wilkesiana* (Alade and Irobi 1993); *Calendula officinalis* (Kasiram et al. 2000); *Salvia mirzayanii* Rech. F. and Esfand. (Portillo et al. 2005); *Solanum dulcamara* (Kumar and Bhadauria 2009); *Ocimum* spp. (Janseen et al. 1989); *O. sanctum* (Gangrade et al. 1989); *Ocimum gratissimum* L. (Silva et al. 2005); *Lawsonia inermis* (Bhakuni et al. 1971; Bhatnagar et al. 1961; Misra and Dixt 1979); *Psoralea corylifolia* (Sharma and Singh 1979; Grover and Rao 1979; Gupta et al. 1962); *Curcuma longa* rhizome (Vaijayanthimala et al. 2004); *Pongamia pinnata* seed oil (Kesari et al. 2010); and lemon and lantana (Jain et al. 2004) were reported to have antifungal activity. The effective antifungal compounds isolated from different medicinal plants against pathogenic fungi are given in Tables 5.1 and 5.2.

Table 5.1 Plant-derived antifungal compounds from different plant parts

Sl. no	Name of the plant	Part used	Active compounds	Name of the compound class	Targeted microorganism studied	Ref. no.
1	*Aesculus pavia* L., Sapindaceae	Leaves	S-6-[2-(Hydroxy methyl) butoxy]-7-hydroxy-4-methyl-2H-chromen-2-one, named Pavietin	Prenylated coumarin	*Guignardia aesculi*	Curir et al. (2007)
2	*Ailanthus excelsa* Roxb., Simaroubaceae	Stem bark	Methanol extract of the plant was partitioned with chloroform	Extract	*A. niger*, *A. fumigatus*, *Penicillium frequentans*, *P. notatum*, *Botrytis cinerea*	Deepa et al. (2004)
3	*Ajania fruticulosa* (Ledeb.) Poljak, Asteraceae	Aerial parts	(I) 1E,2E-Epoxy-3E,4D,8E,10D-titrahydroxyguaia-11(13)en-12-Olide	Guaianolides	*Candida albicans*	Meng et al. (2001)
			(II) 1E,2E-Epoxy-3E,4D,9D,10D-tetrahydroxyguaia-11(13)-en-12,6D-olide			
			(III) 1E,2E-Epoxy-10D-hydroperoxy-3E,4D,8E-trihydroxyguaia-11(13)-en-12,6D-olide			
4	*Asterella angusta* (Steph.), Pandé et al., Aytoniaceae		Asterelin A, asterelin B, 11-O-demethyl marchantin, dihydroptychantol A	Bis(benzyl)	*C. albicans*	Qu et al. (2007)
5	*Azadirachta indica* A. Juss., Meliaceae	Green leaves	Nimonol, isomoldenin	Limonoids		Suresh et al. 1997
6	*Ballota glandulosissima*	Aerial part	Kumatakenin, pachypodol, 5-hydroxy-7,3′,4′-trimethoxy, flavone, velutin, retusin	Flavonoids	*C. albicans*, *C. krusei*, *C. glabrata*	Citoglu et al. (2003)
7	*Bridelia retusa* (Linn.) Spreng., Euphorbiaceae	Stem bark	(E)-4-(1,5,Dimethyl-3-oxo-1-hexenyl) benzoic acid	Bisabolane sesquiterpenes	*Cladosporium cladosporioides*	Jayasinghe et al. 2003
8	*Calophyllum Caledonicum* Vieill. ex Planch. and Triana, Clusiaceae	Stem bark	Caledonixanthone E	Xanthone	*A. fumigatus*	Larcher et al. (2004)
9	*Centaurea*, Asteraceae		(+) Enicin, salonitenolide, (+) costunolide, (−) dehydro-custuslactone, lychnopholideermantholide C	Sesquiterpene lactones	*Cunninghamella echinulata*	Barrero et al. (2000)

10	*Chenopodium procerum* Hochst. ex Moq., Chenopodiaceae	Aerial part	Irilin A, irilin B, dihydrowogonin, pygmol	Isoflavone, flavanone, sesquiterpene	*Cladosporium cucumerinum*	Bergeron et al. (1995)
11	*Clematis tangutica* (Maxim.) Korsh., Ranunculaceae	Aerial parts	(1) 3-o-D-L-Arabinopyranosyl-hederagenin-28-0-D-L-rhamnopyranosyl ester	Terpene saponins	*Saccharomyces cerevisiae*	Du et al. (2003)
			(2) 3-0-E-D-Glucopyranosyl-(1o4)-D-L-arabinopyranosyl-hederagenin-28-o-D-L-rhamnopyranosyl ester			
12	*Clytostoma ramentaceum*		Ursolic acid, 2-(3,4-dihydroxophenyl) ethanol		*A. niger*, *F. oxysporum*	Lorimer and Perry (1994)
13	*Colutea arborescens* L., Fabaceae	Root bark	(1) Coluteol (3,5-dihydroxy-7,2,4-trimethoxyisoflavan)	Isoflavan, isoflavan quinone	*Saccharomyces cerevisiae*	Grosvenor and Gray David (1998)
			(2) Colutequinone B (7,4 6-trimethoxyisoflavan-2 5-quinone)			
14	*Commiphora wightii* (Arn.) Bhandari, Burseraceae		Muscanone	Flavanone	*C. albicans*	Fatope et al. (2003)
15	*Coptis japonica* (Thunb.) Makino, Ranunculaceae		Berberine	Alkaloid	*B. berengeriana*, *G. cingulata*, *P. expansum*	Chung and Paik (1997)
16	*Cordia alliodora* (Ruiz and Pav.) Oken, Boraginaceae	Root bark	(1) 1-(3-Methoxypropanoyl)-2,4,5-trimethoxy benzene	Phenylpropanoid derivative, prenylated hydroquinone	*Cladosporium cucumerinum*	Ioset et al. (2000)
			(2) 2-2(Z)-(3-Hydroxy-3-7-dimethylocta-2,6-dienyl)-1,4-benzenediol			
17	*Cryptomeria japonica* (L. f.) D. Don, Cupressaceae	Bark	12-Methoxy-6D,11-dihydroxyabieta-8,11,13-triene	Diterpenoids	*A. alternata*, *P. oryzae*, *R. solani*, *F. oxysporum*	Kofujita et al. (2006)
18	*Cuban propolis*		A novel polyisoprenylated benzophenone	Ketone	*C. albicans*, *C. tropicalis*	Rubio et al. (1999)
19	*Cudrania cochinchinensis* Lour., Moraceae	Roots	Cudraxanthones S	Xanthone 7	*Cryptococcus neoformans*, *A. fumigatus*, *A. nidulans*, *C. glabrata*	Fukai et al. (2003)
			(1) [1,3,5,6-Tetrahydroxy-2-(1,1-dimethyl-2-propanyl)xanthone]			
			(2) Toxyloxanthone			
			(3) Wightone			

(continued)

Table 5.1 (continued)

Sl. no	Name of the plant	Part used	Active compounds	Name of the compound class	Targeted microorganism studied	Ref. no.
20	*Delphinium denudatum* Wall., Ranunculaceae	Roots	8-Acetylheterophyllisine, vilmarrianone, panicutin	Diterpene, alkaloid, alkaloid	*Allescheria boydii*, *E. floccosum*, and *A. niger*, *A. boydii*, *Stachybotrys atra*, *Pleurotus ostreatus*	Rahman et al. (1997)
21	*Detarium microcarpum* Guill. and Perr., Leguminosae	Fruit pulp	(1) 3,4-Epoxyclerodan-13E-en-15-oic acid	Clerodane diterpenes	*Cladosporium cucumerinum*	Cavin et al. (2006)
			(2) 5D,8D-(2-Oxokolavenic acid)			
			(3) 4-Dihydroxyclerodan-1 3Z-en-15-oic acid			
22	*Eriosema tuberosum* Hochst., Leguminosae	Roots	Eriosemaone A, eriosemaone B, eriosemaone C, eriosemaone D, flemichin D	Polyphenols	*Cladosporium cucumerinum*, *C. albicans*	Ma et al. (1995)
23	*Eugenia dysenterica* DC., Myrtaceae	Leaves	E-Caryophyllene, D-humeelene, limonene, D-thujine	Sesquiterpenes, monoterpene hydrocarbon	*C. albicans*	Costa et al. (2000)
24	*Ferula diversivittata* Rgl. et Schmalh., Apiaceae	Roots	Diversolides A–G, diversolide E	Sesquiterpene, lactone derivative	Antifungal	Iranshahi et al. (2008)
25	*Ganoderma annulare* (Lloyd) Boedijn, Ganodermataceae		5D-ergost-7-en-3E-o1, applanoxidic acids A, C and F	Sterol, triterpenes	*Microsporum canis*, *T. mentagrophytes*	Smania et al. (2003)
26	*Garcinia mangostana*, Linn., Guttiferae	Fruit	Mangostin, 8-desoxygartanin, BR, xanthone, gartanin, beta-mangostin, garcinone-D	Xanthones	*F. oxysporum* f. sp. *vasinfectum*, *Alternaria tenuis*, *Drechslera oryzae*	Gopalakrishnan et al. (1997)
27	*Glaucium oxylobum* Boiss and Buhse, Papaveraceae	Aerial parts	Dicentrine, glaucine, protopine D, allocryptopine	Alkaloids	*M. gypseum*, *M. canis*, *T. mentagrophytes*, *Epidermophyton*, *Floccosum*	Morteza et al. (2003)
28	*Gmelina arborea* Roxb. Lamiaceae	Heartwood	Catalpol	Iridoid glycoside	*Trametes versicolor*	Kawamura and Ohara (2005)
29	*Guarea macrophylla* Vahl., Meliaceae		Caryophyllene oxide, guai-6-en-10E-ol	Sesquiterpenes	*C. cladosporioides*, *C. sphaerospermum*	Costa et al. (2000)
30	*Hildegardia barteri* (Mast.) Kosterm., Sterculiaceae	Root	Hilde gardiol, 2-hydroxy maackiain, farrerol	Isoflavan, flavonoid	*C. albicans*	Meragelman et al. (2005)

31	*Hypoestes serpens* (Vahl) R. Br. Acanthaceae	Leaves	Fusicoserpenol, dolabeserpenic acid	Diterpenes	*C. cucumerinum*, *C. albicans*	Rasoamiaranjanahary et al. (2003)
32	*Khaya senegalensis* (Desr.) A. Juss., Meliaceae	Fruits	Seneganolide A, 2-hydroxyseneganolide A, 2-acetoxyseneganolide A, two limonoids,	Limonoid	*Botrytis cinerea*	Abdelgaleil et al. (2004)
33	*Licania intrapetiolaria*, Chrysobalanaceae	Roots	Intrapetacin B	Clerodane diterpenoid	*A. niger*	Oberlies et al. (2001)
34	*Lycium chinense* Mill., Solanaceae	Roots, barks	Dihydro-N-caffeoyltyramine, trans-N-fereeloyloctopamine, trans-N-caffeoyltyramine, cis-N-caffeoyltyramine	Phenolic amides	*C. albicans*	Lee et al. (2004)
35	*Magnolia grandiflora* L., Magnoliaceae	Stem bark	Costunolide, 1,10 epoxyparthenolide	Sesquiterpene lactone	*Nigrospora* spp., *R. solani*, *Helminthosporium*, *Alternaria alternata*, *Fusarium culmorum*	Ahmed and Abdelgaleil (2005)
36	*Melicope borbonica* (Bory), Rutaceae	Leaves	Xanthoxylin, scoparone, limettin	Acetophenone coumarins	*C. albicans*, *Penicillium expansum*	Simonsen et al. (2004)
37	*Monnina obtusifolia* H. B. and K., Polygalaceae	Aerial parts	Two xanthones	Xanthone	*Trichophyton mentagrophytes*	Pinto et al. (1994)
38	*Nepeta clarkei* Hook. F., Lamiaceae	Aerial part	Actinidine	Terpene alkaloid	*A. flavus*	Saxena Jyoti and Mathela (1996)
39	*Nepeta leucophylla* Benth, Lamiaceae	Aerial part	Iridodial-E-monoenol acetate	Iridoid	*A. flavus*, *A. ochraceus*, *Penicillium citrinum*, *P. viridicatum*	Saxena Jyoti and Mathela (1996)
40	*Peperomia serpens* (*Sw.*) Loudon., Piperaceae	Leaves	Two chromenes	Chromene	*C. cladosporioides*, *C. sphaerospermum*	Li et al. (1999)
41	*Phyllanthus piscatorum* Kunth, Euphorbiaceae		Justicidin B	Arylnapthalide lignan	*A. fumigatus*, *A. flavus*, *C. albicans*	Gutierrez et al. (2002)
42	*Phytolacca tetramera* Hauman, Phytolaccaceae	Berries	Phytolaccoside B, phytolaccoside E	Monodesmosidic triterpenoid saponins	*Trichophyton mentagrophytes*	Kariba et al. (2002)
43	*Piper arboretum*, Piperaceae	Leaves	(1) N-[10-(13,14-Methylenedioxy-phenyl-7(E),9(Z)-pentadienoyl]-pyrolindine	Amide	*Cladosporium sphaerospermum*	Sartori et al. (2003)
			(2) Arboreumine			

(continued)

Table 5.1 (continued)

Sl. no	Name of the plant	Part used	Active compounds	Name of the compound class	Targeted microorganism studied	Ref. no.
44	*Piper crassinervium*	Leaves	(1) 1,4-Dihydroxy-2-(3,7-dimethyl-1-oxo-2-E,6-octadienyl)benzene (2) 1,4-Dihydroxy-2-(3,7-dimethyl-1′ oxo-2-Z,6-octadienyl)benzene (3) 1,4-Dihydroxy-2(7-methyl-3-methylene-1-oxo-4,7-peroxide-octyl) benzene	Prenylated hydroquinone	*C. cladosporioides*, *C. sphaerospermum*	Cavin et al. (2006)
45	*Piper hispidum* Sw., Piperaceae	Leaves	N-[7-(3,4-Methylenedioxyphenyl)-2 (z),4(Z)-heptadienoyl] pyrrolidine	Pyrrolidine amide	*Cladosporium sphaerospermum*	Alecio et al. (1998)
46	*Piper hispidum* Sw., Piperaceae		Two new amides	Amide	*Cladosporium sphaerospermum*	Navickiene et al. (2000)
47	*Piper lanceaefolium* Kunth, Piperaceae	Leaves	Four new benzoic acid derivatives together with tabogenic acid, pinocembrin, and pinocembrin chalcone. Lanceaefolic acid methyl ester and pinocembrin chalcone	Benzoic acid derivative	*C. albicans*	Lopez et al. (2002)
48	*Piper umbellatum* Linn., Piperaceae	Branch	*Piper umbellatum* (A–D)	Alkaloids	*C. albicans*	Borris (1996)
49	*Piscidia erythrina* L., Fabaceae	Root bark	Isopiscerythrone, allolicosi flavone A, piscisoflavone A, piscisoflavone B	Isoflavones	*Cladosporium herbarum*	Monga and Mohapatra (1980)
50	*Plagiochila fasciculata* Lindenb., Plagiochilaceae		2-Hydroxy-4,6-dimethoxyacetophenone, 2-hydroxy-3,4,6-trimethoxyacetophenone	Acetophenone	*T. mentagrophytes*, *Cladosporium resinae*	Alecio et al. (1998)
51	*Pogostemon cablin* Benth, Lamiaceae	Whole plant	5-Hydroxy-7,3,4′trimethoxyflavanone, 5-hydroxy-7,4-dimethoxyflavanone, 3,5-dihydroxy-7,4′dimethoxyflavanone	Flavonoids		Cornet et al. (2002)
52	*Prismatomeris fragrans* Geddes, Rubiaceae	Roots and stem	Nordamnacanthal, damnacanthal	Anthraquinones	*C. albicans*	Saxena et al. (1996)
53	*Rudgea viburnoides* (Cham.) Benth., Rubiaceae	Ripe fruit	Viburgenin	Triterpene	*Cladosporium cladosporioides*	Claudia et al. (1998)
54	*Schizozygia coffaeoides* Baill., Apocynaceae	Leaves and roots	6,7-Dehydro-19E-hydroxy schizozygine	Alkaloid	*Cladosporium cladosporioides*	Kariba et al. (2002)

55	*Selaginella tamariscina* (Beauv.), Selaginellaceae	Whole plant	Amentoflavone	Biflavonoid	*C. albicans*	Jung et al. (2006)
56	*Smallanthus sonchifolius* (Poepp. and Endl.) H. Robinson, Asteraceae	Leaf extract	Sonchifolin	Melampolide	*Pyricularia oryzae*	Hashidoko et al. (1995)
57	*Smilax medica* Schltdl. and Cham. Smilacaceae	Roots	Two new steroidal saponins and the known disporoside A	Steroidal saponins	*C. albicans*, *C. glabrata*, *C. tropicalis*	Sauton et al. (2005)
58	*Sorocea muriculata* Miq., Moraceae	Root extract	Sorocenol H	Oxygen heterocyclic, Diels-Alder-type adduct	*C. albicans*, *Cryptococcus neoformans*, *A. fumigatus*	Samir et al. (2008)
59	*Spathodea campanulata* P. Beauv., Bignoniaceae	Roots, peels	Iridoid glucoside (ajugol) and two phenols *p*-hydroxybenzoic acid, methyl *p*-hydroxybenzoate	Iridoid glycoside, phenols	*Cladosporium herbarum*	Pianara et al. (2007)
60	*Taxus baccata* L., Taxaceae	Needles	Bilobetin	Flavones	*A. alternata*, *Cladosporium oxysporum*, *F. culmorum*	Krauze-Baranowska (2003)
61	*Thonningia sanguinea* Vahl., Balanophoraceae		Extracts	Saponins, quinines, polyphenols	*A. fumigatus*, *A. flavus*, *C. albicans*	Carpano et al. (2003)
62	*Toona ciliata* M. Roem., Meliaceae	Leaves	Toonaciliatin	Pimaradiene-type diterpenoid	*T. rubrum*	Chen et al. (2009)
63	*Toona ciliata* M. Roem., Meliaceae		Cedrelone, azadiradione, limonin, nomilinic acid	Tetranotri terpenoids	*Puccinia arachidis*	Govindachari et al. (2000)
64	*Tordylium apulum* L., Apiaceae	Aerial parts	Flavonoid diglycoside, a coumarin	Coumarin, Flavonoid	*Cladosporium cucumerinum*	Kofinas et al. (1998)
65	*Vochysia divergens* Pohl, Vochysiaceae	–	Sericic acid		*C. albicans*, *C. tropicalis*, *Penicillium*	Hess, et al. (1995)
66	*Wedelia paludosa* DC., Asteraceae	Flower	Kaurenoic acid, luteolin	Flavone	*Epidermophyton floccosum*, *T. mentagrophytes*	Sartori et al. (2003)
67	*Zuccagnia punctata* Cav., Fabaceae	Fruits, aerial parts	(i) 2,4-Dihydroxy-3-methoxy chalcone (ii) 2,4-Dihydroxy chalcone	Chalcone	*M. gypsum*, *T. rubrum*, *T. mentagrophytes*	Svetaz et al. (2007)

Table 5.2 Antifungal compounds from essential oils of medicinal plants

Sl. no.	Botanical name	Compound	Effective against	Reference(s)
1	*Haplophyllum tuberculatum* (Forsskal) A. Juss	α- and β-Phellandrene, limonene, β-ocimene, β-caryophyllene and myrcene	*Curvularia lunata* and *Fusarium oxysporum*	Al-Burtamani et al. (2005)
2	*Thymbra capitata* (L.) Cav.	γ-Terpinene and p-cymene	*Candida*, *Aspergillus*, and dermatophyte	Salgueiro et al. (2004)
3	*Lavandula stoechas* L.	Fenchone, limonene, and myrtenol	*Rhizoctonia solani*	Angioni et al. (2006)
4	*Rosmarinus officinalis* L.	α-Pinene, borneol, camphene, camphor, verbenone, and bornyl acetate	*Candida albicans*	Dalleau et al. (2007)
5	*Lavandula angustifolia*	Linalool and linalyl acetate	*Candida albicans*	D'Auria et al. (2005)
6	*Mentha piperita* L.	Menthol	*Aspergillus flavus* and *A. parasiticus*	Farag et al. (1989)
7	*Thymus vulgaris*	Thymol	*Aspergillus flavus* and *A. parasiticus*	Farag et al. (1989)
8	*Salvia mirzayanii* Rech. F. and Esfand.	Linalool, linalyl acetate, α-terpinyl acetate, 1,8-cineole, α-cadinol, and δ-cadinene	*Fusarium solani* and *Candida albicans*	Portillo et al. (2005)
9	*Calocedrus formosana* Florin.	α-Cadinol and murolol	*Lenzites betulina*, *Pycnoporus coccineus*, *Trametes versicolor*, and *Laetiporus sulphureus*	Cavaleiro et al. (2006)
10	*Trachyspermum ammi* (L.) Sprague	Thymol, p-cymene, γ-terpinene, β-pinene, and terpinen-4-ol	*Aspergillus niger*, *Fusarium moniliforme*, and *Curvularia lunata*	Nigam and Rao (1977)
11	*Cuminum cyminum* L.	β-Pinene, γ-terpinene, and cuminaldehyde	*Aspergillus*	Patra et al. (2002); Kawther (2007); Hammer et al. (2002)
12	*Foeniculum vulgare* Mill.	Trans-anethole	*Alternaria alternata*, *Fusarium oxysporum*, *Aspergillus flavus*, and *A. parasiticus*	Aggarwal et al. (2000); Shabnam et al. (2012).
13	*Bupleurum gibraltaricum* Lamarck.	Sabinene, α-pinene, and 2,3,4-trimethylbenzaldehyde	*Plasmopara halstedii*	Komiya et al. (2006)
14	*Zingiber officinale* Roscoe	Zingiberene	*Aspergillus flavus*, *A. parasiticus*, and *F. oxysporum*	Pandey et al. (2010); Farag et al. (1989)
15	*Curcuma longa* L.	Terpinolene, α-phellendren, and terpinene-4-ol	Dermatophyte	Apisariyakul et al. (1995); Saha R and Bhupendar (2011)
16	*Nigella sativa* L.	Thymoquinone	*Candida albicans*, *Aspergillus flavus*	Mashhadian and Rakhshandeh (2005); Khan et al. (2003); Aggrawal et al. (1979); Amrouche et al. (2011); Singh et al. (2005); Al-Jabre et al. (2005); Zhang et al. (2006)
17	*Syzygium aromaticum* (L.) Merrill and Perry	Eugenol	*Candida*, *Aspergillus*, and dermatophyte	Eugenia et al. (2009)

(continued)

Table 5.2 (continued)

Sl. no.	Botanical name	Compound	Effective against	Reference(s)
18	*Lavandula angustifolia* Mill.	Linalool and linalyl acetate	*Candida albicans*	D'Auria et al. (2005)
19	*Rosmarinus officinalis* L.	α-Pinene, borneol, camphene, camphor, verbenone, and bornyl acetate	*Fusarium graminearum*	Angioni et al. (2004)
20	*Tagetes patula* L.	Piperitone and piperitenone	*Botrytis cinerea* and *Penicillium digitatum*	Romagnoli et al. (2005)
21	*Calocedrus formosana*	α-Cadinol and murolol	*Lenzites betulina*, *Pycnoporus coccineus*, *Trametes versicolor*, and *Laetiporus sulphureus*	Cheng et al. (2004)
22	*Trachyspermum ammi* (L.) Sprague	Thymol, *p*-cymene, γ-terpinene, β-pinene, and terpinen-4-ol	*Aspergillus niger*, *Fusarium moniliforme*, and *Curvularia lunata*	Singh et al. (2004)

5.1.9 Future Prospectus of Plant-Derived Antifungal Drug

Fungal pathogens, namely, the species of *Trichophyton*, *Microsporum*, *Epidermophyton*, *Aspergillus*, *Candida*, etc., have become prevalent in causing diseases in human beings and animals. The increase in fungal infections in human beings and emergence of resistance in pathogens are serious problems in the healthcare system. The unscientific way of using antifungal drugs for prophylactic or therapeutic purposes or for agricultural purposes has provided selective pressure on the survival and spread of resistant organisms. To overcome this emerging global problem, the policy makers, pharmaceutical industries and funding agencies have to create a platform for the scientists to develop new strategies and new antifungal drugs for the management of fungal diseases. As plants have proved to have very strong properties of inhibiting fungal growth without any residual side effect on the human body, the traditionally used medicinal plants could be the best source for the development of new antifungal drugs.

References

Abdelgaleil SAM, Doe M, Iwagawa T, Nakatani M (2004) Antifungal limonoids from the fruits of *Khaya senegalensis*. Fitoterapia 75(6):566–572

Adam KP (1999) Phenolic constituents of the fern *Phegopteris connectilis*. Phytochemistry 52(5):929–934

Aggarwal KK, Ahmed A, Santha TRK, Jain N, Gupta SK, Kumar V, Khanuja SPS (2000) Antimicrobial activity spectra of *Pelargonium graveolens* L. and *Cymbopogon winterrianus* Jowitt. oil constituents and acyl derivatives. J Med Aromat Plant Sci 22:544–548

Aggrawal R, Kharya MD, Srivastava R (1979) Antimicrobial and anthelmintic activities of the essential oil of *Nigella sativa* Linn. Indian J Exp Biol 17 (11):1264–1265

Agrawal RB, Rangari VD (2003) Anti-inflammatory and antiarthritic activities of lupeol and 19 α-H lupeol isolated from *Strobilanthus callosus* and *S. ixiocephala* roots. Ind J Pharmacol 35:384–387

Ahmed SM, Abdelgaleil SAM (2005) Antifungal activity of extracts and sesquiterpene lactones from magnolia grandiflora L. (Magnoliaceae). Int J Agric Biol 07 (4):638–642

Alade PI, Irobi ON (1993) Antimicrobial activities of crude leaf extracts of *Acalypha wilkesiana*. J Ethnopharmacol 39:171–174

Alanis-Garza BA, Gonzalez-Gonzalez GM, Salazar-Aranda R, Waksman de Torres N, Rivas-Galindo VM (2007) Screening of antifungal activity of plants from the northeast of Mexico. J Ethnopharmacol 114:468–471

Al-Burtamani SKS, Fatope MO, Marwah RG, Onifade AK, Al-Saidi SH (2005) Chemical composition, antibacterial and antifungal activities of the essential oil of *Haplophyllum tuberculatum* from Oman. J Ethnopharmacol 96:107–112

Alecio AC, Bolzani, da Silva V, Young MCM, Kato MJ, Furlan (1998) Maysa. J Nat Prod 61(5):637–693

Ali-Shtayeh MS, Abu Ghdeib SI (1999) Antifungal activity of plant extracts against dermatophytes. Mycoses 42:665–672

Aljabre SH, Randhawa MA, Akhtar N, Alakloby OM, Alqurashi AM, Aldossary A (2005) Anti-dermatophyte activity of ether extract of *Nigella sativa* and its principle, thymoquinone. J Ethnopharmacol 101:116–119

Al-Jabre S, Al-Akloby OM, Al-Qurashi AR, Akhtar N (2005) Thymoquinone, an active principle of *Nigella sativa*, inhibited *Aspergillus niger*. Pak J Med Res 42 (3):102–104

Amrouche A, Benmehdi H, Moussaoui A, Mebarki K, Chaoufi A, Saneba A (2011) Evaluation of antifungal activity of some oils from Algerian medicinal plants against *Aspergillus flavus* strain produced aflatoxins. J Appl Pharm Sci 01(08):48–53

Ando H (1983) Use of bryophytes in China 2 Mosses indispensable to the production of Chinese gallnuts. Proc Bryol Soc Jpn 3:124–125

Ando H, Mastuo A (1984) Applied bryology. Adv Bryol 2:133–224

Angioni A, Barra A, Cereti E, Barile D, Coisson JD, Arlorio M, Dessi S, Coroneo V, Cabras P (2004) Chemical composition, plant genetic differences, antimicrobial and antifungal activity – investigation of the essential of *Rosmarinus officinalis* L. J Agric Food Chem 252(11):3530–3535

Angioni A, Barra A, Coroneo V, Dessi S, Cabras P (2006) Seasonal plant part chemical variability and antifungal activity investigation of *Lavandula stoechas* L ssp. stoechas essential oils. J Agric Food Chem 54:4364–4370

Apisariyakul A, Vanitanakom N, Bunddhasukh D (1995) Antifungal activity of turmeric oil extracted from *Curcuma longa*. J Ethanopharmacol 49:163–169

Baek NI, Kennelly EJ, Kardono LBS, Tsauri S, Padmawinata K, Soejarto DD, Kinghorn AD (1994) Flavonoids and a proanthocyanidin from rhizomes of *Selliguea feei*. Phytochemistry 36(2):513–518

Banerjee RD, Sen SP (1979) Antibiotic activity of bryophytes. Bryologist 82:141–153

Barrero AF, Oltra JE, Alvarez M, Raslan DS, Sande DA, Akssira M (2000) Fitoterapia 71(1):60

Beck-Sague CM, Jarvis WR (1993) National nosecomial infections surveillance system. Secular trends in the epidemiology of nosocomial fungal infections in the United states 1980–1990. J Infect Dis 167:1247–1251

Behl PN, Srivastava G (2002) Herbs useful in dermatological therapy, 2nd edn. CBS Publishers, New Delhi

Bergeron C, Marston A, Hakizamungu E, Hostettmann K (1995) Antifungal constituents of Chenopodium procerum. Pharm Biol 33(2):115–119

Bhagavathy S, Sumathi P, Bell JS (2011) Green algae *Chlorococcum humicola* anew source of bioactive compounds with antimicrobial activity. Asian Pac J Trop Biomed 1:S1–S7

Bhakuni DS, Dhar ML, Dhar MM, Dhawan BN, Gupta B, Srimal RC (1971) Screening of Indian plants for biological activity. Indian J Exp Biol 9(1):91–102

Bhatnagar SS, Santapau H, Fernandes F, Kamat VN, Dastoor NJ, Ras TSN (1961) Physiological activity of Indian medicinal plants. J Sci Ind Res (Suppl) 20 (8):1–22

Borchardt JK (2002) The beginnings of drug therapy ancient Mesopotamian medicine. Drug News Perspect 15:187–192

Borris RP (1996) Natural products research perspectives from a major pharmaceutical company. J Ethnopharmacol 51:29–38

Bouhlal R, Haslin C, Chermann JC, Colliec-Jouault S, Sinquin C, Simon G, Cerantola S, Riadi H, Bourgougnon N (2011) Antiviral activities of sulfated polysaccharides isolated from *Sphaerococcus coronopifolius* (Rhodophytha, Gigartinales) and *Boergeseniella thuyoides* (Rhodophyta, Ceramiales). Mar Drugs 9:1187–1209

Brazil (2006) Pharmaceuticals DDA, editor. National policy of medicinal plants and herbal medicine. Ministry of Health SdC, Technology and Estrategic Inputs, Brasilia

Burkill IH (1935) A dictionary of the economic products of the Malay peninsula. Edited by Ministry of agriculture (Malaysia) Crown Agents for the Colonies London 59:839–843

Butler MS (2005) Natural products to drugs: natural product derived compounds in clinical trials. Nat Prod Rep 22:162–195

Cardenas NC, Zavala MA, Aguirre JR, Perez C, Perez S (2005) Chemical composition and antifungal activity of essential oil of *Chrysactinia mexicana* Gray. J Agric Food Chem 53(11):4347–4349

Carnachan SM, Harris PJ (2000) Ferulic acid is bound to the primary cell walls of all gymnosperm families. Biochem Syst Ecol 28(9):865–879

Carpano, Stella M, Spegazzini, Etile D, Rossi, Javier S, Castro, Maria T, Debenedetti Silvia L (2003) Fitoterapia 74(3):294

Cavaleiro C, Pinto E, Gonçalves MJ, Salgueiro L (2006) Antifungal activity of Juniperus essential oils against dermatophyte, *Aspergillus* and *Candida* strains. J Appl Microbiol 100:1333–1338

Cavin AL, Hay A-E, Marston A, Stoeckli Evans H, Scopelliti R, Diallo D, Hostettmann K (2006) J Nat Prod 69:768

Chang YC, Hung WT, Chang HC et al (2011) Pathogenic VCP/TER94 alleles are dominant actives and contribute to neurodegeneration by altering cellular ATP level in *Drosophila* IBMPFD model. PLoS Genet 7(2):1000–1288

Charchari S, Dahoun A, Bachi F, Benslimani A (1996) In vitro antimicrobial activity of essential oils of *Artemisia* herba-*alba* and *Artemisia judaica* from Algeria. Riv Ital EPPOS 18:3–6

Chen HD, Yang SP, Wu Y, Dong L, Yue J-M (2009) Terpenoids from *Toona ciliate*. J Nat Prod 72:685–689

Cheng SS, Wu CL, Chang HT, Kao YT, Chang ST (2004) Antitermitic and antifungal activities of essential oil of *Calocedrus formosana* leaf and its composition. J Chem Ecol 30:1957–1967

Chopra RN, Nayar SL, Chopra (1956) Glossary of Indian medicinal plant. Publication and Information Directorate Council of Scientific and Industrial Research, New Delhi, pp 256–257

Choudhary MI, Naheed N, Abbaskhan A, Musharraf SG, Siddiqui H, Rahman AU (2008) Phenolic and other constituents of fresh water fern *Salvinia molesta*. Phytochemistry 69(4):1018–1023

Chung IM, Paik SB (1997) Anal Sci Technol 10 (2):153–159

Çitoglu GS, Sever B, Antus S, Baitz-Gacs E, Altanlar N (2003) Antifungal flavonoids from *Ballota glandulosissima*. Pharm Biol 41(7):483–486

Claudia M, Young M, Araju AR, da Silvapes Christhian A, Lopes Marcia N, Trevisan Ligia MV, da Bolzani Vanderlan S (1998) J Nat Prod 61:936

Cornet M, Fleury L, Maslo C, Bernard JF, Brucker G (2002) Epidemiology of invasive aspergillosis in France: a six-year multicentric survey in the Greater Paris area. J Hosp Infect 51:288–296

Costa TR, Fernandes OFL, Santos SC, Oliveira CMA, Lia LM, Ferri PH, Paula JR, Ferreira HD, Sales BHN, Silva MRR (2000) Antifungal activity of volatile constituents of *Eugenia dysenterica* leaf oil. J Ethnopharmacol 72:111–117

Curir P, Galeotti F, Dolci M, Barile E, Lanzotti V (2007) Nat Prod 70(10):1668

D'Auria FD, Tecca M, Strippoli V, Salvatore G, Battinelli L, Mazzanti G (2005) Antifungal activity of *Lavandula angustifolia* essential oil against *Candida albicans* yeast and mycelial form. Med Mycol 43 (5):391–396

Dagne E (1996) Phytochemical studies of Ethiopian medicinal plants. In: Proceedings of the workshop on development and utilization of herbal remedies in Ethiopia, Nazareth. pp 40–41

Dai Y, Sun LR (2008) Advances in studies on bibenzyls naturally occurred in plant. Chin Tradit Herb Drugs 39:1753–1756

Dalleau S, Cateau E, Berges T, Berjeaud J, Limbert C (2007) In vitro activity of essential oils and their major components against *Candida albicans* yeasts growing planktonically and as biofilms. Int J Antimicrob Agents 29(2):147

Dayong S, Jing L, Shuju G, Lijun H (2008) Antithrombotic effect of bromophenol, the alga-derived thrombin inhibitor. J Biotechnol 136:577–588

De Felicio R, De Albuquerque S, Young MCM, Yokoya NS, Debonsi HM (2010) Trypanocidal, leishmanicidal and antifungal potential from marine red alga *Bostrychia tenella* J. Agardh (*Rhodomelaceae, Ceramiales*). J Pharm Biomed Anal 52:763–769

Deepa MA, Narmatha Bai V, Basker S (2004) Fitoterapia 75(6):581

Dev S (1999) Ancient-modern concordance in ayurvedic plants some examples. Environ Health Perspect 107:783–789

Devi GK, Manivannan K, Thirumaran G, Rajathi F, Anantharaman (2011) In vitro antioxidant activities of selected seaweeds from Southeast coast of India. Asian Pac J Trop Med 4:205–211

Ding H (1982) Zhong guo Yao yun Bao zi Zhi w. Kexue Jishu Chuban She, Shanghai1, p 409

Douglas KA (1987) Biologically active compounds from plants with reputed medicinal and sweetening properties. J Nat Prod 50(6):1009–1024

Du Z, Zhu N, Ze-Ren-Wang-Mu N, Shen Y (2003) Two new antifungal saponins from the Tibetan herbal medicine Clematis tangutica. Planta Med 69:547–551

Dubey NK, Yadav P, Joshi VK (1998) Screening of some essential oils against dermatophytes. Philipp J Sci 127 (2):139–147

Ernavitha (2008) bioassay-guided fractionation and identification of antioxidant and antimicrobial compounds from *Callistemon viminalis* (Gaertn) G Don. Thesis, School of Biological Sciences

Ernst E (1999) Efficacy of herbal medicine: do they work and are they safe? In: Pharmaceutical news, focus on contemporary medicine, special issue. p 17

Erturk O, Tas B (2011) Antibacterial and antifungal effects of some marine algae. Kafkas Univ Vet Fak Derg 17(Suppl A):S121–S124

eSouza LKH, Alves de Oliveira CM, Ferri PH, Santos SC, de Oliveira JG Jr, Miranda ATB, Liao LM, Silva MdRR (2002) Antifungal properties of Brazilian cerrado plants. Braz J Microbiol 33 (3):247–249

Eugenia P, Vale-Silva L, Cavaleiro C, Salgueiro L (2009) Antifungal activity of the clove essential oil from Syzygium aromaticum on *Candida*, *Aspergillus* and dermatophyte species. J Med Microbiol 58 (11):1454–1462

Farag RS, Daw ZY, Abo- Raya SH (1989) Influence of some spice essential oils on *Aspergillus paraciticus* growth and production of alfatoxins in synthetic medium. J Food Sci 54:74–76

Fatope MO, Al-Burtomani SKS, Ochei JO, Abdulnour AO, Al-Kindy SMZ, Takeda Y (2003) Muscanone: a 3-O-(1″, 8″, 14″-trimethylhexadecanyl) naringenin from Commiphora wightii. Phytochemistry 62(8):1251–1255

Favel A, Kemertelidze E, Benidze M, Fallague K, Regli P (2005) Antifungal activity of steroidal glycosides from Yucca gloriosa L. Phytother Res 19(2):158–161

Fenner R, Sortino M, Rates SM, Dall'Agnol R, Ferraz A, Bernardi AP, Albring D, Nor C, Van Poser G, Schapoval E, Zacchino S (2005) Phytomedicine 12:236

Flowers S (1957) Ethnobryology of the Gosuite Indians of Utah. Bryologist 60:11–14

Frahm JP (1994) Moose-lebende Fossilien. Biol Unserer Zeit 24:120–124

Frahm JP (2001) Biologie der Moose, 1st edn. Spektrum Verlag, Berlin

Frahm JP (2004) New frontiers in bryology and lichenology: Recent developments of commercial products from bryophytes of China, Shanghai. Bryologist 107:277–283. 409pp

Fukai T, Yonekawa M, Hou A-J, Nomura T, Sun H-D, Uno J (2003) Antifungal agents from the roots of Cudrania cochinchinensis against Candida, Cryptococcus, and Aspergillus species. J Nat Prod 66 (8):1118–1120

Gangrade SK, Shrivastava RD, Sharma OP, Jain NK, Trivedi KC (1989) Evaluation of antifungal properties of essential oils of Ocimum species. Indian Perfumer 33:97–101

Garg SC, Jain R (1999) Antifungal activity of Luvunga scadens against some keratinophilic fungi. Indian J Pharm Sci 61:248–249

Genovese G et al (2013) The mediterranean red alga *Asparagopsis taxiformis* has antifungal activity against *Aspergillus* species. Mycoses 56:516–519

Ghahfarokhia MS, Goodarzia M, Abyanehb R, Al-Tiraihic T, Seyedipourd G (2004) Morphological evidences for onion-induced growth inhibition of *Trichophyton rubrum* and *Trichophyton mentagrophytes*. Fitoterapia 75:645–655

Gopalakrishnan G, Banumathi B, Suresh G (1997) J Nat Prod 60:519–524

Gopalakrishnan V, Rao KNV, Loganathan V, Shanmuganathan S, Bollu VK, Sharma TB (2000) Antimicrobial activity of extracts of *Acalypha indica* Linn. Ind J Pharmacol Sci 62:347–350

Govindachari TR, Suresh G, Gopalakrishnan G, Masilamani S, Banumathi B (2000) Antifungal activity of some tetranortriterpenoids. Fitoterapia 71 (3):317–320

Grosvenor PW, Gray David O (1998) J Nat Prod 61 (1):99–101

Grover GS, Rao J (1979) In vitro antimicrobial studies of essential oil of *psoralea corylifolia*. Indian Perfum 23:135

Guarro J, Abdullah SK, Gené J, Al-Saadoon AH (1997) A new species of Preussia from submerged plant debris. Mycol Res 101:305–308

Guedes E, de Azevedo Prata MC, dos Reis ES, Cançado PHD, Leite RC (2012) Comparative efficiency of two models of CO2 traps in the collection of free-living stages of ixodides. Parasitol Res 111(6):2325–2328

Gupta KC, Bhatia MC, Chopra CL, Amar N, Chopra IC (1962) Bull Reg Res Lab Jammu 1:59

Gurgel LA, Sidrim JJ, Martins DT, Cechinel V, Rao VS (2005) J Ethnopharmacol 97:409

Gutierrez, Delle Monache A, Gonzalez Sierra M, Zacchino SA, Hammer KA, Carson CF, Riley TV (2002) Antifungal activity of the components of *Melaleuca alternifolia* (tea tree) oil. J Appl Microbiol 95:853–860

Hammer KA, Carson CF, Riley TV (2002) In vitro activity of Melaleuca alternifolia oil against dermatophytes and other filamentous fungi. J Antimicrob Chem 50:195–199

Hashidoko Y, Inoue A, Tamogami S, Kato H, Nakazato Y, Akiyama M, Kodama O, Akatsuka T (1995) Antifungal melampolides from leaf extracts of *Smallanthus sonchifolius*. Phytochemistry 39(4):845–848

Heinrichs J, Groth H, Gradstein SR, Rycroft DS, Cole WJ, Anton H (2001) *Plagiochila rutilans* (Hepaticae) a poorly known species from tropical America. Bryologist 104:350–361

Hess SC, Lima EO, Cechinel Filho V, Yunes RA (1995) Fitoterapia 66(6):549–550

Huang KC (1999) The pharmacology of chinese herbs. CRC Press, Boca Raton, p 2

Huffman MA (2001) Self-medicative behavior in the African great apes: an evolutionary perspective into the origins of human traditional medicine. Biol Sci 51 (8):651–661

Ibrahim D, Osman H (1995) Antimicrobial activity of *Cassia alata* from Malaysia. J Ethnopharmacol 45 (3):151–156

Ioset JR, Marston A, Gupta MP, Hostettmann K (2000) Antifungal and larvicidal compounds from the root bark of *Cordia alliodora*. J Nat Prod 63:424–426

Iranshahi M, Hosseini ST, Shahverdi AR, Molazade K, Khan SS, Ahmad VU (2008) Phytochemistry 69 (15):2753

Irobi ON, Darambola SO (1993) Antifungal activities of crude extracts of *Mitracarpus villosus* (Rubiaceae). J Ethnopharmacol 40:137–140

Isabel, Arnason JT, Smith ML (2008) Antifungal and antioxidant activities of the phytomedicine pipsissewa, *Chimaphila umbellata*. Phytochemistry 69 (3):738–746

Jain N, Sharma M, Kumar P (2004) Regulatory effect of some plant extracts on the growth of dermatophytic fungi. Ind J Microbiol 44:59–62

Janssen AM, Scheffer JJC, Ntezurubanza L, Baerheim Svendsen A (1989) Antimicrobial activities of some *Ocimum* species grown in Rwanda. J Ethnopharmacol 26:57–63

Javad H, Peyman Z et al (2014) The association of coagulation factor V (Leiden) and factor II (Prothrombin) mutations with stroke. Iran Red Crescent Med J 16 (11):11548

Jayasinghe L, Kumarihamy Mallika BM, Jayarathna Nishantha KHR, Udishani Gayathri NWM, Bandara Ratnayake BM, Hara N, Fujimoto Y (2003) Phytochemistry 62(4):637

Jennings OE (1926) Mosses immune to molds. Bryologist 29:75–76

Jung HJ, Sung WS, Yeo SH, Kim HS, Lee I- S, Woo E-R, Lee, Dong G (2006) Arch Pharma Res 29(9):746

Kannan PB, Shanmugavadivu C, Petchiammal W, Hopper (2006) In vitro antimicrobial activity of *Wrightia tinctoria* leaf extracts against skin microorganisms. Acta Bot Hungarica 48(3–4):323–329

Kariba RM, Houghton PJ, Yenesew A (2002) Antimicrobial activities of a new schizozygane indoline alkaloid from *Schizozygia coffaeoides* and the revised structure of isoschizogaline. J Nat Prod 65:566–569

Karunyal Samuel J, Andrews B, Shyla Jebashree H (2001) In vitro evaluation of the ant fungal activity of *Allium sativum* bulb extract against *Trichophyton rubrum*, a human skin pathogen. World J Microbiol Biotechnol 16(7):617–620

Kasiram K, Sakharkar PR, Patil AT (2000) Antifungal activity of *Calendula officinalis*. Ind J Pharmacol Sci 62:464–466

Kawamura F, Ohara S (2005) Antifungal activity of iridoid glycosides from the heartwood of Gmelina arborea. Holzforschung 59(2):153–155

Kawther FA (2007) Antimicrobial activity of essential oils of some medicinal plants from Saudi Arabia. Saudi J Biol Sci 14(1):53–60

Kesari V, Das A, Rangan L (2010) Physico-chemical characterization and antimicrobial activity from seed oil of Pongamia pinnata a potential biofuel crop. Biomass Bioenergy 34:108–115

Khan MA, Ashfaq MK, Zuberi HS, Mahmood MS, Gilani AH (2003) The in vivo antifungal activity of the aqueous extract from *Nigella sativa* seed. Phytother Res 17(2):183–186

Kholia BS, Punetha NN (2005) Useful Pteriophytes of kumaon central Himalaya India. Ind Fern J 22:1–6

Kim SK, Karadeniz F (2011) Anti-HIV activity of extracts and compounds from marine algae. Adv Food Nutr Res 64:255–265

Kirtikar KR, Basu BD (1975) Indian medicinal plants, vol II, periodical experts, D- 42. Vivek Vihar, Delhi, p 877

Kirtikar KR, Basu BO, ICS (1935) Indian medicinal plants, vol 4, 2nd edn. Bishen Singh Mahendra Pal Singh, Dehra Dun

Kishore N, Mishra AK, Chansouria JP (1993) Fungal toxicity of essential oils against dermatophytes. Mycoses 36:211–225

Kofinas C, Chinou I, Loukis A, Harvala C, Maillard M, Hostettmann K (1998) Flavonoids and bioactive coumarins of *Tordylium apulum*. Phytochemistry 48:637

Kofujita H, Fujino Y, Ota M, Takahashi K (2006) Antifungal diterpenes from the bark of *Cryptomeria japonica* D. Don. Holzforchung 60(1):20

Komiya M, Takeuchi T, Harada E (2006) Lemon oil vapor causes an anti-stress effect via modulating the 5-HT and DA activities in mice. Beha Brain Res 172 (2):240–249

Krauze-Baranowska M (2003) Antifungal activity of biflavones from Taxus baccata and Ginkgo biloba. Z Naturforsch C J Biosci 58(1/2):65

Kulandairaj D, De Britto J (2000) Antibacterial and antifungal activity of secondary metabolites from some medicinal and other common plant species. J Ecol Taxon Bot 24:21

Kumar P, Bhadauria S (2009) Bioactive nature of an alkaloid and flavonoid from *Solanum dulcamara*. Res Environ Life Sci 2:11–12

Kumar A, Kaushik P (1999) Antibacterial effect of *Adiantum capillus veneris* Linn. Indian Fern J 16:72–74

Lachoria R, Jain PC, Agrawal SC (2000) *In vitro* antimycotic activity of some essential oils against dermatophytes. MicroBiotech. AMI 41st annual conference

Lahlou EH, Hashimoto T, Asakawa Y (2000) Chemical constituents of the liverworts *Plagiochasma japonica* and *Marchantia tosana*. J Hattori Bot Lab 88:271–275

Lamidi M, DiGiorgio C, Delmas F, Favel A, Eyele Mve-Mba C, Rondi ML, Ollivier E, Nze-Ekekang L, Balansard G (2005) In vitro cytotoxic, antileishmanial and antifungal activities of ethnopharmacologically selected *Gabonese* plants. J Ethnopharmacol 102:185–190

Larcher G, Morel C, Tronchin G, Landreau A, Seraphin D, Richomme P, Bouchara JP (2004) Investigation of the antifungal activity of caledonixanthone E and other xanthones against *Aspergillus fumigatus*. Planta Med 70:569–571

Lau KM, Fu LH, Cheng L, Wong CW, Wong Y, Lau CP, Han SQB, Chan PKS, Fung KP, Lau CBS, Hui M, Leung PC (2010) Two antifungal components isolated from *Fructus Psoraleae* and *Folium Eucalypti Globuli* by bioassay-guided purification. Am J Chin Med 38:1005–1014

Lee DG, Park Y, Kim M-R, Jung HJ, Seu YB, Hahm (2004) Anti-fungal effects of phenolic amides isolated from the root bark of *Lycium chinense*. Biotechnol Lett 26(14):1125–1130

Li X-C, Elsholy HN, Nimrod AC, Clark AM (1999) Antifungal Jujubogenin Saponins from *Colubrina retusa*. J Nat Prod 62(5):674–677

Lillykutty L, Santhakumari G (1969) Antimicrobial activity of *Cassia-fistula*. J Res Ind Med Proc Int Seed Test Ass 33:167

Lima R, Diaz RF, Castro A, Fievez V (2011) Digestibility methane production and nitrogen balance in sheep fed ensiled or fresh mixtures of sorghum-soybean forage. Livest Sci 141(1):36–46

Lopez A, Ming DS, Towers GH (2002) Antifungal activity of benzoic acid derivatives from *Piper lanceaefolium*. J Nat Prod 65:62–64

Lorimer SD, Perry NB (1993) An antifungal bibenzyl from the New Zealand liverwort *Plagiochila stevensoniana*. J Nat Prod 56:1444–1450

Lorimer SD, Perry NB (1994) Antifungal hydroxy acetophenones from the New Zealand liverwort. *Plagiochila fasciculate*. Planta Med 60:386–387

Ma W, Guang Fuzzati N, Li QS, Yang CR, Evans HS, Hostettmannn K (1995) Polyphenols from *Eriosema tuberosum*. Phytochemistry 39(5):1049–1061

Machado KE, Cechinel Filho V, Cruz RC, Meyre-Silva C, Cruz AB (2009) Antifungal activity of *Eugenia umbelliflora* against dermatophytes. Nat Prod Commun 4:1181–1184

Manickam VS, Irudayaraj V (1992) Pteridophyte flora of the western Ghats. South India B I Publications, New Delhi

Mansuang WU, Wardee G, Omboon L, Wanda C (2002) Antifungal activity of Curcuma long grown in Thailand Mahidol University. Annual research abstracts 29(492)

Maoz M, Neeman I (1998) Antimicrobial effects of aqueous plant extracts on the fungi *M. canis* and *Trichophyton rubrum* and on three bacterial species. Lett Appl Microbiol 26:61–63

Mashhadian NV, Rakhshandeh H (2005) Antibacterial and antifungal effects of *Nigella sativa* extracts against *S. aureus*, *P. aeruginosa* and *C. albicans*. Pak J Med Sci 21(1):47–52

Matsuo A, Yuki S, Nakayama M, Hayashi S (1982a) Three new sesquiterpene phenols of the *ent*-herbertane class from the liverwort *Herberta adunca*. Chem Lett 11:463–466

Matsuo A, Yuki S, Higashi R, Nakayama M, Hayashi S (1982a) Structure and biological activity of several sesquiterpenoids having a novel herbertane skeleton from the liverwort Herberta adunca. In: 25th Symposium on chemistry of natural products. Symposium papers, pp 242–249

Matsuo, A, Yuki, S, Nakayama, M (1983) -(−)-Herbertenediol and (−)-herbertenolide, two new sesquiterpenoids of the ent-herbertane class from the liverwort Herberta adunca. Chem Lett:1041–1042

McCleary A, Walkington DL (1966) Mosses and antibiosis. Rev Bryol Lichnol 34:309–314

Meng JC, Hu YF, Chen JH, Tan RX (2001) Antifungal highly oxygenated guaianolides and other constituents from *Ajania fruticulosa*. Phytochemistry 58:1141–1145

Meragelman TL, Tucker KD, McCloud TG, Cardellina JH, Shoemaker RH (2005) Antifungal flavonoids from *Hildegardia barteri*. J Nat Prod 68:1790–1792

Miller HA (1982) Bryophyte evolution and geography. Biol J Linn Soc 18:145–196

Mishra R, Verma DL (2009) Flavonol glycosides of Cheilanthes anceps Roxb. J Am Sci 5(4):183–188

Misra SS, Dixt SN (1979) Acta Bot Indic 7:147

Monga DP, Mohapatra LN (1980) A compilation of published reports of mycoses in animals in India. Mycopathologia 72(1):3–11

Morteza Semnani K, Amin G, Shidfar MR, Hadizadeh H, Shafiee A (2003) Antifungal activity of the methanolic extract and alkaloids of *Glaucium oxylobum*. Fitoterapia 74(5):493–496

Mourice MI, Angela RD, Chris OO (1999) In: Janick J (ed) New antimicrobials of plant origin in perspectives on new crops and new uses. ASHS Press, Alexanderia, pp 457–462

Mukherjee PK, Saha K, Giri SN, Paland M, Saha BP (1995) Ant fungal screening of *Nelumbo nucifera* (Nympheaceae) rhizome extract. Ind J Microbiol 35:327–330

Na HJ, Moon PD, Lee HJ, Kim HR, Chae HJ, Shin T, Seo Y, Hong SH, Kim HM (2005) Regulatory effect of atopic allergic reaction by Carpopeltis affinis. Ethnopharmacol 101:43–48

Nadkarni BK (1954) Indian materia medica with ayurvedic, unantibii, siddha, allopathic, homeopathic, naturopathic and home remedies. Popular Book Depot, Bombay

Natarajan PV, Venugopal T, Menon (2003) Effect of *azadirachta indica* (neem) on the growth pattern of dermatophytes. Ind J Med Microbiol 21(2):98–101

Navickiene HMD, Alecio AC, Kato MJ, Bolzani VS, Young MCM, Cavalheiro AJ, Furlan (2000) Antifungal amides from *piper hispidum and piper tuberculatum*. Phytochemistry 55(6):621–626

Nayar BK (1959) Medicinal ferns of India. Bull Nat Bot Gard Lucknow 29:1–36

Newman DJ, Cragg GM (2007) Natural products as sources of new drugs over the last 25 years. J Nat Prod 3:461–477

Ngono Ngane A, Biyiti L, Bouchet P, Nkegfack A, Amvam Zollo PH (2003) Antifungal activity of *Piper guineense* of Cameroon. Fitoterapia 74:464–468

Nigam SS, Rao TST (1977) Antimicrobial efficacy of some Indian essential oils. In: Kapoor LD, Ramakrishnan (eds) Advance in essential oil industry. Today's and Tomorrow's Printer and Publishers, New Delhi, pp 177–180

Oberlies NH, Burgess JP, Navarro HA, Pinos RE, Soejarto DD, Farnsworth NR, Kinghorn A, Douglas Wani Mansukh C, Wall MEJ (2001) Nat Prod 64:497

Pandey KP, Mishra RK, Kamran A, Mishra P, Bajaj AK, Dikshit A (2010) Studies on antidermatophytic activity of waste leaves of *Curcuma longa* L. Physiol Mol Biol Plants 16:177–185

Pandian P, Selvamuthukumar S, Manavalan R, Parthasarathy V (2011) Screening of antibacterial and antifungal activities of red marine algae *Acanthaphora spicifera (Rhodophyceae)*. J Biomed Sci Res 3(3):444–448

Pankajalakshmi VV (1994) Antidermatophytic activity of Neem (*Azadirachta indica*) leaves In vitro. Indian J Pharmacol 25:141–143

Pant G, Tewari SD (1989) Various human uses of bryophytes in the Kumaun region of Northwest Himalaya. Bryologist 92:120–122

Pant G, Tewari SD (1990) Bryophytes and mankind. Ethnobotany 2:97–103

Pant N, Garg HS, Madhusudhanan KP, Bhakuni DS (1986) Sulpharous compounds from Azadirachta indica leaves. Fitoterapia 57:302–304

Pargavi B, Sivakumar (2014) Phytochemical screening and antimicrobial activity of drynaria quercifolia J. SMITH. Int J Curr Res Chem Pharm Sci 1 (5):68–73

Parihar P, Bohra A (2002) Screening of some ferns for their Antimicrobial activity against *salmonella typhi*. Adv Plant Sci 15(2):365–367

Parihar P, Bohra A (2003) Effect of some Pteridophytic plant extracts on human pathogenic bacteria – *Salmonella typhi*. Indian Fern 20:39–41

Parihar P, Parihar L (2006) Some pteridophytes of medicinal importance from Rajasthan. Nat Prod Radiat 5 (4):297–301

Pates AL, Madsen GC (1955) Occurrence of antimicrobial substances in chlorophyllose plants growing in Florida. II Bot Gaz 116:250–261

Patra M, Shahi SK, Midgely G, Dikshit A (2002) Utilization of essential oil as natural antifungal against nail-infective fungi. Flav Frag 17:91–94

Patra M, Shahi S, Dikshit A (2003) Utilization of pericarp of *Citrus sinensis* oil for the development of natural antifungal against nail infection. Curr Sci 84:1512–1515

Pedersen JA, Ollgaard B (1982) Phenolic acids in the genus *Lycopodium*. Biochem Syst Ecol 10 (3):90044–90048

Pianaro A, Pinto J, Pereira Ferreira DT, Ishikawa NK, Filho RB, Semina (2007) Ciencias Agrarias. Londrina 28(2):251

Pinheiro da Silva MF, Lisboa CL, de Vasconcelos Brazao R (1989) Contribucao ao estudo de briofitas como fonts de antibióticos. Acta Amazon 19:139–145

Pinto DC, Fuzzati N, Pazmino XC, Hostettmann K (1994) Xanthone and antifungal constituents from *Monnina obtusifolia*. Phytochemistry 37(3):875–878

Plaza MA, Cifuentes, Ibanez E (2008) In the search of new functional food ingredients from algae. Trends Food Sci Technol 19:31–39

Ponnusamy S et al (2010) sphingolipids and cancer: ceramide and sphingosine -1-1 phosphate in the regulation of cell death and drug resistance. Future Oncol 6:1603–1624

Portillo MC, Viramontes S, Muñoz LN, Gastelum MG, Nevarez GV (2005) Antifungal activity of Mexican oregano (*Lippia berlandieri* Shauer.). J Food Prot 68 (12):2713–2717

Pryce RJ (1972) Metabolism of lunularic acid to a new plant *stilbene* by *Lunularia cruciata*. Phytochemistry 11:1355–1364

Pyun MS, Shin S (2006) Antifungal effects of the volatile oils from *Allium* plants against *Trichophyton* species and synergism of the oils with ketoconazole. Phytomedicine 13(6):394–400

Qu J, Xie C, Huaifang G, Wentao Y, Lou H (2007) Phytochemistry 68:13

Rahman A, Nasreen A, Akhtar F, Shekhani MS, Clardy J, Parvez M, Choudhary MI (1997) Antifungal diterpenoid alkaloids from Delphinium denudatum. J Nat Food Prod 60:472–474

Rai MK, Acharya D (1999) Screening of some *Asteraceous* plants for antimycotic activity. My J 34:37–43

Rai MK, Upadhyay S (1998) Screening of medicinal plants of Chhindwara district against *T. mentagrophytes*. A causal organism of Tinea pedis. Hindustan Antibiot Bull 30:33–36

Raja A, Vipin C, Aiyappan A (2013) Biological importance of marine algae – an overview. Int J Curr Microbiol Appl Sci 2(5):222–227

Ranganathan S (1996) Antidermatophytic activities of *Azadirachta indica* an in vitro and in vivo studies. Ind J Derm 41(4):113–177

Rasoamiaranjanahary L, Marston A, Guilet D, Schenk K, Randimbivololona F, Hostettmann K (2003) Antifungal diterpenes from *Hypoestes serpens* (Acanthaceae). Phytochemistry 62(3):333–337

Richardson JS (1993) Limits to productivity in streams: evidence from studies of macroinvertebrates. Can Spec Publ Fish Aquat Sci 118:09–15

Rojas R, Doroteo V, Bustamante B, Baver J, Lock O (2004) Antimicrobial and free radical scavenging activity of Gentianella nitida. Fitoterapia 75:754–757

Romagnoli C, Bruni R, Andreotti E, Rai MK, Vicentini CB, Mares D (2005) Chemical characterization and antifungal activity of essential oil of *capitula* from wild Indian *Tagetes patula* L. Protoplasma 225:57–65

Rubio OC, Cuellar AC, Rojas N, Castro HV, Rastrelli L, Aquino RA (1999) A polyisoprenylated benzophenone from Cuban propolis. J Nat Prod 62:1013–1015

Saha R, Bhupendar K (2011) Pharmacognosy and pharmacology of *Nigella sativa* – a review. Int Res J Pharmcol 2(11):36–39

Sahayaraj K, Borgio JAF, Raju G (2009) Antifungal activity of three fern extracts on causative agents of groundnut early leaf spot and rust diseases. J Plant Prot Res 49:2

Saidani K, Bedjou F, Benabdesselam F, Touati N (2012) Antifungal activity of methanolic extracts of four Algerian marine algae species. Afr J Biotechnol 11:9496–9500

Salgueiro LR, Piato E, Goncalves MJ, Pina-Vaz C, Cavaleiro C, Rodriques AG (2004) Chemical composition and antifungal activity of the essential oil of *Thymbra capitata*. Planta Med 70(6):572–575

Samant SS, Dhar U, Palni LMS (1998) Medicinal plants of Indian Himalaya, diversity, distribution and potential value. Gyannodya Prakasha, Nainital

Samir RA, Guzman RR, Radwan MM, Jacob M, Ding Y, Li XC, Ferreira D, Manly SP (2008) Sorocenols G and H, Anti-MRSA Oxygen Heterocyclic DielsAlder-Type Adducts from Sorocea muriculata Roots. J Nat Prod 71:1764–1767

Sartori MRK, Pretto JB, Cruz AB, Bresciani LFV, Yunes RA, Sortino M, Zacchino SA (2003) Filho V Cechinel. Pharmazie 58(8):567

Sasidharan S, Darah I, Jain K (2007) In vitro and In vivo antifungal activity of the methanol extract from Gracilaria changii. Int J Pharmacol 6:1–10

Saxena J, Mathela CS (1996) Antifungal activity of new compounds from *Nepeta leucophylla and Nepeta clarkei*. Appl Environ Microbiol 62(2):702–704

Saxena G, Farmer SW, Hancock REW, Towers GHN (1996) Chlorochimaphilin: a new antibiotic from *Moneses uniflora*. J Nat Prod 59:62–65

Schuster RM (1966) The hepaticae and anthocerotae of North America. Columbia University Press, New York, pp 1–1344

Shabnam J, Sobia M, Ibatsam K, Rauf A, Saleem MH (2012) Comparative antimicrobial activity of clove and fennel essential oils against food borne pathogenic fungi and food spoilage bacteria. Afr J Biotechnol 11 (94):16065–16070

Shams M, Shokoohamiri MR, Amirrajab N, Moghadasi B, Ghajari A, Zeini F, Sadeghi G (2006) In vitro antifungal activities of Allium cepa, Allium sativum and ketoconazole against some pathogenic yeasts and dermatophytes. Fitoterapia 77(4):321–323

Sharma SK, Singh VP (1979) The antifungal activity of some essential oils. Indian Drugs Pharmcol Ind 14 (1):3–6

Shizadiann A, Azad H, Khalghani J (2009) Introductional study of antifungal activities of bryophyte extracts. Appl Entomol Phytopathol 77(1):1–22, 88

Silva MR, Oliveira JG, Fernändez OF Jr, Passos XS, Costa CR, Souza LK, Lemos JA, Paula JR (2005) Antifungal activity of *Ocimum gratissimum* towards dermatophytes. Mycoses 48:172–175

Silveira GP, Nome F, Gesser JC, Sá MM, Terenzi T (2006) The strategies used to combat bacterial resistance. Quim Nova 29:844–855

Simonsen HT, Adsersen A, Bremner P, Heinrich M, Wagner Smitt U, Jaroszewski JW (2004) Antifungal constituents of *Melicope borbonica*. Phytother Res 18 (7):542–545

Singh HB (1999) Potential medicinal pteridophytes of India and their chemical constituents. J Econ Taxon Bot 23(1):63–78

Singh CJ, Singh BG (1997) Antifungal activity of some plant extracts against dermatophytes and related keratinophilic fungi. Adv Plant Sci 10:249–251

Singh S, Singh SK, Tripathi SC (1988) Fungitoxic properties of essential oil of *Eucalyptus rostrata*. Indian Perfum 32:190–193

Singh G et al (2004) Chemical constituents, antifungal and antioxidative effects of ajwain essential oil and its acetone extract. J Agri Food Chem 52:3292–3296

Singh G, Marimuthu P, Heluani C, Catalan C (2005) Chemical constituents and antimicrobial and antioxidant potentials of essential oil and acetone extract of *Nigella sativa* seeds. J Sci Food Agric 85:2297–2306

Singh M, Singh N, Khare PB, Rawat AKS (2008) Antimicrobial activity of some important *Adiantum* species used traditionally in indigenous system of medicine. J Ethnopharmacol 115(2):327–329

Smania EFA, Delle MF, Smania A, Yunes RA, Cuneo RS (2003) Antifungal activity of sterols and triterpenes isolated from *Ganoderma annulare*. Fitoterapia 74 (4):375–377

Smith AR, Pryer KM, Schuettpelz E, Korall P, Schneider H, Wolf PG (2006) A classification for extant ferns. Taxon 55:705–731

Sonia Pereira Leite, Jeyesson Raphael Cardoso Vieira, Palom Lys de Medeiros, Roberta Maria Pereira Leite, Vera Lucia de Menezea Lima, Haroudo Satiro Xavier, Edeltrudes de Oliveira Lima (2006) Antimicrobial activity of *Indigofera suffruticosa*. eCAM Advance access, pp 1–5

Souza TM, Morais-Braga MFB, Costa JGM, Saraiva AAF, Coutinho HDM (2012) Enhancement of antimicrobial activity of antibiotics and antifungals by the use of natural products from pityrogramma calomelanos (L.) LINK. Arch Biol Sci Belgrade 64(1):43–48

Stein AC, Sortino M, Avancini C, Zacchino S, Von Poser G (2005) Ethnoveterinary medicine in the search for antimicrobial agents: antifungal activity of some species of *Pterocaulon* (Asteraceae). J Ethnopharmacol 99(2):211–214

Subhisha S, Subramoniam A (2005) Antifungal activities of a steroid from *Pallavicinia lyellii*, a liverwort. Indian J Pharmacol 37(5):304–308

Subramoniam A, Subhisha S (2005) Bryophytes of India: a potential source of antimicrobial agents. In: Khan IA, Khanum A (eds) Role of biotechnology in medicinal and aromatic plants, vol 11. Ukaaz Publications, Hyderabad

Sumathi P, Parvathi A (2010) Antimicrobial activity of some traditional medicinal plants. J Med Plant Res 4 (4):316–321

Sun L, Sun S, Cheng A, Wu X, Zhang Y et al (2009) In vitro activities of retigeric acid B alone and in combination with azole antifungal agents against Candida albicans. Antimicrob Agents Chemother 53:1586–1591

Suresh G, Narasimha NS, Masilamani S, Partho PD, Gopalakrishnan, Goetha (1997) Antifungal fractions and compounds from uncrushed green leaves of *Azadirachta indica*. Phytoparasitica 25(1):33–39

Svetaz L, Aguero MB, Alvarez S, Luna L, Feresin G, Derita M, Tapia A, Zacchino S (2007) Antifungal activity of *Zuccagnia punctata* Cav evidence for the mechanism of action. Planta Med 73(10):1074–1080

Tadeg H, Mohammed E, Asres K, Gebre-Mariam T (2005) Antimicrobial activities of some selected traditional ethiopian medicinal plants used in the treatment of skin disorders. J Ethnopharmacol 100:168–175

Taguchi Y, Ishibashi H, Takizawa T, Inoue S, Yamaguchi H, Abe S (2005) Protection of oral or intestinal candidiasis in mice by oral or intragastric administration of herbal food, clove(syzygium aromaticum). Nipoun Ishiukin Gakkai Zasshi 46:27–33

Takahashi T, Kokubo R, Sakaino M (2004) Antimicrobial activities of eucalyptus leaf extracts and flavonoids from Eucalyptus maculata. Lett Appl Microbiol 39:60

Tarfa FD, Obodozic OO, Mshelin E, Ibrahim K, Temple V (2004) Evaluation of phytochemical and antimicrobial properties of leaf extract of Tapinanthus sessilifolius (P. Beauv) van Tiegh. J Indian J Exp Biol 42:326–329

Tiwari TN, Chansouria JPN, Dubey NK (2003) Antimycotic potency of some essential oils in the treatment of induced dermatomycosis of an experimental animal. Pharm Biol 41(5):351–356

Turney CSM, Kershaw AP, Lowe JJ, van der Kaars S, Johnston R, Rule S, Moss P, Radke L, Tibby J, McGlone MS, Wilmshurst JM, Vandergoes MJ, Fitzsimons SJ, Bryant C, Branch NP, Jacobsen G, Fifield LK (2006) Climatic variability in the southwest Pacific during the last termination. Quat Sci Rev 25:886–903

Uddin MG, Mirza MM, Pasha MK (1998) The medicinal uses of pteridophytes of Bangladesh. Bangladesh J Plant Taxon 5(2):29–41

University of Michigan, Dearborn (2003) Native American ethnobotany, a database of plants used as drugs, foods, dyes, fibers, and more, by native Peoples of North America. http://herb.umd.umich.edu. Accessed 8 Jan 2004

Vaijayanthimala J, Rajendra Prasad N, Anandi C, Pugalendi KV (2004) Anti-dermatophytic activity of some Indian medicinal plants. J Nat Remedies 4(1):26–31

Vasudeva SM (1999) Economic importance of Pteridophytes. Indian Fern J 16:130–152

Weideman L (2005) An investigation into antibacterial activities of medicinal plants traditionally used in the Eastern Cape to treat secondary skin infections associated with burn wounds; Thesis submitted in Biomedical Technology at the Nelson Mandela Metropolitan University, Cape town, South Africa

Wren RW (1956) Potters new encyclopedia of botanical drugs and preparations, 7th edn. Harper & Row, London

Wu PC (1977) *Rhodobryum giganteum* (Schwaegr.) Par can be used for curing cardiovascular disease. Acta Phytotax Sin 15:93

Wyatt R (1982) Population ecology of bryophytes. J Hattori Bot Lab 52:179–198

Xu Y (2004) Dermatology in traditional Chinese medicine. Donica Publishing, St. Albans

Yadav P, Dubey NK (1994) Screening of some essential oils against ringworm fungi. Indian J Pharmaceut Sci 2:27–230

Zaidi MA, Crow SA Jr (2005) Biologically active traditional medicinal herbs from Balochistan. Pakistan J Ethnopharmacol 96:331–334

Zaika LL (1975) Spices and herbs their antimicrobial activity and its determination. J Food Saf 9:97–118

Zakaria Z, Sanduran PS (2010) Antifungal activity of the edible ferns application for public health. Int J Humanit 8(8):113–118

Zhang ZZ, Li YB, Qi L, Wan XC (2006) Antifungal activities of major tea leaf volatile constituents toward *Colletorichum camelliae* Massea. J Agric Food Chem 54(11):3936–3940

Recent Advancements in Combinational Antifungal Therapy and Immunotherapy

6

Sudarshan Singh Rathore, Jayapradha Ramakrishnan, and Thiagarajan Raman

Abstract

The most predominant opportunistic fungal pathogens Candida sp, Aspergillus and Cryptococcus sp are emerging as a global problem of great concern as it is associated with increased morbidity and mortality. Although there are different classes of approved antifungal agents like polyenes, pyrimidines, azoles, and echinocandins, the morbidity and mortality rate due to systemic infection remains high, due to its toxicity, narrow spectrum of activity and tolerability. The continuous usages of these drugs are associated with resistance development and thus this is another area of emerging global problem. In addition, biofilm formation by fungal pathogens are proven to be an important virulence factor and are resistant to host defense mechanism and antifungal agents. Combination therapy has shown to be a promising choice to overcome the emergence of drug resistance, minimizing the toxic effects by reducing the drug dosage, and to increase the spectrum of killing. In this chapter, we have reviewed the last 5 years' developments made in combinational antifungal therapy with special emphasize on biofilm related infections and have presented a reappraisal of the current practices associated with immunotherapy to treat fungal infections and future challenges.

Keywords

Combinational therapy • Immunotherapy • Fungal infections • Biofilm

S.S. Rathore • J. Ramakrishnan (✉)
Actinomycetes Bioprospecting Lab, Centre for Research in Infectious Diseases (CRID), SASTRA University, Tirumalaisamudram, Thanjavur, Tamil Nadu, India
e-mail: sdsrathore2511@gmail.com; jayapradha@scbt.sastra.edu

T. Raman (✉)
Actinomycetes Bioprospecting Lab, Centre for Research in Infectious Diseases (CRID), SASTRA University, Tirumalaisamudram, Thanjavur, Tamil Nadu, India

Laboratory for Inflammation Research, Centre for Research in Infectious Diseases (CRID), SASTRA University, TirumalaisamudramThanjavur, Tamil Nadu, India
e-mail: raman@biotech.sastra.edu

A. Basak et al. (eds.), *Recent Trends in Antifungal Agents and Antifungal Therapy*,
DOI 10.1007/978-81-322-2782-3_6

6.1 Introduction

The improvement in medical advances incidentally has led to an increase in the incidence of opportunistic fungal infections. The most predominant opportunistic fungal pathogens are *Candida* sp., *Aspergillus* sp., and *Cryptococcus* sp. (Paramythiotou et al. 2014). During the last decade, there has been an emergence of other pathogens like *Zygomycetes*, *Fusarium*, etc., (Pagano et al. 2011). These pathogens take the chance to cause disease when the immune system of an individual is altered due to a plethora of reasons like organ transplantation, cancer therapies, diabetes mellitus, immature birth, dietary factors, and nutrient deficiency (Riddell and Shuman 2012). The other contributing factor for the increase of fungal infections includes surgical problems with prolonged antibiotic treatment (Miceli et al. 2011).

Although there are different classes of approved antifungal agents like polyenes, pyrimidines, azoles, and echinocandins, the morbidity and mortality rate due to systemic infection remains high, due to its toxicity, narrow spectrum of activity, and tolerability (Bal 2010). The continuous treatment with similar kind of drugs evolves different drug resistance mechanisms in fungi, and the emergence of drug-resistant strains is a growing global problem (Allen 2010). The common drug resistance mechanisms established in fungi are (I) transport alteration (ATP-binding cassette transporters or major facilitator superfamily transporter), (II) target alteration (by mutation or structural change in target protein), (III) gene upregulation (overexpression of target protein) and utilization of compensatory and catabolic pathway, (IV) biofilm formation, (V) efflux of antifungal drug, and (VI) decreased uptake of antifungal drug (Kontoyiannis and Lewis 2002; Sanglard et al. 2009).

Combination therapy has shown to be a promising choice to overcome the emergence of drug resistance, minimizing the toxic effects by reducing the drug dosage, and to increase the spectrum of killing. The drug combinations are always selected based on the two different modes of actions of the antifungals. These are inhibitors of cell wall synthesis and inhibitors of protein/DNA synthesis or inhibitors of the cell membrane. The drug interactions in the combinational therapy are classified as synergistic, antagonistic, and additive (Chen et al. 2014). Some proposed mechanisms for synergistic drug interactions are inhibition of biochemical pathway at different stages, for example, ergosterol biosynthesis and weakening of the fungal cell membrane by combinational drug therapy using terbinafine and azoles (Barchiesi et al. 1998; Johnson et al. 2004); increased penetration of combined drugs by impairing the cell wall or cell membrane activity, for example, combination of fluconazole or amphotericin B with quinolones and rifampin allows these antibacterial agents to penetrate fungal cell to target DNA synthesis (Medoff 1983); transport interaction of antifungal drug with the help of another antifungal drug, for example, the cell membrane degradation by amphotericin B via oxidative decay allows flucytosine into the cell (Beggs and Sarosi 1982); and instantaneous inhibition of multiple fungal cell targets (Bartizal et al. 1997). Antagonism among antifungal agents might occur in one of several ways. Administration of antifungal drug at the same site and time will decrease the efficacy of drug, for example, amphotericin B and azole: amphotericin B binds with ergosterol in the cell membrane and usually blocks the synthesis of ergosterol (Cosgrove et al. 1978). Adsorption of one antifungal agent to the cell surface inhibits binding of another antifungal agent to target site of activity, for example, itraconazole and ketoconazole are lipophilic azole that bind with the cell surface and inhibit binding of amphotericin B (Johnson et al. 2004). Modification of a drug target upon previous exposure to an antifungal agent reduces the susceptibility, for example, exposure of pathogen to an azole compound results in alteration of membrane ergosterol with methylated sterol derivative, which will reduce the activity of amphotericin B (Currie et al. 1995). The selection of drugs in combination should accomplish these criteria for better treatment option.

Considering these facts, the therapy should also incorporate the appropriate dosage for eradication of fungal biofilms. Biofilm formation is an important contributing factor for virulence and an emerging global problem of great concern. The biofilms are resistant to host defense mechanism and antifungal agents (Martinez and Fries 2010). The three main drug resistance mechanisms proposed in biofilms are (I) incomplete or slow penetration of antimicrobials due to exopolysaccharide polymers; (II) resistant phenotype, known as persister cells characterized by a slower growth rate and greater resistance to antimicrobials; and (III) an altered environment that may affect the antibiotic action such that fungal biofilm persists even when appropriate antifungal treatment is given (Chatzimoschou et al. 2011). For instance, fluconazole and amphotericin combination therapy was unsuccessful in the treatment of *Cryptococcus neoformans*-infected hip joints due to poor penetration of drugs through the biofilm (Johannsson and Callaghan 2009). Catheter-related fungal infections are serious infectious complications that worsen the patient outcomes. The guidelines of the European Society of Clinical Microbiology and Infectious Diseases (ESCMID) have suggested removing the catheter, but a recent report, by analyzing 842 patients from two randomized clinical trials, reveals that the removal of the catheter does not improve the patient outcome (Nucci et al. 2010). Hence, development of new strategies for fungal infection management is required. One such approach is identifying an antiquorum sensing molecule which comprises targets of biofilm, virulence factors, and hyphal formation. The most studied quorum sensing molecule in fungi is farnesol, which has a control over growth, morphogenesis, and biofilm formation. The fungus quorum sensing research is still in infancy, and further research could lead to the development of new antifungal agents (Albuquerque and Casadevall 2012). Immunotherapy is based on the premise that most opportunistic fungal infections are a result of compromised immune system. And thus, any means of modulating the immune system using immune or immunogenic components could prove to be useful in achieving immune activation in the patient that will enable pathogen clearance. Such strategy minimizes the toxic effects of conventional antifungal drugs (and in some cases even synergizes with the drug or potentiates the activity of the drug), thereby opening up a new way of combating multidrug-resistant fungal pathogens.

In this chapter, we have reviewed the last 5 years' developments made in combinational antifungal therapy with special emphasis on biofilm-related infections and have presented a reappraisal of the current practices associated with immunotherapy and future challenges.

6.2 *Cryptococcus* Meningitis

Cryptococcus meningitis (CM) is an opportunistic neurological infection in immunocompromised individuals. It has emerged as the leading cause of morbidity and mortality in HIV-positive and HIV-negative patients (Muzoora et al. 2012). Recently, the Center for Disease Control estimated approximately one million new cases of cryptococcal meningitis each year, resulting in the death of 625,000 people worldwide. The standard therapy for CM consists of three phases: induction, consolidation, and maintenance (Vaidhya et al. 2015). The current standard initial therapy consists of amphotericin B and flucytosine (5-FC), but the availability of 5-FC is an issue in the countries of Asia and Africa, where CM is common (Perfect et al. 2010). Hence the probability of effective antifungal combination therapies is being investigated by the researchers.

The primary objectives of the therapies are to reduce the high infectious dose in the lungs and brain tissues which in turn increase the intracranial pressure and tend to have a worse prognosis. The induction therapy with amphotericin B deoxycholate (0.7–1.0 mg/-kg) combined with flucytosine (100 mg/kg/day for 2 weeks), followed by further consolidation therapy with fluconazole, is considered a standard management regime in the developed world (Denning and Hope 2010). An alternate but promising treatment strategy includes administering amphotericin B and fluconazole (both at 800 mg/14 days), followed by administration of fluconazole alone (800 mg/56 days). However, validation of this strategy is required through randomized phase III clinical trial. Fluconazole resistance is not common if it is used as a part of amphotericin B induction therapy. But continuous use of fluconazole has been shown to result in other resistant yeast infections, for example, fluconazole resistant (Govender et al. 2013). Toward the end of these antibiotic therapies, protein abnormalities may persist for years, which would cause adverse effects including nephrotoxicity (Fine and Atta 2007). Hence the combination therapy of fluconazole and flucytosine was recommended and has greater antifungal activity compared with fluconazole monotherapy, even though fluconazole has high penetration in CSF (Yao et al. 2014). However, the effect of this combination in comparison with amphotericin B is unknown (Denning and Hope 2010). Although the lipid-based formulations of amphotericin B have excellent efficiency by reducing its nephrotoxicity, the major drawback is its cost. To execute significant measures to overcome drug resistance, few studies have attempted to search for an alternative antifungal drug (Nikapitiya 2012). In search of new antifungal agents, in vivo efficacy of miltefosine, alkylphosphocholine, was evaluated for its usage as monotherapy and in combination with fluconazole and amphotericin. The monotherapy does not significantly improve survival of mice. Combination with miltefosine revealed neither synergy nor antagonism in both models, suggesting caution with the utility of miltefosine for anticryptococcal therapy (Wiederhold et al. 2013).

Apart from cost, the lipid formulation of amphotericin B is unable to pass through the blood-brain barrier. In CNS fungal infections, the permeability of lipid amphotericin B is enhanced by the inflammatory process. However, the inflammation is associated with the risk of intracranial pressure. Hence drugs directly administered into the CNS could be a suitable therapy for CN infections, especially in those cases where conventional therapy fails or a relapse of the infection occurs. Gazzoni et al. (2012) compared the in vivo efficacy of intrathecal administration of L-AmB alone and in combination with voriconazole in a murine model of cryptococcal meningitis. The animals were treated with liposomal amphotericin B administered intrathecally at 0.006 mg/kg weekly or intravenously at 10 mg/kg daily or with voriconazole administered orally at 30 mg/kg per dose twice daily or with combinations of both drugs, i.e., L-AmB i.t.c. + voriconazole or L-AmB i.v. + voriconazole at the same doses as used in the monotherapies. All treatments significantly increased the survival and reduced fungal burden in comparison with the control group, with voriconazole being less effective in comparison with all other treatments. The combination of L-AmB i.t.c. + voriconazole showed a synergistic effect in the reduction of fungal load that was significantly superior to any tested therapy (Gazzoni et al. 2012).

The increase of fungal burden in the CSF is a major issue to be focused to increase the survival. Studies were conducted to determine the therapeutic value of amphotericin B and voriconazole (alone and in combination) that would support reduction of fungal burden. Preclinical studies were conducted using BALB/c-SCID mice infected with fluconazole-resistant strains of *Cryptococcus neoformans* var. *grubii*. The survival curves and yeast quantifications indicate that amphotericin B was effective in increasing the survival of the animal without reducing the fungal burden in the lung and brain tissue (Silva et al. 2011). The rate of clearance of *C. neoformans* from cerebrospinal fluid is facilitated by amphotericin B and flucytosine as first-line therapy. But flucytosine is not available in many resource-limited countries (Bisson et al. 2013). Thus, combinations with different modes of action and different doses were tried with AmB + fluconazole (800 mg/day), AmB + fluconazole (600 mg twice per day), and AmB + voriconazole (300 mg twice per day) that provides a means to compare the antifungal activity of alternate regimens in small phase II studies. The combination therapy comprising rifampin and amphotericin B for treating fungal diseases is of past interest to eradicate the toxic effect of conventional AMB. In vivo studies were not conducted due to the lack of laboratory testing standards, of *Cryptococcus* sp. (Loyse et al. 2012).

Gamaletsou et al. (2012) experimented salvage therapy for two patients with HIV-related, refractory *Cryptococcus* meningitis as there are no potential comparative studies for salvage therapy. The salvage regimen provides a successful treatment, consisting of the combination of liposomal amphotericin B (3 mg/kg), intravenous voriconazole, and subcutaneous recombinant interferon γ-1b (200 μg thrice weekly). And to overcome interactions with co-administered ritonavir, voriconazole was given at an increased dose (5 mg/kg, twice daily). Oral fluconazole was given as maintenance therapy. The salvage therapy eliminates the fungal burden in the CSF after 10 weeks of therapy, also no adverse side effects, and no recurrence of infections has been observed even after 4 years of follow-up. Further validation is needed to use this strategy as initial therapy for HIV-related CM.

Recently, Yao et al. (2014) conducted a meta-analysis for prospective cohort studies to compare the survival benefits between fluconazole and 5-flucytosine given in combination with AmB as the initial therapy for CM. The survival rate was similar in case of AmB in combination with high-dose fluconazole, and therefore this combination is alternative for 5-flucytosine for the management of HIV-associated CM therapy, which is not available in resource-limited countries (Yao et al. 2014). Clinical data disclose that high-dose fluconazole shows no adverse liver function abnormalities (Vaidhya et al. 2015). However, high-dose fluconazole might lead to the emergence of drug-resistant strains. For patients without CNS involvement with limited pulmonary cryptococcosis and non-pulmonary cryptococcal infections in immunocompetent patients, initial therapy with fluconazole (400 mg per day) could be considered. The standard therapy of amphotericin B and flucytosine is not recommended for disseminated cryptococcosis (Probst et al. 2010).

Cryptococcus neoformans biofilms are becoming common due to the increased use of brain valves and other medical devices such as ventriculoatrial shunt catheters, peritoneal dialysis fistula, cardiac valves, and prosthetic joints. The knowledge on the mechanism of biofilm formation by *Cryptococcus* is still in infancy; hence, there are very limited reports on the evaluation of antifungals on *C. neoformans* biofilms. An in vitro study reveals that amphotericin B and caspofungin are effective against the mature biofilms of *C. neoformans*; however, the efficacy profile is not similar for melanized *Cryptococcus* biofilms. Recently, antibiofilm activity of antifolate drugs (sulfamethoxazole-trimethoprim and sulfadiazine-pyrimethamine) alone and in combinations with amphotericin B was studied (Liao et al. 2015). However clinical study and combination therapy for cryptococcal shunt infections are lacking.

6.3 Candidiasis

Candida albicans is the most prevalent infectious species among the 20 different species of *Candida* (Centre for Disease Control 2014). It is the normal commensal of the alimentary tract, vagina, and oral cavity. The emergence of *Candida* sp. as the important nosocomial pathogen is owing to the increased use of medical devices such as central venous catheters and various implanted devices (Kim et al. 2011). Although *C. albicans* is the most prevalent, non-albican species, *C. glabrata*, *C. tropicalis*, *C. parapsilosis*, *C. guilliermondii*, *C. krusei*, *C. lusitaniae*, and *C. kefyr* are also increasing recently, which is evident in a recent 5-year prospective study and population-based surveys of candidemia (Alexander et al. 2013). For candidemia management, the most frequently administered antifungal agent is fluconazole and echinocandins. The effectiveness of the antifungal agent varies with the species of *Candida* isolated from different clinical samples (Pfaller et al. 2012; Sardi et al. 2013).

Andes et al. (2012) conducted an individual patient-level quantitative review of randomized trials consisting of 1,915 patients for treatment of invasive candidiasis. They identified that the use of an echinocandin was associated with reduced mortality when compared to the use of a drug from either the triazole or polyene classes. And their study also supports the greater toxic effects associated with amphotericin B than triazole or echinocandin therapy. These findings lend support to recent treatment guidelines that recommend echinocandin as a first-line choice for invasive candidiasis particularly for patients who are critically ill and those with previous exposure to triazole (Andes et al. 2012).

The treatment for *Candida* infections varies with patient's underlying conditions. For instance, the delayed treatment of candidemia in septic shock is shown to be an important factor of patient outcome. Septic shock due to invasive candidiasis is a near-fatal condition. Retrospective cohort studies of septic shock patients with *Candida* infections reveal the risk of death is exceptionally high among patients with septic shock associated with *Candida* infection. The antifungal therapy regime is needed to improve the clinical outcomes (Kollef et al. 2012). Invasive candidiasis is also a leading cause of mortality and morbidity in immune-compromised neonates. A comparative outcome of antifungal treatment of candidiasis in infants was provided by a large multicenter cohort study. Infants treated with L-AmB had a higher mortality when compared to infants treated with either amphotericin B deoxycholate or fluconazole. This finding may be related to inadequate penetration of L-AmB into the kidneys. There are no recommendations for combination therapy to treat life-threatening infection and deep tissue infection as the comparative outcome data are limited (Tragiannidis et al. 2015). In case of candidiasis in hematologic malignancies, echinocandins are recommended as initial therapy, in non-neutropenic patients with candidial infections, initial therapy with the echinocandin followed by fluconazole, is recommended. There are no reports on combination therapy for invasive candidiasis in leukemic patients (Walsh and Gamaletsou 2013). The summary of recent treatment recommendations for different types of candidiasis is given in Table 6.1.

The important factor for drug resistance in *Candida* sp. is their ability to grow and adhere as biofilms on various surfaces including the medical devices. *Candida* biofilms are the leading cause for intravascular catheter-related infections. The presence of biofilms on intravascular devices makes the treatment of CR-BSI difficult and resists the action of clinically important drugs including amphotericin B and azole (Chatzimoschou et al. 2011). Also a recent randomized clinical trial reveals the higher toxicity effects of amphotericin B with fluconazole (Sanati et al. 1997). Thus, many researchers have turned their focus to evaluate the antifungal therapy comprising different combinations including antimicrobial peptides, natural products, and non-antifungal agents. And many studies are being carried out to demonstrate the antimicrobial lock therapy to treat catheter-related blood infections. Antimicrobial lock therapy involves the instillation of a small volume of the highly concentrated antibiotic solution, which is allowed to stay in the lumen of the catheter for 12–72 h, to eradicate biofilm formed on the inner surface of the device. Ku et al. (2010) determined the efficacy of antibiofilm and persistence of the activity for the lock solutions of caspofungin, micafungin, and posaconazole in in vitro model of *Candida* biofilm on silicone catheter. Micafungin and caspofungin had significant inhibitory activity against mature *C. albicans* and *C. glabrata* biofilms, but caspofungin displayed less persistence in *C. glabrata* biofilms. Even though posaconazole was less effective against *C. albicans* biofilms, its activity was maintained. They had suggested that echinocandin lock therapy may be useful to control candidiasis in catheterized patients (Cateau et al. 2011). Until recently, the guidelines of the Infectious Diseases Society of America (IDSA) had suggested the use of antibiotic lock therapy only under special circumstances as a salvage treatment. For the management of catheter-

Table 6.1 Summary of recent treatment recommendations for different types of candidiasis

Type	Organism	Symptoms	Estimated statistics	Treatment	References
Oral candidiasis	*Candida* sp., *Candida albicans* (69.23 %)	Soreness and redness in infected area, difficulty in swallowing, dryness at the corners of mouth	5–7 %, babies less than 1 month; 9–31 %, AIDS patients; ~20 %, cancer patients	Clotrimazole with fluconazole or itraconazole	Pieralisi et al. (2014), Jensen and Peterson (2014), Center for Disease Control and Prevention
Vulvovaginal candidiasis	*Candida albicans* (87.5 %), *C. glabrata* (6.8 %), other *Candida* species less than 3 %	Woman (itching, burning, and vaginal discharge)	75 % adult woman	Oral medication (clotrimazole with fluconazole), suppositories, and creams	Mendling (2015), Center for Disease Control and Prevention
Invasive candidiasis	*Candida albicans* (70 %), *C. glabrata* (11 %)	Fever, chills, and failure of antibiotics	8–10 cases per 100,000 person, 10–20 % in new born child (less than 1 month)	Neonates (amphotericin B), child and adults (fluconazole, micafungin, caspofungin, anidulafungin, and echinocandin)	Cornely et al. (2011), Kobayashi et al. (2015), Laua et al. (2015), Center for Disease Control and Prevention
Genital candidiasis	*Candida albicans* (87.5 %), *C. glabrata* (6.8 %), other *Candida* species less than 3 %	Man (itchy rash), woman (itching, burning, and vaginal discharge)	75 % adult woman, men (rare occasion)	Oral medication, suppositories, and creams	Mendling (2015), Center for Disease Control and Prevention

related bloodstream infections, tigecycline, an antibacterial agent, was evaluated alone and in combination with fluconazole, amphotericin B, or caspofungin against *Candida* biofilms. It has been found that high-dose tigecycline is highly active against in vitro *Candida* biofilms; however, the combination with antifungals is not significant. Further investigation is required to utilize the antimicrobial lock therapy containing tigecycline for the management of CR-BSI (Ku et al. 2010). The antibiofilm activity of amphotericin B was found to be significantly enhanced when used in combination with polyamine biosynthesis inhibitors, 1,4-diamino-2-butanone (DAB) and difluoromethylornithine (DFMO). Further studies have shown that DAB and DFMO also enhanced the antibiofilm activities of several other antifungal agents by enhancing the formation of reactive oxygen species in biofilm cells.

In vitro, the combination of caspofungin or anidulafungin with antimicrobial peptides is very effective in killing *Candida* sp.; however, the efficacy was not improved in a murine model of systemic candidiasis, wherein the combination of caspofungin and antimicrobial peptide, ranalexin, was used. The enhanced efficacy was observed in the same animal model when the combination therapy of amphotericin B and lactoferrin was used. There is no literature support for the synergism between antimicrobial peptides and echinocandins. Rossignol et al. (2011) investigated the tryptophan-rich, ApoEdpL-W peptide, derived from human

ApoE apolipoprotein shown to have less antifungal activity against mature biofilms and partially inhibited biofilms on medical devices.

Synergism of natural products with antibiotics is the currently developing area of phytopharmaceuticals. Many natural products including essential oils, peptides, and polyphenols are active against different shapes of *Candida* biofilms. Recently the effectiveness of the antifungal activity of the essential oils of *Ocimum sanctum* was evaluated for the fluconazole-resistant *Candida* sp. Besides its selective cytotoxicity, the synergism with the fluconazole and ketoconazole showed a better antifungal activity in in vitro fluconazole-resistant *Candida* strains (Amber et al. 2010). Similar synergistic interactions were observed in farnesol with micafungin, fluconazole, and amphotericin B against in vitro *C. albicans* biofilms (Katragkou et al. 2015). The in vitro effect of essential oil components of eugenol and methyleugenol alone and in combination was checked against fluconazole-sensitive and fluconazole-resistant strains of clinical *Candida* sp. The synergistic interactions of eugenol and methyleugenol with fluconazole are very effective killing fluconazole-resistant strains (Ahmad et al. 2010a). Mandal et al. (2011) demonstrated the activity of Tn-AFP1 from *Trapa natans* against biofilm inhibition of *C. tropicalis* in vitro. Similarly, Delattina et al. (2014) identified OSIP108, a nontoxic decapeptide from the model plant *Arabidopsis thaliana*, that inhibits *C. albicans* biofilm. These recent reports support the synergism of natural products with antifungal agents in controlling the *Candida* biofilms. However, the major drawback of natural products' utility is the unestablished toxicity profiles. One such well-established compound, fulvic acid, was proven to be nontoxic in an epithelial cell line and rat model alone (Nett 2014). The information from several studies indicating synergistic interactions of non-antifungal drugs in combination with fluconazole are summarized in Table 6.2

Due to the high cost and side effects of combination therapy, the focus of research recently turned to examine the combinations with non-antifungal agents, including tetracyclines (Loreto et al. 2011), quinolones (Wang et al. 2012), calcineurin inhibitors (Chen et al. 2011), heat shock protein 90 inhibitors, calcium homeostasis regulators, and traditional medicines (Muñoz et al. 2012).

One such combination is azole with calcineurin inhibitor. The calcineurin signaling pathway is a key mediator of stress responses in *C. albicans* as a potential drug target for systemic candidiasis and candidal keratitis. Calcineurin inhibitors when given as a monotherapy were not suitable for other candidial infections of oropharyngeal, pulmonary, and vaginal tract. The virulence pathway inhibitor in combination with azole would increase the efficiency of the antifungal drug. The establishment of virulence pathway varies with the species. Hence, identifying suitable drug targets and finding its synergism with the antibiotics would be potential therapeutic options (Reedy et al. 2010). However, drugs targeting the inhibition of virulence expression would be suitable targets to enhance the current antifungal therapy and to reduce the evolution of drug-resistant strains. The synergism shown by calcineurin inhibitors with the antifungal agents amphotericin B and caspofungin is potentially good as the combination lock therapy in the rat catheter infection model; however, its utility is limited as the drugs have immunosuppressive effects that prevents the systemic administration.

Diclofenac, aspirin in combination with caspofungin and amphotericin B, was evaluated to treat infections of *Candida albicans* and *C. parapsilosis*. Bink et al. (2012) screened 1,600 compounds of a drug-repositioning library that enhance antibiofilm activity against *C. albicans*. They have found out that hexachlorophene, pyrvinium pamoate, and artesunate can act synergistically with miconazole in eradicating C. *albicans* biofilms. They suggested miconazole and artesunate as the potential combination therapy to treat oral and vulvovaginal candidiasis. In vivo toxicity studies are required to execute the potentiality of the miconazole and artesunate (De Cremera et al. 2015). Finasteride, a

Table 6.2 Synergism of fluconazole with non-antifungal agents

Combined drug	Clinical *Candida* spp.	Methods and results	Antifungal mechanisms/targets
Antimicrobial agents			
Minocycline	Azole-susceptible *C. albicans* strains	FICI, E method, indifference	Minocycline significantly enhances the amount of FLC penetrating *C. albicans* biofilms as well as interrupting the cellular calcium balance
	Azole-resistant *C. albicans* strains	FICI, E method, synergy	
Doxycycline	*C. albicans* (SC5314), FLC-resistant strain	FICI, synergy	Interference with iron homeostasis
	C. albicans biofilm	FICI, synergy	Inhibition of biofilm formation
	C. albicans biofilm	FICI, E method, synergy	Inhibition of biofilm formation is associated with calcium
Ciprofloxacin, trovafloxacin	*C. albicans*, *C. tropicalis*, *C. glabrata*	FICI, in vivo, prolonged the survival of mice	Possibly inhibits fungal topoisomerase
	Calcineurin inhibitors		
FK506 (tacrolimus)	Azole-susceptible *C. albicans* isolates	FICI, E method, time-kill curves, synergy, or indifference	CsA and FK506 bind to cyclophilin A and FKBP12, respectively, to form protein–drug complexes that inhibit the calcineurin-mediated stress response, cause damage to the cell membrane, and increase fungal cell membrane permeability, thus increasing the concentration of azoles in fungal cells to achieve a fungicidal effect
	Azole-resistant *C. albicans* isolates	FICI, E method, time-kill curves, synergy	
	C. albicans wild type	FICI, synergy	
	C. albicans mutants	FICI, synergy	
	C. albicans wild-type biofilm	FICI, SEM, synergy	
	C. albicans biofilms (mutants: cnb1/cnb1, crz1/crz1)	FICI, indifference, or antagonism	
Cyclosporine A(CsA)	*C. albicans* (standard strain, ATCC90028) planktonic cells	FICI, synergy	
	C. albicans (standard strain, ATCC90028) biofilm cells	FICI, synergy	
	C. albicans wild type	FICI, synergy, or indifference	
	C. albicans mutants	FICI, synergy, or indifference	
	C. albicans biofilm (wild-type SC5314)	FICI, synergy	
	C. albicans (mutants: cnb1/cnb1,crz1/crz1)	FICI, synergy	
	C. albicans biofilms (mutants: cnb1/cnb1, crz1/crz1)	FICI, indifference, or antagonism	
Heat shock protein 90 (Hsp90) inhibitors			
Geldanamycin	*C. albicans*	FICI, synergy	Inhibit the stress responses induced by Hsp90 and reverse fungal azole resistance

(continued)

Table 6.2 (continued)

Combined drug	Clinical *Candida* spp.	Methods and results	Antifungal mechanisms/targets
Calcium homeostasis regulators			
Amiodarone (AMD)	FLC-susceptible *C. albicans* strains	FICI, indifference	AMD elicits an immediate, dose-dependent hyperpolarization of the membrane, causing Ca2+ influx and release from internal stores, inducing a disruption of calcium homeostasis
	FLC-resistant *C. albicans* strains	FICI, synergy	
	FLC-resistant *C. albicans*	FICI, E method, synergy	
	FLC-susceptible *C. albicans* strains	FICI, E method, indifference, or antagonism	
	C. albicans SC5314	FICI, synergy	
Traditional Chinese medicine			
Pseudolaric acid B	*C. albicans*	FICI, synergy	–
	C. krusei, *C. tropicalis*, *C. dubliniensis*	FICI, indifference	–
	C. glabrata, *C. guilliermondii*		–
	FLC-resistant *C. albicans*	FICI, E method, time-kill curves, synergy	–
	FLC-sensitive *C. albicans*	FICI, E method, time-kill curve, synergy, or indifference	–
Berberine chloride	FLC-resistant *C. albicans*	FICI, time-kill curves, synergy	–
Baicalein	FLC-resistant clinical *C. albicans*	FICI, time-kill curves, synergy	Inhibition of efflux pumps thus may increase the susceptibility of cells to antifungals. Exerts a synergistic effect with FLC by affecting some other factors on resistance, which requires further research
Eugenol, methyleugenol	FLC-sensitive *Candida* isolates	FICI, synergy, or indifference	Target the sterol biosynthetic pathway, altering fluidity and permeability
	FLC-resistant *Candida* isolates	FICI, synergy, or indifference	
	FLC-sensitive *Candida* isolates	FICI, synergy, or indifference	
	FLC-resistant *Candida* isolates	FICI, synergy, or indifference	

Source: Liu et al. (2014)

5-α-reductase inhibitor, commonly used for the treatment of benign prostatic hyperplasia was tested alone and in combination with amphotericin B and fluconazole against *C. albicans* urinary biofilms. It exhibits synergistic activity in combination with fluconazole. However, further investigation is required to explore the clinical utility of finasteride in the prevention of candidiasis (Chavez-Dozal et al. 2014). Synergistic interactions were observed during in vitro combination studies of allicin and flucytosine against *C. albicans*, *C. glabrata*, and *C. tropicalis* (Alexander et al. 2013). L-AmB in combination with heat shock protein, HSP 90 inhibitor (Mycograb), also provides promising therapy; however, the antibody-based inhibitor is not available for clinical trials (Pachl et al. 2006). The synergistic interactions of fluconazole with non-antifungal agents are summarized in Table 6.2.

The present review recommends the necessity for standard guidelines for the management of fungal biofilm infection.

6.4 Aspergillosis

Aspergillus exposure to human is quite common due to their ubiquitous nature. In immune-compromised host, it causes aspergillosis. Invasive aspergillosis (IA) has become increasingly common in recent days resulting in a mortality ranging from 40 % to 80 % (Ascher et al. 2012; Aguado et al. 2015). The predominant causative agent is *A. fumigatus* which also finds concurrence in a recent literature study reported from 1985 to 2013. The study comprised of 116 liver transplant recipients with invasive aspergillosis. The most common infecting species are *Aspergillus fumigatus* (73 %), *A. flavus* (14 %), and *A. terreus* (8 %). Voriconazole improves the survival among the commonly used AMB and itraconazole (Barchiesi et al. 2015; Ogawa et al. 2015). However, an increase in drug resistance restricts the usage of azole for treatment and necessitates the combinational therapy approach.

A randomized, double-blind, placebo-controlled, multicenter trial comprising 454 patients with hematologic malignancies and hematopoietic stem cell transplantation suspected with invasive aspergillosis were treated with voriconazole and anidulafungin. The study showed the improvement of survival when compared with monotherapy (Marr et al. 2015). In another retrospective study that was conducted between 1993 and 2008, the combination of L-AmB and echinochandins for the management of IA in hematological malignancies was analyzed, and the results suggested that the combination therapy offered no improvement in terms of response and mortality (Mihu et al. 2010).

Though sinus aspergillosis was believed to be a rare disease for many years, recently there has been an increase in the number of cases. Chronic invasive aspergillosis of the sinus occurs in mildly immunodeficient patients. Two of the most important treatment regimens for invasive aspergillosis of the sinus are surgical debridement and antifungal therapy. The pharmacodynamic interactions of amphotericin B and voriconazole were assessed using conventional in vitro tests that mimic human serum concentrations. The combination was more effective than monotherapy regimens (Siopi et al. 2015).

Various formulations of L-AmB differ in terms of toxicity and efficacy which concurs with a recent finding by Olson et al. (2015) while comparing, AmBisome and Lambin, in *Aspergillus fumigatus* mice model. Lambin was more toxic than the other, emphasizing the adequacy requirement to compare liposomal formulations. The in vitro combination studies of nystatin-intralipid with caspofungin demonstrated effective synergistic interaction against *A. terreus* when compared with voriconazole and 5-fluorocytosine (Semis et al. 2015; Shadomy et al. 1975). In a murine aspergillosis model, the combinations of L-AmB, micafungin, and deferasirox improved the fungal clearance and patient survival. The in vitro efficacy of PTX3, a glycoprotein produced by monocytes in combination with voriconazole, improved patient survival and clearance of fungal burden (Giudice et al. 2012). Erythropoietin in combination with AMB improved the outcome of in vivo disseminated aspergillosis.

The synergistic interactions vary with the species and resistance mechanism. In a similar kind of study, the synergistic interactions of posaconazole and caspofungin varied with 20 different wild and resistant types of *A. fumigatus*, with ten different resistant mechanisms. The interactions are insignificant when compared with wild isolates (Mavridou et al. 2015). Similarly, while investigating the efficacy of the combination of voriconazole and anidulafungin in an in vitro model of invasive pulmonary mice infected with triazole-resistant *Aspergillus*, the addition of anidulafungin does not significantly improve the voriconazole activity, whereas another study suggests the combination of triazole and echinocandin as the treatment strategy for triazole-resistant isolates (Lepak et al. 2013). The efficacy varied with the type of strain and the drug resistance pattern evolved. A study by Jeans et al. (2012) has proven the efficacy of voriconazole and caspofungin in an in vivo model of pulmonary aspergillosis infected with different species of *Aspergillus*. The synergism between *A. flavus* and *A. niger* was found to be significant while not in the case of *A. fumigatus* (Zhang et al. 2014). In a neutropenic *A. fumigatus* murine model, the synergistic interactions of voriconazole and anidulafungin were observed in voriconazole-susceptible strain, but additive in resistant type, which creates concern for the management of infections by voriconazole-resistant strains.

The inappropriate use of antibiotics is another major factor for the emergence of drug-resistant strains. A recent cross-sectional cohort study states that systemic antifungal therapy was administered irrespective of the presence of invasive fungus infection (Ascher et al. 2012; Olson et al. 2015). Based on this, a recent survey study conducted revealed the weakness in the knowledge on current guidelines of invasive candidiasis and invasive aspergillosis among the European physicians (Valerio et al. 2015).

A. fumigatus biofilms are difficult to manage with conventional antifungal agents; hence, even after antimycotic therapy, these infections are not eradicated and are emerging as life-threatening infections (Mowat et al. 2007). Melanin produced by *A. fumigatus* during infection is required for lung tissue invasion, conferring less susceptibility to drugs (Bowyer et al. 2011). Biofilms of *A. fumigatus* are resistant to itraconazole and, some extent, to caspofungin due to its extracellular polymeric substances (EPS) (Singh et al. 2011). An in vitro finding reveals *A. fumigatus* are not responding to amphotericin B, when they have a prior exposure to voriconazole, suggesting the improper usage of azole will be a hindrance for treatment with amphotericin B. Synergistic interactions were observed in combination with caspofungin and amphotericin B or voriconazole against *Aspergillus* biofilms (Liu et al. 2012). Another in vitro study with combinations of amphotericin B or Lamb with alginate lyase, the degrading enzyme of EPS (Bugli et al. 2013; Rajendran et al. 2015), suggesting the EPS as a suitable target for antibiofilm activity. There are very limited clinical studies that are focused on the management of biofilm-associated *A. fumigatus* infections.

In a recent literature review, data addresses the significance of combination antifungal therapy in the treatment of invasive aspergillosis. The information suggests that evidence is not convincing enough to support the use of combination antifungal therapy (Garbati et al. 2012). Another systematic review comprising animal and clinical studies addressed the efficacy of combination therapy of triazole and echinocandin in treatment of invasive aspergillosis, suggesting a development toward improved overall survival (Day et al. 2013; Marr et al. 2015).

In summary, combination therapies appear to be conflicting in some studies, and the limited clinical data are not convincing to support combination antifungal therapy over monotherapy. In addition, well-sketched clinical trials focusing on management of biofilm-related fungal infections are immediately needed to overcome the emergence of drug resistance in these pathogens.

6.5 Immunotherapy Against Fungal Infections

Opportunistic fungal infections become a major challenge to health care since a majority of them have affected immune-compromised individuals such as HIV, hypersensitive, autoimmune, or transplant patients. Most of these infections in the case of normal persons are easily combated by the immune system, which is not the case with patients whose immune system is suppressed. Added to this are the problems associated with hospital multiple drug-resistant pathogens. Opportunistic fungal infections such as candidiasis, aspergillosis, and cryptococcosis are three of the leading causes of significant mortality in both hospital-derived infections and in those of immune-compromised patients (Truan et al. 1994). Studies also clearly suggest that due to modified lifestyles and exposure, these fungal infections are set to increase and dominate the concerns of medical industry in the future (Kauffman 2001). This creates the need for either combinatorial therapies or rapid discovery of novel antibiotics. However, most such strategies are replete with adverse side effects and may not be favorable, though the benefits far outweigh the risks. One way forward to combat such opportunistic fungal infections, with nil or minimal side effects, is immunotherapy (Ahmad et al. 2010b).

Majority of the immunotherapy strategies rely on immunomodulation of the patient's immune system to prime it against the fungal pathogen. This indirect method has many advantages, including lack of side effects and resistance development (Ahmad et al. 2010b; Owais et al. 2010), and more importantly such treatment to some extent can be catered to suit an individual patient's needs. However, it's important to recognize that host-fungi interaction (Ahmad et al. 2010b), immune responses of the host to fungi, and, importantly, immune escape mechanisms of fungi are quite complicated and the underlying mechanisms of fungal infections are beginning to be unraveled. With better understanding of these components, there is a better hope of developing much more effective immune modulators. In the next few sections, we will primarily focus on the use of immunotherapy as the latest and futuristic treatment for three of the major opportunistic pathogenic fungi: *Candida*, *Aspergillus*, and *Cryptococcus*.

6.5.1 Antimicrobial Proteins

These form one of the first lines of innate immune defense against a variety of pathogens and are widely distributed in animal kingdom. These are usually cationic and amphipathic peptides (Owais et al. 2010) and expressed in cells of the innate immunity such as macrophages, mast cells, and neutrophils, as well as in nonimmune cells such as keratinocytes and epithelial cells. The antimicrobial proteins can either directly defend the pathogens (Braff et al. 2005) or stimulate cells such as neutrophils (Sorensen et al. 1997) to degranulate resulting in inflammatory responses (Hancock and Diamond 2000). These antimicrobial peptides comprise cathelicidins (Sorensen et al. 1997), β-defensins, substance P, RANTES (Braff et al. 2005), LL-37 (Larrick et al. 1995; Nagaoka et al. 2001), etc. Apart from immune cell activation, these peptides can also activate epithelial cells upon microbial invasion leading to proinflammatory cytokine signaling (Van Watering et al. 1997) and toll-like receptor (TLR) activation (Akira et al. 2006). This could end up activating and recruiting immune cells to the site of infection and thus boosting the host immune defense. Thus, components of the innate immunity can be tweaked that could prove to be effective not only in controlling the spread of infection but also end up activating the adaptive immune components. However, the effectiveness of the use of these antimicrobial peptides as a therapy against opportunistic fungal pathogens needs more attention.

6.5.2 Cytokines

Both pro- and anti-inflammatory cytokines are regulators of innate and adaptive immune responses (Iwasaki and Medzhitov 2015), including cell recruitment, immune cell activation, and immune cell differentiation. Nowhere is this more pronounced than in the interplay of cytokines during development of T helper subsets (Romani 2000). Studies have shown the importance of cytokines in not only modulating immune responses against fungal pathogens (Mencacci et al. 1998) but have also demonstrated mechanisms involved in hijacking of the host cytokine network for immune escape by the fungal pathogens (Netea et al. 2006). Cytokines such as proinflammatory TNF-α and IFN-γ are unregulated (Rodriguez-Adrian et al. 1998) upon infection by *Candida* and *Aspergillus*, resulting in the activation of phagocytic cells. Cytokines can modulate immune cells very effectively if properly applied and are one of the potential immunotherapeutic strategies against opportunistic fungal infections. TNF is a proinflammatory cytokine and has been found to be very effective against *A. fumigatus* (Steinbach and Stevens 2003). TNF has pleiotrophic effect on phagocytic cells, including cell activation and phagolysosomal killing of fungal pathogens (Cisalpino et al. 1996). The study by Steinbach and Stevens (2003) has shown the importance of TNF-α in suppressing immunodeficiency induced by dexamethasone, and the mechanism consists of upregulation of TLR and IFN-γ and preventing leukocyte apoptosis (van der Meer et al. 2005). Although both Th1 and Th2 are essential for anticryptococcal immunity, Th1 are very effective (Hoag et al. 1997). Thus, higher levels of Th1 cytokines such as TNF-α and IFN-γ are associated with increased fungal clearance (Milam et al. 2007; Wormley et al. 2007). More importantly, TNF-α is necessary for IFN-γ production (Bekker et al. 2001) which results in macrophage activation. So, it is clear that TNF therapy could potentially stimulate IFN activity leading to better antifungal responses. Apart from this, pathogen-associated molecular patterns (PAMPs) are potent antigens recognized by innate pattern recognition receptors (PRRs) such as TLRs. Both TNF-α and IFN-γ synergize to increase the cell surface expression of TLRs (Bosisio et al. 2002) resulting in better recognition and clearance of *Candida*, *Aspergillus* (Tsiodras et al. 2008), and *Cryptococcus* (Pappas et al. 2004). Thus, TNF-α immunotherapy could be a potential therapeutic strategy against fungal pathogens.

IFN-γ synergizes with TNF-α in a variety of immune functions including Th1 differentiation (Milam et al. 2007), TLR expression (Tsiodras et al. 2008), and phagocyte activation (Bekker et al. 2001). In rodents, the use of IFN-γ has shown good results against systemic cryptococcosis (Lutz et al. 2000), and the same cytokine was also successfully tested in one human patient (Netea et al. 2004). Studies have shown the effectiveness of IFN-γ against a variety of fungal pathogens including *Candida*, *Aspergillus*, and *Cryptococcus* (Rodriguez-Adrian et al. 1998; Shahid et al. 2010). IFN-γ has been shown to stimulate phagocytic activity and intracellular killing of *A. fumigatus* (Steinbach and Stevens 2003; Shahid et al. 2010), through stimulating generation of free radical production (Steinbach and Stevens 2003). In murine models, immunotherapy using TNF-α and IFN-γ was shown to produce significant reduction in mortality due to *Aspergillus* infection (Shahid et al. 2010). However, one issue with IFN-γ usage is that it reduces anticryptococcal activity of human macrophage and thus its use in humans is debatable (Reardon et al. 1996). Nevertheless, in 2010, the Infectious Diseases Society of America has approved recombinant IFN-γ against persistent cryptococcosis (Antachopoulos and Walsh 2012). Gallin et al. (1995) have shown that the use of IFN-γ under clinical settings is very effective in the case of patients suffering from chronic granulomatous disease (CGD) and who are especially prone to *Aspergillus* (Antachopoulos and Roilides 2005). Unlike anticryptococcal action, macrophage anti-*Candida* activity is enhanced by IFN-γ treatment (Redmond et al. 1993; Baltch et al. 2005). But other studies in rodents have shown lack of an effect of IFN-γ (Brummer and Stevens 1987; Marcil et al. 2002) in case of pulmonary macrophages, but peritoneal or peripheral blood PMNs (from both human and murine) showed good anti-*Candida* activity after IFN-γ treatment (Djeu et al. 1986; Kullberg et al. 1993). In spite of the lack of strong evidence for IFN-γ immunotherapy, Dignani et al. (2005) showed in the case of three patients with candidiasis that IFN-γ therapy was actually beneficial. Similarly, IFN-γ immunotherapy of an HIV-infected patient proved to be beneficial (Riddell et al. 2001; Bodasing et al. 2002). Nonetheless, large-scale clinical trials are the need of the hour to better understand the efficacy of this cytokine for immunotherapy.

IL-12, like TNF-α and IFN-γ, is a Th1 cytokine (Milam et al. 2007; Wormley et al. 2007), and Th1 activation and responses are vital for strong antifungal responses against *Candida*, *Aspergillus*, and *Cryptococcus*. IL-12 has been shown to inhibit Th2 responses, thereby increasing anti-candidiasis response of the host (Morrison et al. 1986; Deepe 1997). Other studies have shown the potential of IL-12 in enhancing the antifungal efficacy of fluconazole against *Candida* (Afonso et al. 1994), murine cryptococcosis (Muchmore et al. 1968), and aspergillosis (Allendoerfer et al. 1996). From these studies, it is recognized that one way by which IL-12 could exert its antifungal effect is through enhancement in IFN-γ levels (Antachopoulos and Roilides 2005; Kawakami et al. 1996; Jacobson et al. 2000), synergization with IL-18 in the induction of anticryptococcal activity (Qureshi et al. 1999), and suppression of IL-4 activity (Qureshi et al. 1999). However, IL-12 usage has been shown to exacerbate mucosal disease in mouse (Morein et al. 1984) and produce adverse reactions on the immune system (Murphy et al. 1993). Thus, more studies are warranted to check the efficacy of this cytokine as an immunotherapy agent.

Colony-stimulating factors (CSFs) are essential for phagocyte development (especially in the case of neutropenia) and function and have been approved for humans (Rodriguez-Adrian et al. 1998). Being able to facilitate antifungal activity of phagocytes, CSFs are one of the promising immunotherapy agents. Approved and accepted since 1991 for patient use (Welte et al. 1996), granulocyte colony-stimulating factor (G-CSF) is effective in prolonging phagocyte and respiratory burst activity against *Aspergillus* infection (Polak-Wyss 1991; Liles et al. 1997) and *Candida hyphae* (Gaviria et al. 1999; Kullberg et al. 2004). However, when compared to other CSFs, G-CSF also produces anti-inflammatory cytokines and could be useful against endotoxins in immune-compromised patients (Gomez et al. 1995). G-CSF also favors a Th2 response and so can be beneficial against *Candida* (Condon et al. 1996) and *Aspergillus* (Cisalpino et al. 1996). However, anticryptococcal immunity requires a robust Th1 activation, and thus G-CSF may not prove to be successful here. However, neutrophils from G-CSF-treated HIV patients have been shown to be effective against *Cryptococcus* and *Candida* (Han and Cutler 1995), and this was shown to be due to enhanced superoxide anion generation. Granulocyte-macrophage colony-stimulating factor (GM-CSF) is very important for the development and activation of neutrophils (Rodriguez-Adrian et al. 1998) and macrophage (Steinbach and Stevens 2003). In murine models, Brummer and Stevens (1987) demonstrated the usefulness of GM-CSF in rescuing dexamethasone-induced suppression of macrophage and synergistically working with TNF-α (in potentiating anti-*Aspergillus* function of macrophage (Bodey 1994)). Rowe et al. (1995) have shown that GM-CSF was effective against aspergillosis in clinical trials, but issues remain (Kong and Levine 1967; Bodey et al. 1993; Groll et al. 1996). However, when compared to other CSFs, studies have shown that GM-CSF is more effective (Kirkland and Fierer 1985; Kirkland et al. 1991) and thus is more promising. In combination with antifungal therapy, GM-CSF administration was shown to improve clinical outcome of *Candida* infection (Vazquez et al. 2000), but controlled trials are the need of the hour as far as GM-CSF is concerned.

Macrophage colony-stimulating factor (M-CSF), important for mononuclear phagocyte functions (Shahid et al. 2010), has been demonstrated to be very effective against *Candida* infections (Mencacci et al. 1996). A study employing rabbits (Gonzalez et al. 2001) has shown that M-CSF is very effective against pulmonary aspergillosis and resulted in higher numbers of activated alveolar macrophages. But its usefulness, when compared to G-CSF or GM-CSF, is less clear (Steinbach and Stevens 2003). Currently, human recombinant G-CSF and GM-CSF are available for clinical use but not so in the case of M-CSF (Hubel et al. 2002). Thus, extensive studies are needed to confirm the effectiveness of these three CSFs as potent immunotherapy agents.

6.5.3 Antibodies

Antibody represents one of the best effector arms of the humoral immune response and is important for establishing a robust secondary immune response. Being highly specific in their action, antibodies are receiving a lot of interest as adjuncts for antifungal therapy. Antibody-mediated immunity (AMI) thus is a potential immunotherapeutic strategy and can be activated by direct mechanism or by indirect mechanism

(Casadevall and Pirofski 2012). In the case of direct action, antibodies can interfere with the fungal pathogen and inhibit *Cryptococcus* virulence production (Martinez et al. 2004) and resistance development (Martinez and Casadevall 2005), *Candida* replication (Brena et al. 2007), gene expression in *Cryptococcus* (McClelland et al. 2010), and metabolism in *Candida* (Brena et al. 2011). By their indirect action, antibodies have been shown to enhance antifungal responses against *Cryptococcus* by inducing phagocytic activity (Schlagetter and Kozel 1990), complement activation (Shapiro et al. 2002), NK cell-mediated killing (Nabavi and Murphy 1986), and proinflammatory responses (Feldmesser et al. 2002). Thus, for both *Candida* and *Cryptococcus*, a variety of monoclonal antibodies have been developed (Ullmer et al. 1993; Beno et al. 1995) but only very few against *Aspergillus* (Torosantucci et al. 2009). In most of these studies, it is clear that administration of mAbs produces a variety of antifungal effects including opsonization (Rachini et al. 2007), inhibition of cell growth (Rodrigues et al. 2000), reduction of capsule/cell wall thickness (Rachini et al. 2007), regulating cytokine production and interplay (Beenhouwer et al. 2001), and in general triggering inflammation (Rivera et al. 2005). But, patient response to mAb administration remains inconsistent since their function is affected by dose and immune parameters of the patient (Casadevall and Pirofski 2006). In addition, the actual antigen load can also influence antibody activity (Casadevall and Pirofski 2003). Under these circumstances, studies have started to show the importance of idiotypic antibodies, especially against *Candida* (Van De Veerdonk et al. 2010), primarily because these produce highly specific immunotherapy and affect the so-called "natural antibody" network.

6.5.4 Transfusion of Leukocytes

Opportunistic infection by fungal pathogens such as *Candida*, *Aspergillus*, and *Cryptococcus* is high in the case of immune-compromised patients, including those with neutropenia. The use of blood transfusion, especially using leukocytes, represents an effective strategy for immunotherapy. In spite of early failures, leukocyte transfusion is gaining importance now due to advances in our understanding of leukocyte development and function. Thus, for example, use of CSFs along with leukocyte transfusion increases the effectiveness of immunotherapy (Casadevall 1995; Deepe 1997). Indeed, clinical trials have shown that the use of G-CSF-pretreated leukocytes for transfusion is very effective against *Candida* (Condon et al. 1996) and *Aspergillus* (Cisalpino et al. 1996) infections. In fact phase 1/2 trial using G-CSF primer leukocytes has been reported to be effective (Cox et al. 1988). Bozza et al. (2004) have reported that transfer of antigen-primed dendritic cells results in enhanced adaptive immune response against *Candida*. In vitro priming of leukocytes, by appropriate cytokine or fungal antigen, and subsequent transfer of the patient represent a simple strategy for antifungal immunotherapy (Bacci et al. 2002). It is recognized that neutrophils are effective against disseminated candidiasis, whereas cell-mediated immunity is triggered against mucosal candidiasis (Deepe 1997). In the case of *Aspergillus* infection of mononuclear cells (Deepe 1994) and cryptococcosis, macrophages (Rohatgi and Pirofski 2015) appear to be important. Tramsen et al. (2007) have shown the effectiveness of anti-*Candida* T cells after transfusion, especially against the hyphal forms and synergize with neutrophils to restrict *Candida* infection (Tramsen et al. 2007). In the case of cryptococcosis, Th1 responses are vital (Hoag et al. 1997) and higher levels of Th1 cytokines such as TNF and IFN are associated with increased fungal clearance (Milam et al. 2007; Wormley et al. 2007). In addition to this, a recent work by Muller et al. (2007) has demonstrated the importance of IL-17 and the associated Th-17 response in immune modulation of *C. neoformans*-infected mice. Moreover, such adoptive transfer of leukocyte also has the added advantage of increasing the activity of specific components of either innate or adaptive immune system, resulting in a robust response. Thus, it is clear that transfusion of a specific type of leukocyte after priming, based on the nature of fungal infection, could be an important immunotherapy strategy in the case of immune-compromised patients.

6.5.5 Vaccines

Antigens from fungus have been successfully isolated and have been used as vaccine candidates. An excellent review by Casadevall and Pirofski (2012) shows that some of these antigens include β-glucan (Torosantucci et al. 2009) and cell wall glycoprotein (Chaturvedi et al. 2005) from *A. fumigatus*, Hsp 90 (Matthews et al. 1991), aspartyl proteinase (Sandini et al. 2011), phosphoglycerate kinase (Calcedo et al. 2011) from *C. albicans* and melanin (Rosas et al. 2001), glucosylceramide (Rodrigues et al. 2000), β-glucan (Torosantucci et al. 2009), and glucuronoxylomannan (Beenhouwer et al. 2007) from *C. neoformans*. These studies have shown that most of these antigens elicit excellent antibody-mediated immunity in the host and thus are quite successful (for more information, please refer to specific reviews on antifungal vaccines). However, a successful vaccination requires intact cellular components and thus might be an issue in the case of immune-compromised individuals. However, vaccination strategy has worked well in murine models of both systemic and mucosal candidiasis (Torosantucci et al. 2005), using diphtheria toxoid CRM197 and laminarin conjugate. Similarly, vaccine conjugates containing mannan too have been successful against disseminated candidiasis and vaginal infection by *Candida* in mice (Han et al. 1999). The use of inert antigens from the fungal pathogens appears to be preferred since it is safer when compared to attenuated or live fungus. But, such a strategy may not be very effective in terms of the antigenicity or purity of the antigen preparation. Under these circumstances recombinant technology provides an excellent opportunity to precisely select and express the desired epitope to be used as an effective vaccine. Nevertheless, the use of adjuvants, such as cytokines, has opened up huge possibilities in antifungal vaccine development (Deepe 1997), which allows us to modulate the cellular response postvaccination. Another way forward is DNA vaccines, as has been shown for virulent bacterial infections such as *Mycobacterium* (Tascon et al. 1996). But antifungal DNA vaccines are yet to be reported for major fungal pathogens. In addition, extensive studies are needed to analyze antifungal vaccines in humans to better understand the mechanisms involved and the therapeutic effectiveness reached. This will enable us to develop an effective antifungal vaccination for use in the general population.

References

Afonso LCC, Scharton TM, Viera LQ, Wysocka M, Trinchieri G, Scott P (1994) The adjuvant effect of interleukin-12 in a vaccine against Leishmania major. Science 263:235–237

Aguado JM, Vázquez L, Fernández-Ruiz M, Villaescusa T, Ruiz-Camps I, Barba P, Silva JT, Batlle M, Solano C, Gallardo D, Heras I, Polo M, Varela R, Vallejo C, Olave T, López-Jiménez J, Rovira M, Parody R, Cuenca-Estrella M (2015) Serum galactomannan versus a combination of Galactomannan and Polymerase chain reaction – based *Aspergillus* DNA detection for early therapy of invasive aspergillosis in high-risk hematological patients: a randomized controlled trial. Clin Infect Dis 60(3):405–414

Ahmad A, Khan A, Khan LA, Manzoor N (2010a) In vitro synergy of eugenol and methyleugenol with fluconazole against clinical *Candida* isolates. J Med Microbiol 59(10):1178–1184

Ahmad I, Khan MSA, Zahin M, Owais M, Shahid M, Mehmood Z, Pant AB et al (2010b) Combinational antifungal therapy and recent trends in drug discovery. In: Ahmad I (ed) Combating fungal infections. Springer-Verlag, Berlin. doi:10.1007/978-3-642-12173-9_10

Akira S, Uematsu S, Takeuchi O (2006) Pathogen recognition and innate immunity. Cell 124:783–801

Albuquerque P, Casadevall A (2012) Quorum sensing in fungi – a review. Med Mycol 50(4):337–345

Alexander BD, Johnson MD, Pfeiffer CD, Jiménez-Ortigosa C, Catania J, Booker R, Castanheira M, Messer SA, Perlin DS, Pfaller MA (2013) Increasing echinocandin resistance in *Candida glabrata*: clinical failure correlates with presence of FKS mutations and elevated minimum inhibitory concentrations. Clin Infect Dis 56(12):1724–1732

Allen UD (2010) Antifungal agents for the treatment of systemic fungal infections in children. Paediatr Child Health 15(9):603

Allendoerfer R, Maresca B, Deepe GS (1996) Cellular immune responses to recombinant heat shock protein 70 from Histoplasma capsulatum. Infect Immun 64:4123–4128

Amber K, Aijaz A, Immaculata X, Luqman KA, Nikhat M (2010) Anticandidal effect of Ocimum sanctum essential oil and its synergy with fluconazole and ketoconazole. Phytomedicine 17(12):921–925

Andes DR, Safdar N, Baddley JW, Playford G, Reboli AC, Rex JH, Sobel JD, Pappas PG, Kullberg BJ (2012) Impact of treatment strategy on outcomes in patients with candidemia and other forms of invasive candidiasis: a patient-level quantitative review of randomized trials. Clin Infect Dis 54:1110–1122, cis021

Angela Bacci*, Claudia Montagnoli*, Katia Perruccio†

Antachopoulos C, Roilides E (2005) Cytokine and fungal infections. Br J Haematol 129:583–596

Antachopoulos C, Walsh TJ (2012) Immunotherapy of *Cryptococcus* infections. Clin Microbiol Infect 18:126–133

Ascher SB, Smith B, Watt K, Benjamin DK, Cohen-Wolkowiez M, Clark RH, Benjamin DK, Moran C (2012) Antifungal therapy and outcomes in infants with invasive *Candida* infections. Pediatr Infect Dis J 31(5):439

Bacci A, Montagnoli C, Perruccio K, Bozza S, Gaziano R, Pitzurra L, Velardi A, D'Ostiani CF, Cutler JE, Romani L (2002) Dendritic cells pulsed with fungal RNA induce protective immunity to *Candida albicans* in hematopoetic transplantation. J Immunol 168:2904–2913

Bal AM (2010) The echinocandins: three useful choices or three too many? Int J Antimicrob Agents 35(1):13–18

Baltch AL, Bopp LH, Smith RP, Ritz WJ, Carlyn CJ, Michelsen PB (2005) Effects of voriconazole, granulocyte-macrophage colony-stimulating factor, and interferon gamma on intracellular fluconazole-resistant *Candida glabrata* and *Candida krusei* in human monocyte-derived macrophages. Diagn Microbiol Infect Dis 52:299–304

Barchiesi F, Di Francesco LF, Compagnucci P, Arzeni D, Giacometti A, Scalise G (1998) In-vitro interaction of terbinafine with amphotericin B, fluconazole and itraconazole against clinical isolates of *Candida albicans*. J Antimicrob Chemother 41(1):59–65

Barchiesi F, Mazzocato S, Mazzanti S, Gesuita R, Skrami E, Fiorentini A, Singh N (2015) Invasive aspergillosis in liver transplant recipients: epidemiology, clinical characteristics, treatment, and outcomes in 116 cases. Liver Transpl 21(2):204–212

Bartizal K, Gill CJ, Abruzzo GK, Flattery AM, Kong L, Scott PM, Smith JG, Leighton CE, Bouffard A, Dropinski JF, Balkovec J (1997) In vitro preclinical evaluation studies with the echinocandin antifungal MK-0991 (L-743,872). Antimicrob Agents Chemother 41:2326–2332

Beenhouwer DO, Shapiro S, Feldmesser M, Casadevall A, Scharff MD (2001) Both Th1 and Th2 cytokines affect the ability of monoclonal antibodies to protect mice against *Cryptococcus neoformans*. Infect Immun 69:6445–6455

Beenhouwer DO, Yoo EM, Lai CW, Rocha MA, Morrison SL (2007) Human immunoglobulin G2 (IgG2) and IgG4, but not IgG1 or IgG3, protect mice against *Cryptococcus neoformans* infection. Infect Immun 75:1424–1435

Beggs WH, Sarosi GA (1982) Further evidence for sequential action of amphotericin B and 5-fluorocytosine against *Candida albicans*. Chemotherapy 28:341–344

Bekker LG, Freeman S, Murray PJ, Ryffel B, Kaplan G (2001) TNF-α controls intracellular mycobacterial growth by both inducible nitric oxide synthase-dependent and inducible nitric oxide synthase-independent pathways. J Immunol 166:6728–6734

Beno DW, Stover AG, Mathews HL (1995) Growth inhibition of *Candida albicans* hyphae by CD81 lymphocytes. J Immunol 154:5273–5281

Bink A, Kucharíková S, Neirinck B, Vleugels J, Van Dijck P, Cammue BP, Thevissen K (2012) The nonsteroidal antiinflammatory drug diclofenac potentiates the in vivo activity of caspofungin against *Candida albicans* biofilms. J Infect Dis 206(11):1790–1797

Bisson GP, Molefi M, Bellamy S, Thakur R, Steenhoff A, Tamuhla N, Rantleru T, Tsimako I, Gluckman S, Ravimohan S, Weissman D, Tebas P (2013) Early versus delayed antiretroviral therapy and cerebrospinal fluid fungal clearance in adults with HIV and cryptococcal meningitis. Clin Infect Dis 56(8):1165–1173

Bodasing N, Seaton RA, Shankland GS, Pithie A (2002) Gamma-interferon treatment for resistant oropharyngeal candidiasis in an HIV-positive patient. J Antimicrob Chemother 50:765–766

Bodey GP (1994) The potential role of granulocyte-macrophage colony stimulating factor in therapy of fungal infections: a commentary. Eur J Clin Microbiol Infect Dis 13:363–366

Bodey GP, Anaissie E, Gutterman J, Vadhan-Raj S (1993) Role of granulocyte macrophage colony-stimulating factor as adjuvant therapy for fungal infections in patients with cancer. Clin Infect Dis 17:705–707

Bosisio D, Polentarutti N, Sironi M, Bernasconi S, Miyake K, Webb GR, Martin MU, Mantovani A, Muzio M (2002) Stimulation of toll-like receptor 4 expression in human mononuclear phagocytes by interferon-γ: a molecular basis for priming and synergism with bacterial lipopolysaccharide. Blood 99:3427–3431

Bowyer P, Moore CB, Rautemaa R, Denning DW, Richardson MD (2011) Azole antifungal resistance today: focus on Aspergillus. Curr Infect Dis Rep 13 (6):485–491

Bozza S, Montagnoli C, Gaziano R, Rossi G, Nkwanyuo G, Bellocchio S, Romani L (2004) Dendritic cell-based vaccination against opportunistic fungi. Vaccine 22:857–864

Braff MH, Bardan A, Nizet V, Gallo RL (2005) Cutaneous defense mechanisms by antimicrobial peptides. J Invest Dermatol 125:9–13

Brena S, Omaetxebarria MJ, Elguezabal N, Cabezas J, Moragues MD, Ponton J (2007) Fungicidal monoclonal antibody C7 binds to *Candida albicans* Als3. Infect Immun 75:3680–3682

Brena S, Cabezas-Olcoz J, Moragues MD, de Fernandez LI, Dominguez A, Quindos G, Ponton J (2011) Fungicidal monoclonal antibody C7 interferes with iron acquisition in *Candida albicans*. Antimicrob Agents Chemother 55:3156–3163

Brummer E, Stevens DA (1987) Activation of pulmonary macrophages for fungicidal activity by gamma-interferon or lymphokines. Clin Exp Immunol 70:520–528

Bugli F, Posteraro B, Papi M, Torelli R, Maiorana A, Sterbini FP, Posterarod P, Sanguinettia M, De Spirito M (2013) In vitro interaction between alginate lyase and amphotericin B against Aspergillus fumigatus biofilm determined by different methods. Antimicrob Agents Chemother 57(3):1275–1282

Calcedo R, Ramirez-Garcia A, Abad A, Rementeria A, Pontón J, Hernando FL (2011) Phosphoglycerate kinase and fructose bisphosphate aldolase of *Candida albicans* as new antigens recognized by human salivary IgA. Rev Iberoam Micol 29:172–174

Casadevall A (1995) Antibody immunity and invasive fungal infections. Infect Immun 63:4211–4218

Casadevall A, Pirofski LA (2003) Antibody-mediated regulation of cellular immunity and the inflammatory response. Trends Immunol 24:474–478

Casadevall A, Pirofski LA (2006) A reappraisal of humoral immunity based on mechanisms of antibody-mediated protection against intracellular pathogens. Adv Immunol 91:1–44

Casadevall A, Pirofski LA (2012) Immunoglobulin in defense, pathogenesis and therapy of fungal diseases. Cell Host Microbe 11:447–456

Cateau E, Berjeaud JM, Imbert C (2011) Possible role of azole and echinocandin lock solutions in the control of *Candida* biofilms associated with silicone. Int J Antimicrob Agents 37(4):380–384

Centre for Disease Control (CDC) (2014) http://www.cdc.gov/fungal/diseases/candidiasis/genital/index.html

Chaturvedi AK, Kavishwar A, Shiva Keshava GB, Shukla PK (2005) Monoclonal immunoglobulin G1 directed against *Aspergillus fumigatus* cell wall glycoprotein protects against experimental murine aspergillosis. Clin Diagn Lab Immunol 12:1063–1068

Chatzimoschou A, Katragkou A, Simitsopoulou M, Antachopoulos C, Georgiadou E, Walsh TJ, Roilides E (2011) Activities of triazole-echinocandin combinations against *Candida* species in biofilms and as planktonic cells. Antimicrob Agents Chemother 55(5):1968–1974

Chavez-Dozal AA, Lown L, Jahng M, Walraven CJ, Lee SA (2014) In vitro analysis of Finasteride activity against *Candida albicans* urinary biofilm formation and filamentation. Antimicrob Agents Chemother 58 (10):5855–5862

Chen SC, Slavin MA, Sorrell TC (2011) Echinocandin antifungal drugs in fungal infections: a comparison. Drugs 71(1):11–41

Chen X, Ren B, Chen M, Liu MX, Ren W, Wang QX, Zhang LX, Yan GY (2014) ASDCD: antifungal synergistic drug combination database. PLoS One 9(1): e86499

Cisalpino CS, Puccia R, Yamauchi LM, Cano MI, da Silva JF, Travassos LR (1996) Cloning, characterization, and epitope expression of the major diagnostic antigen of Paracoccidioides brasiliensis. J Biol Chem 271:4553–4560

Condon C, Watkins SC, Celluzzi CM, Thompson K, Falo LD (1996) DNA-based immunization by in vivo transfection of dendritic cells. Nat Med 2:1122–1127

Cornely OA, Pappas PG, Young JAH, Maddison P, Ullmann AJ (2011) Accumulated safety data of micafungin in therapy and prophylaxis in fungal diseases. Expert Opin Drug Saf 10(2):171–183

Cosgrove RF, Beezer AE, Miles RJ (1978) In vitro studies of amphotericin B in combination with the imidazole antifungal compounds clotrimazole and miconazole. J Infect Dis 138:681–685

Cox RA, Kennell W, Boncyk L, Murphy JW (1988) Induction and expression of cell-mediated immune responses in inbred mice infected with Coccidioides immitis. Infect Immun 56:13–17

Currie B, Sanati H, Ibrahim AS, Edwards JE, Casadevall A, Ghannoum MA (1995) Sterol compositions and susceptibilities to amphotericin B of environmental *Cryptococcus neoformans* isolates are changed by murine passage. Antimicrob Agents Chemother 39:1934–1937

Day JN, Chau TTH, Wolbers M, Mai PP, Dung NT, Mai NH, PhuNghia HD, Phong ND, Thai CQ, Thai LH, Chuong LV, Sinh DX, Duong VA, Hoang TN, Campbell JI, Sieu TPM, Baker SG, Chau NVV, Hien TT, Lalloo DG, Farrar JJ (2013) Combination antifungal therapy for cryptococcal meningitis. N Engl J Med 368(14):1291–1302

De Cremera K, Lanckackerc E, Coolsa TL, Baxa M, Bruckera KD, Cosc P, Cammuea BPA, Thevissen K (2015) Artemisinins, new miconazole potentiators resulting in increased activity against *Candida albicans* biofilms. Antimicrob Agents Chemother 59 (1):421–426

Deepe GS (1994) Role of CD81 T cells in host resistance to systemic infection with Histoplasma capsulatum in mice. J Immunol 152:3491–3500

Deepe GS (1997) Prospects for the development of fungal vaccines. Clin Microbiol Rev 10:585–596

Delattina N, De Bruckera K, Craikb DJ, Chenevalb O, Fröhlichc M, Veberc M, Girandonc L, Davisd TR, Weeksd AE, Kumamotod CA, Cose P, Coenyef T, Conincka BD, Cammuea BPA, Thevissen K (2014) Plant-derived decapeptide OSIP108 interferes with *Candida albicans* biofilm formation without affecting cell viability. Antimicrob Agents Chemother 58 (5):2647–2656

Denning DW, Hope WW (2010) Therapy for fungal diseases: opportunities and priorities. Trends Microbiol 18(5):195–204

Dignani MC, Rex JH, Chan KW, Dow G, deMagalhaes-Silverman M, Maddox AM, Walsh T, Anaissie E (2005) Immunomodulation with interferon-gamma and colony-stimulating factors for refractory fungal infections in patients with leukemia. Cancer 104:199–204

Djeu JY, Blanchard DK, Halkias D, Friedman H (1986) Growth inhibition of *Candida albicans* by human polymorphonuclear neutrophils: activation by interferon-g and tumor necrosis factor. J Immunol 137:2980–2984

Feldmesser M, Mednick A, Casadevall A (2002) Antibody-mediated protection in murine *Cryptococcus neoformans* infection is associated with subtle pleotrophic effects on the cytokine and leukocyte response. Infect Immun 70:1571–1580

Fine DM, Atta MG (2007) Review: kidney disease in the HIV-infected patient. AIDS Patient Care STDs 21 (11):813–824

Gallin JI, Farber JM, Holland SM, Nutman TB (1995) Interferon-gamma in the management of infectious diseases. Ann Intern Med 123:216–222

Gamaletsou MN, Sipsas NV, Kontoyiannis DP, Tsiakalos A, Kontos AN, Stefanou I, Kordossis T (2012) Successful salvage therapy of refractory HIV-related cryptococcal meningitis with the combination of liposomal amphotericin B, voriconazole, and recombinant interferon-γ. Diagn Microbiol Infect Dis 74(4):409–411

Garbati MA, Alasmari FA, Al-Tannir MA, Tleyjeh IM (2012) The role of combination antifungal therapy in the treatment of invasive aspergillosis: a systematic review. Int J Infect Dis 16(2):e76–e81

Gaviria JM, Van Burik JA, Dale DC, Root RK, Liles WC (1999) Modulation of neutrophil-mediated activity against the pseudohyphal form of *Candida albicans* by granulocyte colony-stimulating factor (G-CSF) administered in vivo. J Infect Dis 179:1301–1304

Gazzoni AF, Capilla J, Mayayo E, Guarro J (2012) Efficacy of intrathecal administration of liposomal amphotericin B combined with voriconazole in a murine model of cryptococcal meningitis. Int J Antimicrob Agents 39(3):223–227

Giudice PL, Campo S, De Santis R, Salvatori G (2012) Effect of PTX3 and voriconazole combination in a rat model of invasive pulmonary aspergillosis. Antimicrob Agents Chemother 56(12):6400–6402

Gomez FJ, Allendoerfer R, Deepe GS (1995) Vaccination with recombinant heat shock protein 60 from Histoplasma capsulatum protects mice against pulmonary histoplasmosis. Infect Immun 63:2587–2595

Gonzalez CE, Lyman CA, Lee S, del Guercio C, Roilides E, Bacher J, Gehrt A, Feuerstein E, Tsokos M, Walsh TJ (2001) Recombinant human macrophage colony-stimulating factor augments pulmonary host defences against *Aspergillus fumigatus*. Cytokine 15:87–95

Govender NP, Meintjes G, Bicanic T, Dawood H, Harrison TS, Jarvis JN, Karstaedt AS, Maartens G, McCarthy KM, Rabie H, Variava E, Venter WDF, Boulware DR, Chiller T, Meya DB, Scriven J (2013) Guideline for the prevention, diagnosis and management of cryptococcal meningitis among HIV-infected persons: 2013 update. S Afr J HIV Med 14(2):76–86

Groll AH, Shah PM, Mentzel C, Schneider M, Just-Nuebling G, Huebner K (1996) Trends in the postmortem epidemiology of invasive fungal infections at a university hospital. J Infect 33:23–32

Han Y, Cutler JE (1995) Antibody response that protects against disseminated candidiasis. Infect Immun 63:2714–2719

Han Y, Ulrich MA, Cutler JE (1999) *Candida albicans* mannan extract-protein conjugates induce a protective immune response against experimental candidiasis. J Infect Dis 179:1477–1484

Hancock RE, Diamond G (2000) The role of cationic antimicrobial peptides in innate host defences. Trends Microbiol 8(9):402–410

Hoag KA, Lipscomb MF, Izzo AA, Street NE (1997) IL-12 and IFN-gamma are required for initiating the protective Th1 response to pulmonary cryptococcosis in resistant C.B-17 mice. Am J Respir Cell Mol Biol 17:733–739

http://www.cdc.gov/fungal/diseases/candidiasis/thrush/index.html

Hubel K, Dale DC, Liles WC (2002) Therapeutic use of cytokines to modulate phagocyte function for the treatment of infectious diseases: current status of granulocyte colony-stimulating factor, granulocyte-macrophage colony-stimulating factor, macrophage colony-stimulating factor, and interferon-gamma. J Infect Dis 185:1490–1501

Iwasaki A, Medzhitov R (2015) Control of adaptive immunity by the innate immune system. Nat Immunol 16:343–353

Jacobson MA, Hardy D, Connick E, Watson J, DeBruin M (2000) Phase 1 trial of a single dose of recombinant human interleukin—12 in human immunodeficiency virus—infected patients with 100–500 CD4 Cells/μL. J Infect Dis 182(4):1070–1076

Jeans AR, Howard SJ, Al-Nakeeb Z, Goodwin J, Gregson L, Warn PA, Hope WW (2012) Combination of voriconazole and anidulafungin for treatment of triazole-resistant *Aspergillus fumigatus* in an in vitro model of invasive pulmonary aspergillosis. Antimicrob Agents Chemother 56(10):5180–5185

Jensen SB, Peterson DE (2014) Oral mucosal injury caused by cancer therapies: current management and new frontiers in research. J Oral Pathol Med 43 (2):81–90

Johannsson B, Callaghan JJ (2009) Prosthetic hip infection due to *Cryptococcus neoformans*: case report. Diagn Microbiol Infect Dis 64(1):76–79

Johnson MD, MacDougall C, Ostrosky-Zeichner L, Perfect JR, Rex JH (2004) Combination antifungal therapy. Antimicrob Agents Chemother 48(3):693–715

Katragkou A, McCarthy M, Alexander EL, Antachopoulos C, Meletiadis J, Jabra-Rizk MA, Petraitis V, Roilides E, Walsh TJ (2015) In vitro interactions between farnesol and fluconazole, amphotericin B or micafungin against *Candida albicans* biofilms. J Antimicrob Chemother 70 (2):470–478

Kauffman CA (2001) Fungal infections in older adults. Clin Infect Dis 33:550–555

Kawakami K, Tohyama M, Xie Q, Saito A (1996) IL-12 protects mice against pulmonary and disseminated infection caused by *Cryptococcus neoformans*. Clin Exp Immunol 104:208–214

Kim JS, Holtom P, Vigen C (2011) Reduction of catheter-related bloodstream infections through the use of a central venous line bundle: epidemiologic and economic consequences. Am J Infect Control 39 (8):640–646

Kirkland TN, Fierer J (1985) Genetic control of resistance to Coccidioides immitis: a single gene that is expressed in spleen cells determines resistance. J Immunol 135:548–552

Kirkland TN, Zhu S, Kruse D, Hsu L, Seshan KR, Cole GT (1991) Coccidioides immitis fractions which are antigenic for immune T lymphocytes. Infect Immun 59:3952–3961

Kobayashi C, Hanadate T, Niwa T, Hirano Y, Yoshiyasu T, So M, Matsui K (2015) Safety and efficacy of micafungin for prophylaxis against invasive fungal infections in Japanese patients undergoing hematopoietic stem cell transplantation: results of a post-marketing surveillance study. J Infect Chemother 21:438–443

Kollef M, Micek S, Hampton N, Doherty JA, Kumar A (2012) Septic shock attributed to *Candida* infection: importance of empiric therapy and source control. Clin Infect Dis 54(12):1739–1746

Kong Y, Levine HB (1967) Experimentally induced immunity in the mycoses. Bacteriol Rev 31:35–53

Kontoyiannis DP, Lewis RE (2002) Antifungal drug resistance of pathogenic fungi. Lancet 359 (9312):1135–1144

Ku TSN, Palanisamy SK, Lee SA (2010) Susceptibility of *Candida albicans* biofilms to azithromycin, tigecycline and vancomycin and the interaction between tigecycline and antifungals. Int J Antimicrob Agents 36(5):441–446

Kullberg BJ, Van't Wout JW, Hoogstraten C, Hoogstraten C, van Furth R (1993) Recombinant interferon-g enhances resistance to acute disseminated *Candida albicans* infection in mice. J Infect Dis 168:436–443

Kullberg BJ, Oude Lashof AM, Netea MG (2004) Design of efficacy trials of cytokines in combination with antifungal drugs. Clin Infect Dis 39:S218–S223

Larrick JW, Hirata M, Balint RF, Lee J, Zhong J, Wright SC (1995) Human CAP18: a novel antimicrobial lipopolysaccharide-binding protein. Infect Immun 63:1291

Lars Tramsen1, Olaf Beck1, Friedhelm R. Schuster3,4

Laua AF, Kabira M, Chena SCA, Playford EG, Marriottd DJ, Jonese M, Lipmanf J, McBrydeg E, Gottliebh T, Cheungi W, Seppeltj I, Iredella J, Sorrell TC (2015) *Candida* colonization as a risk marker for invasive candidiasis in mixed medical-surgical intensive care units: development and evaluation of a simple, standard protocol. J Clin Microbiol 53(4):1324–1330

Lepak AJ, Marchillo K, VanHecker J, Andes DR (2013) Impact of in vivo triazole and echinocandin combination therapy for invasive pulmonary aspergillosis: enhanced efficacy against Cyp51 mutant isolates. Antimicrob Agents Chemother 57(11):5438–5447

Liao Z, ZhangGuan X, Zhu Z, Yao X, Yang Y, Jiang Y, Cao Y (2015) Enhancement of the antibiofilm activity of amphotericin B by polyamine biosynthesis inhibitors. Int J Antimicrob Agents (in press)

Liles WC, Huang JE, van Burik JA, Bowden RA, Dale DC (1997) Granulocyte colony stimulating factor administered in vivo augments neutrophil-mediated activity against opportunistic fungal pathogens. J Infect Dis 175:1012–1015

Liu W, Li L, Sun Y, Chen W, Wan Z, Li R, Liu W (2012) The interaction of echinocandin caspofungin with amphotericin B or voriconazole against the Aspergillus biofilms in vitro. Antimicrob Agents Chemother 56:6414–6416, AAC-00687

Liu S, Hou Y, Chen X, Gao Y, Li H, Sun S (2014) Combination of fluconazole with non-antifungal agents: a promising approach to cope with resistant Candida albicans infections and insight into new antifungal agent discovery. Int J Antimicrob Agents 43 (5):395–402

Loreto ÉS, Mario DAN, Denardi LB, Alves SH, Santurio JM (2011) In vitro susceptibility of Pythium insidiosum to macrolides and tetracycline antibiotics. Antimicrob Agents Chemother 55(7):3588–3590

Loyse A, Wilson D, Meintjes G, Jarvis JN, Bicanic T, Bishop L, Rebe K, Williams A, Jaffar S, Bekker LG, Wood R, Harrison TS (2012) Comparison of the early fungicidal activity of high-dose fluconazole, voriconazole, and flucytosine as second-line drugs given in combination with amphotericin B for the treatment of HIV-associated cryptococcal meningitis. Clin Infect Dis 54(1):121–128

Lutz JE, Clemons KV, Stevens DA (2000) Enhancement of antifungal chemotherapy by interferon-gamma in experimental systemic cryptococcosis. J Antimicrob Chemother 46:437–442

Mandal SM, Migliolo L, Franco OL, Ghosh AK (2011) Identification of an antifungal peptide from Trapa natans fruits with inhibitory effects on Candida tropicalis biofilm formation. Peptides 32 (8):1741–1747

Marcil A, Harcus D, Thomas DY, Whiteway M (2002) *Candida albicans* killing by RAW 264.7 mouse macrophage cells: effects of *Candida* genotype, infection ratios, and gamma interferon treatment. Infect Immun 70:6319–6329

Marr KA, Schlamm HT, Herbrecht R, Rottinghaus ST, Bow EJ, Cornely OA, Heinz WJ, Jagannatha S, Koh LP, Kontoyiannis DP, Lee DG, Nucci M, Pappas PG, Slavin MA, Queiroz-Telles F, Selleslag D, Walsh TJ, Wingard JR, Maertens JA (2015) Combination antifungal therapy for invasive aspergillosis: a randomized trial. Ann Intern Med 162(2):81–89

Martinez LR, Casadevall A (2005) Specific antibody can prevent fungal biofilm formation and this effect correlates with protective efficacy. Infect Immun 73:6350–6362

Martinez LR, Fries BC (2010) Fungal biofilms: relevance in the setting of human disease. Curr Fungal Infect Reports 4(4):266–275

Martinez LR, Moussai D, Casadevall A (2004) Antibody to *Cryptococcus neoformans* glucuronoxylomannan inhibits the release of capsular antigen. Infect Immun 72:3674–3679

Matthews RC, Burnie JP, Howat D, Rowland T, Walton F (1991) Autoantibody to heat shock protein 90 can mediate protection against systemic candidosis. Immunology 74:20–24

Mavridou E, Meletiadis J, Rijs A, Mouton JW, Verweij PE (2015) The strength of synergistic interaction between Posaconazole and Caspofungin depends on the underlying azole-resistance mechanism of *Aspergillus fumigatus*. Antimicrob Agents Chemother 59:1738–1744, AAC-04469

McClelland EE, Nicola AM, Prados-Rosales R, Casadevall A (2010) Ab binding alters gene expression in *Cryptococcus neoformans* and directly modulates fungal metabolism. J Clin Invest 120:1355–1361

Medoff G (1983) Antifungal action of rifampin. Rev Infect Dis J 5(suppl. 3):S614–S619

Mencacci A, Spaccapelo R, Del Sero G, Enssle KH, Cassone A, Bistoni F, Romani L (1996) CD41 T-helper-cell responses in mice with low-level *Candida albicans* infection. Infect Immun 64:4907–4917

Mencacci A, del Sero G, Cenci E, d'Ostiani CF, Bacci A, Montagnoli C, Kopf M, Romani L (1998) Endogenous IL-4 is required for development of protective CD4+ T helper type 1 cell responses to *Candida albicans*. J Exp Med 187:307–317

Mendling W (2015) Guideline: vulvovaginal candidosis (AWMF 015/072), S2k (excluding chronic mucocutaneous candidosis). Mycoses 58(1):1–15

Miceli MH, Díaz JA, Lee SA (2011) Emerging opportunistic yeast infections. Lancet Infect Dis 11 (2):142–151

Mihu CN, Kassis C, Ramos ER, Jiang Y, Hachem RY, Raad II (2010) Does combination of lipid formulation of amphotericin B and echinocandins improve outcome of invasive aspergillosis in hematological malignancy patients? Cancer 116(22):5290–5296

Milam JE, Herring-Palmer AC, Pandrangi R, McDonald RA, Huffnagle GB, Toews GB (2007) Modulation of the pulmonary type 2 T-cell response to *Cryptococcus neoformans* by intratracheal delivery of a tumor necrosis factor alpha-expressing adenoviral vector. Infect Immun 75:4951–4958

Morein B, Sundquist B, Hoglund S, Dalsgaard K, Osterhaus A (1984) Iscom, a novel structure for antigenic presentation of membrane proteins from enveloped viruses. Nature 308:457–460

Morrison LA, Lukacher AE, Braciale VL, Fan DP, Braciale TJ (1986) Differences in antigen presentation to MHC class I- and class II-restricted influenza virus-specific cytolytic T lymphocyte clones. J Exp Med 163:903–921

Mowat E, Butcher J, Lang S, Williams C, Ramage G (2007) Development of a simple model for studying the effects of antifungal agents on multicellular communities of Aspergillus fumigatus. J Med Microbiol 56(9):1205–1212

Muchmore HG, Felton FG, Salvin SB, Rhoades ER (1968) Delayed hypersensitivity to cryptococcin in man. Sabouraudia 6:285–288

Muller U, Stenzel W, Kohler G, Werner C, Polte T, Hansen G (2007) IL-13 induces disease-promoting type 2 cytokines, alternatively activated macrophages and allergic inflammation during pulmonary infection of mice with *Cryptococcus neoformans*. J Immunol 179:5367–5377

Muñoz A, Marcos JF, Read ND (2012) Concentration-dependent mechanisms of cell penetration and killing by the de novo designed antifungal hexapeptide PAF26. Mol Microbiol 85(1):89–106

Murphy JW, Hidore MR, Wong SC (1993) Direct interactions of human lymphocytes with the yeast-like organism, *Cryptococcus neoformans*. J Clin Invest 91:1553–1566

Muzoora CK, Kabanda T, Ortu G, Ssentamu J, Hearn P, Mwesigye J, Longleya N, Jarvisb JN, Jaffard S, Harrisonb TS (2012) Short course amphotericin B with high dose fluconazole for HIV-associated cryptococcal meningitis. J Infect 64(1):76–81

Nabavi N, Murphy JW (1986) Antibody-dependent natural killer cell-mediated growth inhibition of *Cryptococcus neoformans*. Infect Immun 51:556–562

Nagaoka I, Hirota S, Niyonsaba F, Hirata M, Adachi Y, Tamura H, Heumann D (2001) Cathelicidin family of antibacterial peptides CAP18 and CAP11 inhibit the expression of TNF-α by blocking the binding of LPS to CD14+ cells. J Immunol 167:3329

Netea MG, Brouwer AE, Hoogendoorn EH, Van der Meer JW, Koolen M, Verweij PE, Kullberg BJ (2004) Two patients with cryptococcal meningitis and idiopathic CD4 lymphopenia: defective cytokine production and reversal by recombinant interferon- gamma therapy. Clin Infect Dis 39:e83–e87

Netea MG, Van der Meer JW, Kullberg BJ (2006) Role of the dual interaction of fungal pathogens with pattern recognition receptors in the activation and modulation of host defence. Clin Microbiol Infect 12:404–409

Nett JE (2014) Future directions for anti-biofilm therapeutics targeting *Candida*. Expert Rev Anti-Infect Ther 12(3):375–382

Nikapitiya C (2012) Bioactive secondary metabolites from marine microbes for drug discovery. Adv Food Nutr Res 65(8):363–387

Nucci M, Anaissie E, Betts RF, Dupont BF, Wu C, Buell DN, Kovanda L, Lortholary O (2010) Early removal of central venous catheter in patients with candidemia does not improve outcome: analysis of 842 patients from 2 randomized clinical trials. Clin Infect Dis 51 (3):295–303

Ogawa T, Matsumoto K, Tsujimoto K, Hishiya N, Yamada Y, Uno K, Kasaharaa K, Maedaa K, Nariob K, Mikasaa K, Morita K (2015) Chronic invasive sinus and intracerebral aspergillosis controlled by combination therapy with micafungin and a daily dose of 400 mg itraconazole oral solution. J Infect Chemother 21(2):134–137

Olson JA, Schwartz JA, Hahka D, Nguyen N, Bunch T, Jensen GM, Adler-Moore JP (2015) Toxicity and efficacy differences between liposomal amphotericin B formulations in uninfected and *Aspergillus fumigatus* infected mice. Med Mycol 53(2):107–118

Owais M, Ansari MA, Ahmad I, Zia Q, Pierard G, Chauhan A et al (2010) Innate immunity in pathogenesis and treatment of dermatomycosis. In: Ahmad I (ed) Combating fungal infections. Springer-Verlag, Berlin. doi:10.1007/978-3-642-12173-9_10

Pachl J, Svoboda P, Jacobs F, Vandewoude K, Hoven BV, Spronk P, Masterson G, Malbrain M, Aoun M, Garbino J, Takala J, Drgona L, Burnie J, Matthews R (2006) A randomized, blinded, multicenter trial of lipid-associated amphotericin B alone versus in combination with an antibody-based inhibitor of heat shock protein 90 in patients with invasive candidiasis. Clin Infect Dis 42(10):1404–1413

Pagano L, Akova M, Dimopoulos G, Herbrecht R, Drgona L, Blijlevens N (2011) Risk assessment and prognostic factors for mould-related diseases in immunocompromised patients. J Antimicrob Chemother 66:i5–i14

Pappas PG, Rex JH, Sobel JD, Filler SG, Dismukes WE, Walsh TJ, Edwards JE (2004) Guidelines for treatment of candidiasis. Clin Infect Dis 38:161–189

Paramythiotou E, Frantzeskaki F, Flevari A, Armaganidis A, Dimopoulos G (2014) Invasive fungal infections in the ICU: how to approach, how to treat. Molecules 19(1):1085–1119

Perfect JR, William ED, Francoise D, David LG, John RG, Richard JH, Thomas SH, RA Larsen RA, Lortholary O, Nguyen MH, Pappas PG, Powderly WG, Singh N, Sobel JD, Sorrell TC (2010) Clinical practice guidelines for the management of cryptococcal disease: 2010 update by the Infectious Diseases Society of America. Clin Infect Dis 50(3):291–322

Pfaller M, Neofytos D, Diekema D, Azie N, Meier-Kriesche HU, Quan SP, Horn D (2012) Epidemiology and outcomes of candidemia in 3648 patients: data from the Prospective Antifungal Therapy (PATH Alliance®) registry, 2004–2008. Diagn Microbiol Infect Dis 74(4):323–331

Pieralisi N, Godoy J, Yamada S, Santana R, Svidzinski T (2014) Oral lesions and colonization by yeasts in hemodialysis patients. J Oral Pathol Med 44(8): 585–590

Polak-Wyss A (1991) Protective effect of human granulocyte colony-stimulating factor (hG-CSF) on *Cryptococcus* and *Aspergillus* infections in normal and immunosuppressed mice. Mycoses 34:205–215

Probst C, Pongratz G, Capellino S, Szeimies RM, Scholmerich J, Fleck M, Salzberger B, Ehrenstein B (2010) Cryptococcosis mimicking cutaneous cellulitis in a patient suffering from rheumatoid arthritis: a case report. BMC Infect Dis 10(1):239

Qureshi MH, Zhang T, Koguchi Y, Nakashima K, Okamura H, Kurimoto M, Kawakami K (1999) Combined effects of IL-12 and IL-18 on the clinical course and local cytokine production in murine pulmonary infection with *Cryptococcus neoformans*. Eur J Immunol 29:643–649

Rachini A, Pietrella D, Lupo P, Torosantucci A, Chiani P, Bromuro C, Proietti C, Bistoni F, Cassone A, Vecchiarelli A (2007) An anti-beta-glucan monoclonal antibody inhibits growth and capsule formation of *Cryptococcus neoformans* in vitro and exerts therapeutic, anticryptococcal activity in vivo. Infect Immun 75:5085–5094

Rajendran R, Mowat E, Jones B, Williams C, Ramage G (2015) Prior in vitro exposure to voriconazole confers resistance to amphotericin B in Aspergillus fumigatus biofilms. Int J Antimicrob Agents 46:342–345

Reardon CC, Kim SJ, Wagner RP, Kornfeld H (1996) Interferon-gamma reduces the capacity of human alveolar macrophages to inhibit growth of *Cryptococcus neoformans* in vitro. Am J Respir Cell Mol Biol 15:711–715

Redmond HP, Shou J, Gallagher HJ, Kelly CJ, Daly JM (1993) Macrophage-dependent Candidacidal mechanisms in the murine system. Comparison of murine Kupffer cell and peritoneal macrophage Candidacidal mechanisms. J Immunol 150:3427–3433

Reedy JL, Filler SG, Heitman J (2010) Elucidating the *Candida albicans* calcineurin signaling cascade controlling stress response and virulence. Fungal Genet Biol 47(2):107–116

Riddell J, Shuman EK (2012) Epidemiology of central nervous system infection. Neuroimaging Clin N Am 22(4):543–556

Riddell LA, Pinching AJ, Hill S, Ng TT, Arbe E, Lapham GP, Ash S, Hillman R, Tchamouroff S, Denning DW, Parkin JM (2001) A phase III study of recombinant human interferon gamma to prevent opportunistic infections in advanced HIV disease. AIDS Res Hum Retroviruses 17:789–797

Rivera J, Zaragoza O, Casadevall A (2005) Antibody-mediated protection against *Cryptococcus neoformans* pulmonary infection is dependent on B cells. Infect Immun 73:1141–1150

Rodrigues ML, Travassos LR, Miranda KR, Franzen AJ, Rozental S, De Souza W, Alviano CS, Barreto-Bergter E (2000) Human antibodies against a purified glucosylceramide from *Cryptococcus neoformans* inhibit cell budding and fungal growth. Infect Immun 68:7049–7060

Rodriguez-Adrian LJ, Grazziutti ML, Rex JH, Anaissie EJ (1998) The potential role of cytokine therapy for fungal infection in patients with cancer: is recovery from neutropenia all that is needed? Clin Infect Dis 26:1270–1278

Rohatgi S, Pirofski LA (2015) Host immunity to Cryptococcus neoformans. Future Microbiol 10:565–581

Romani L (2000) Innate and adaptive immunity in *Candida albicans* infections and saprophytism. J Leukoc Biol 68:175–178

Rosas AL, Nosanchuk JD, Casadevall A (2001) Passive immunization with melanin-binding monoclonal antibodies prolongs survival in mice with lethal *Cryptococcus neoformans* infection. Infect Immun 69:3410–3412

Rossignol T, Kelly B, Dobson C, d'Enfert C (2011) Endocytosis-mediated vacuolar accumulation of the human ApoE apolipoprotein-derived ApoEdpL-W antimicrobial peptide contributes to its antifungal activity in *Candida albicans*. Antimicrob Agents Chemother 55(10):4670–4681

Rowe JM, Andersen JW, Mazza JJ, Bennett JM, Paietta E, Hayes FA, Oette D, Cassileth PA, Stadtmauer EA, Wiernik PH (1995) A randomized, placebo-controlled phase III study of granulocyte–macrophage colony-stimulating factor in adult patients (15 to 70 years of age) with acute myelogenous leukemia: a study of the Eastern Cooperative Oncology Group (E1490). Blood 86:457–462

Sanati H, Ramos CF, Bayer AS, Ghannoum MA (1997) Combination therapy with amphotericin B and fluconazole against invasive candidiasis in neutropenic-mouse and infective-endocarditis rabbit models. Antimicrob Agents Chemother 41(6):1345–1348

Sandini S, La VR, Deaglio S, Malavasi F, Cassone A, De BF (2011) A highly immunogenic recombinant and truncated protein of the secreted aspartic proteases family (rSap2t) of *Candida albicans* as a mucosal anticandidal vaccine. FEMS Immunol Med Microbiol 62:215–224

Sanglard D, Coste A, Ferrari S (2009) Antifungal drug resistance mechanisms in fungal pathogens from the perspective of transcriptional gene regulation. FEMS Yeast Res 9(7):1029–1050

Sardi JCO, Scorzoni L, Bernardi T, Fusco-Almeida AM, Giannini MM (2013) *Candida* species: current epidemiology, pathogenicity, biofilm formation, natural antifungal products and new therapeutic options. J Med Microbiol 62(1):10–24

Schlagetter AM, Kozel TR (1990) Opsonization of *Cryptococcus neoformans* by a family of isotype-switch variant antibodies specific for the capsular polysaccharide. Infect Immun 58:1914–1918

Schlamm HT, Marr KA (2015) Combination therapy for invasive aspergillosis: controversies and conclusions. Curr Fungal Infect Rep 9(2):130–134

Semis R, Nahmias M, Lev S, Frenkel M, Segal E (2015) Evaluation of antifungal combinations of nystatin-intralipid against *Aspergillus terreus* using checkerboard and disk diffusion methods. J Mycol Méd/J Med Mycol 25(1):63–70

Shadomy S, Wagner G, Espinel-Ingroff E, Davis BA (1975) In vitro studies with combinations of 5-fluorocytosine and amphotericin B. Antimicrob Agents Chemother 8:117–121

Shahid M, Tripathi T, Khardori N, Huq A, Ahmad I et al (2010) Cytokine therapy: possible tools in management of fungal infection. In: Ahmad I (ed) Combating fungal infections. Springer, Berlin. doi:10.1007/978-3-642-12173-9_16

Shapiro S, Beenhouwer DO, Feldmesser M, Taborda C, Carroll MC, Casadevall A, Scharff MD (2002) Immunoglobulin G monoclonal antibodies to *Cryptococcus neoformans* protect mice deficient in complement component C3. Infect Immun 70:2598–2604

Silva EG, Paula CR, Dias ALT, Chang MR, da Silva Ruiz L, Gambale V, Prates RA, Ribeiro MS (2011) Combination efficacy of voriconazole and amphotericin B in the experimental disease in immunodeficient mice caused by fluconazole-resistant *Cryptococcus neoformans*. Mycopathologia 171 (4):261–266

Singh R, Shivaprakash MR, Chakrabarti A (2011) Biofilm formation by zygomycetes: quantification, structure and matrix composition. Microbiology 157 (9):2611–2618

Siopi M, Siafakas N, Vourli S, Zerva L, Meletiadis J (2015) Optimization of polyene-azole combination therapy against aspergillosis using an in vitro pharmacokinetic-pharmacodynamic model. Antimicrob Agents Chemother 59:3973–3983, AAC-05035

Sorensen OE, Arnljots K, Cowland JB, Bainton DF, Borregaard N (1997) The human antibacterial cathelicidin, hCAP-18, is synthesized in myelocytes and metamyelocytes and localized to specific granules in neutrophils. Blood 90:2796

Steinbach WJ, Stevens DA (2003) Review of newer antifungal and immunomodulatory strategies for invasive aspergillosis. Clin Infect Dis 37:S157–S187

Tascon RE, Colston MJ, Ragno S, Stavropoulos E, Gregory D, Lowrie D (1996) Vaccination against tuberculosis by DNA injection. Nat Med 2:888–898

Torosantucci A, Bromuro C, Chiani P, De Bernardis F, Berti F, Galli C, Norelli F, Bellucci C, Polonelli L,

Costantino P, Rappuoli R, Cassone A (2005) A novel glyco-conjugate vaccine against fungal pathogens. J Exp Med 202:597–606

Torosantucci A, Chiani P, Bromuro C, De BF, Palma AS, Liu Y, Mignogna G, Maras B, Colone M, Stringaro A, Zamboni S, Feizi T, Cassone A (2009) Protection by anti-beta-glucan antibodies is associated with restricted beta-1,3 glucan binding specificity and inhibition of fungal growth and adherence. PLoS One 4: e5392

Tragiannidis A, Tsoulas C, Groll AH (2015) Invasive candidiasis and Candidaemia in neonates and children: update on current guidelines. Mycoses 58(1):10–21

Tramsen L, Beck O, Schuster FR, Hunfeld KP, Latgé JP, Sarfati J, Röger F, Klingebiel T, Koehl U, Lehrnbeche T (2007) Generation and characterization of anti-*Candida* T cells as potential immunotherapy in patients with *Candida* infection after allogeneic hematopoietic stem-cell transplant. J Infect Dis 196:485–492

Truan G, Epinat JC, Rougeulle C, Cullin C, Pompon D (1994) Cloning and characterization of a yeast cytochrome b5 encoding gene which suppresses ketoconazole hypersensitivity in a NADPH P-450 reductase deficient strain. Gene 142:123–127

Tsiodras S, Samonis G, Boumpas DT (2008) Fungal infections complicating tumor necrosis factor α blockade therapy. Mayo Clin Proc 83:181–194

Ullmer JB, Donnelly JJ, Parker SE, Rhodes GH, Felgner PL, Dwarki VJ, Gromkowski SH, Deck RR, DeWitt CM, Friedman A, Hawe LA, Leander KR, Martinez D, Perry HC, Shiver JW, Montgomery DL, Liu MA (1993) Heterologous protection against influenza by injection of DNA encoding a viral protein. Science 259:1745–1749

Vaidhya SA, Gupta BB, Jha RK, Kumar R (2015) Combination versus Monotherapy for the treatment of HIV associated cryptococcal meningitis. J Clin Diagn Res: JCDR 9(2):14

Valerio M, Vena A, Bouza E, Reiter N, Viale P, Hochreiter M, Giannella M, Muñoz P (2015) How much European prescribing physicians know about invasive fungal infections management? BMC Infect Dis 15(1):80

Van De Veerdonk FL, Netea MG, Joosten LA, Van Der Meer LWM, Kullberg BJ (2010) Novel strategies for the prevention and treatment of *Candida* infections: the potential of immunotherapy. FEMS Microbiol Rev 34:1063–1075

van der Meer JW, Popa C, Netea MG (2005) Side effects of anticytokine strategies. Neth J Med 63:78–80

Van Wetering S, Mannesse-Lazeroms SP, Van Sterkenburg MA, Daha MR, Dijkman JH, Hiemstra PS (1997) Effect of defensins on interleukin-8 synthesis in airway epithelial cells. Am J Physiol 272:L888

Vazquez JA, Hidalgo JA, De Bono S (2000) Use of sargramostim (rh-GM-CSF) as adjunctive treatment of fluconazole-refractory oropharyngeal candidiasis in patients with AIDS: a pilot study. HIV Clin Trials 1:23–29

Walsh TJ, Gamaletsou MN (2013) Treatment of fungal disease in the setting of neutropenia. ASH Educ Program Book 1:423–427

Wang Y, Damu GL, Lv JS, Geng RX, Yang DC, Zhou CH (2012) Design, synthesis and evaluation of clinafloxacin triazole hybrids as a new type of antibacterial and antifungal agents. Bioorg Med Chem Lett 22(17):5363–5366

Welte K, Gabrilove J, Bronchud MH, Platzer E, Morstyn G (1996) Filgrastim (r-metHuG-CSF): the first 10 years. Blood 88:1907–1929

Wiederhold NP, Najvar LK, Bocanegra R, Kirkpatrick WR, Sorrell TC, Patterson TF (2013) Limited activity of miltefosine in murine models of cryptococcal meningoencephalitis and disseminated cryptococcosis. Antimicrob Agents Chemother 57(2):745–750

Wormley FL, Perfect JR, Steele C, Cox GM (2007) Protection against cryptococcosis by using a murine gamma interferon-producing Cryptococcus neoformans strain. Infect Immun 75(3):1453–1462

Yao ZW, Lu X, Shen C, Lin DF (2014) Comparison of flucytosine and fluconazole combined with amphotericin B for the treatment of HIV-associated cryptococcal meningitis: a systematic review and meta-analysis. Eur J Clin Microbiol Infect Dis 33(8):1339–1344

Zhang M, Su X, Sun WK, Chen F, Xu XY, Shi Y (2014) Efficacy of the combination of voriconazole and caspofungin in experimental pulmonary Aspergillosis by different *Aspergillus* species. Mycopathologia 177 (1–2):11–18

Nanocarriers of Antifungal Agents

7

Sevgi Güngör and M. Sedef Erdal

Abstract

Fungal infections are still one of the major healthcare problems worldwide. Particularly, the treatment of invasive fungal infections has become more difficult in high-risk patient groups such as AIDS patients, organ transplant recipients, and cancer patients undergoing immunosuppressive chemotherapy. Superficial fungal infections have a less tendency to develop to systemic infection, but they could be serious due to the inefficiency of the treatment. In addition, the controlling of drug release is also important to minimize systemic absorption and to decrease the toxicity of these agents. Therefore, delivery of antifungal agents used for the treatment of superficial and systemic infections has a great impact in terms of both therapeutic aspect and safety. The novel nano-sized drug carriers play a key role to overcome the limitations of conventional dosage forms to improve the efficacy and to ensure safety of the treatment. In the present chapter, the limitations of conventional drug carriers of fungal agents are concisely emphasized, and the impact of nano-sized carriers in delivering antifungal agents has been addressed, paying particular attention toward the studies performed in recent literature.

7.1 Introduction

Nanotechnology has gain great interest in many areas of scientific research during last decades due to its potential benefits. The pharmaceutical field has also taken the advantage of nanotechnology, and the advances in that area have led to the optimization of different types of nano-sized materials for various biomedical applications. In this context, nano-drug delivery systems are increasingly being explored to deliver drugs, cosmetic compounds, or vaccines or for therapeutic purposes. Nano-drug carriers have superiorities of improving aqueous solubility of hydrophobic drugs, controlling/prolonging the drug release, decreasing/minimizing side effects of the drugs such as irritation, keeping the drugs away from environmental factors, and targeting of drugs

S. Güngör (✉) • M.S. Erdal
Faculty of Pharmacy, Department of Pharmaceutical Technology, Istanbul University, 34116 Istanbul, Turkey
e-mail: sgungor@istanbul.edu.tr

A. Basak et al. (eds.), *Recent Trends in Antifungal Agents and Antifungal Therapy*,
DOI 10.1007/978-81-322-2782-3_7

into specific tissues. Therefore, a great range of nanomaterials including colloidal carriers, vesicles, nanocrystalline, and polymeric nanoparticles has been widely investigated, while some of them are already commercially available; there are also some remarkable data which promise a commercialization in the future.

Fungal infections are widely observed and remain as one of the major healthcare problems worldwide. They can be classified as superficial fungal infections which affect the skin, nails, hair, or mucous membranes and systemic infections affecting the whole body. The prevalence and the incidence of either systemic or superficial fungal infections have been increased in the last three decades. Particularly, the treatment of invasive fungal infections has become more difficult in high-risk patient groups such as AIDS patients, organ transplant recipients, and cancer patients undergoing immunosuppressive chemotherapy. Superficial fungal infections have a less tendency to develop to systemic infection, but they could be serious due to the inefficiency of the treatment (Zhang et al. 2007; Kumar et al. 2014).

Overall, it has a great impact to assess the function of the pathogenetic fungi that cause disease in order to choose the most appropriate therapy and antifungal agent in the treatment of fungal infections. Besides, the delivery system of these agents may play a major role for penetration of drugs into target sites of the disease in terms of improving therapeutic aspect and diminishing or preventing the possible adverse effects (Zhang et al. 2007; Kaur and Kakkar 2010).

The antifungal drugs are administered by topical, oral, or parenteral routes. Topical therapy is the most common approach to the management of mucosal and cutaneous fungal infections because of its advantages including targeting of drugs to the site of disease and reducing of systemic side-effect risk of drugs. The conventional dosage formulations including cream, powder, gels, etc. are widely used in the topical treatment of superficial fungal infections. Some adverse reactions may occur at the site of application, but are less than those of systemic agents. However, in some cases topical therapy could lead to side effects like redness, burning, or swelling, which decrease patient compliance. Furthermore, the topical treatment of deep-seated fungal infections like invasive *candidiasis* could be inefficient due to the low penetration capacity of the drug into the target site and inadequate deposition in the skin, resulting in low topical bioavailability. As well as the barrier characteristics of the skin, the physicochemical properties of the drugs including lipophilic character and low aqueous solubility affect the penetration of compounds across the membranes. Noticeably, almost all of antifungal agents currently available in the market are highly lipophilic, and they have poor aqueous solubility, which lead to low release rate of drugs into the skin (Zhang et al. 2007; Kaur and Kakkar 2010; Güngör et al. 2013). Likewise, in case of fungal infection of nail and eye, conventional formulations may show problem of less bioavailability.

Oral treatment of fungal infections is preferred when the disease shows extensive harmful characteristics. But, the main drawbacks of the oral therapy are the serious adverse effects like hepatotoxicity and drug interactions (Kumar et al. 2014). The parenteral administration of antifungal agents in the treatment of systemic infections may also lead to severe toxicity reactions. For example, amphotericin B, a broad antimycotic agent, is widely used in the very serious systemic infections. However, this drug has limitations due to its severe nephrotoxicity, which may lead to kidney failure. Therefore, liposome formulation of amphotericin B for parenteral administration has been first marketed under the name AmBisome in 1990. Then, other products containing lipidic carriers of amphotericin B have come into market in the commercial names of Abelcet® (lipid complex) and Amphotec® (colloidal lipid dispersion). The toxicity of the antifungal agent has been reduced with these liposome-based and submicronic colloidal systems of amphotericin B (Torrado et al. 2008; Kaur and Kakkar 2010). In this context, the formulation of antifungal agents into a suitable delivery system is crucial for either topical or systemic treatment of fungal infections.

The solubility and permeability of drugs, which are affecting parameters for the enhancement of therapeutic efficacy, should be taken into account. The controlling of drug release also has a great impact to minimize systemic absorption and to decrease the toxicity of these agents. Therefore, the development of novel drug delivery systems is a great challenge to overcome the limitations of conventional dosage forms in terms of improving the efficacy and safety of the treatment. For this aim, different nano-sized drug carriers including liposomes, niosomes, ethosomes, microemulsions, nanoparticles, microspheres, and micelles are intensively investigated. Further, they also provide better penetration of drugs into the deep skin to treat invasive fungal infections. Thus, nano-sized carriers could be considered as a promising tool to improve the efficacy and safety of the treatment of fungal infections. Some comprehensive reviews on novel drug delivery approaches for the treatment of fungal infections have already been published (Kaur and Kakkar 2010; Carneiro et al. 2012; Güngör et al. 2013; Madhu et al. 2012). In the present chapter, the limitations of conventional drug carriers of fungal agents are concisely emphasized and the impact of nano-sized carriers in delivering antifungal agents has been addressed paying particular attention toward the studies performed in recent literature.

7.2 Fungal Infections and Current Treatment Options

Fungal infections of the hair, skin, and nails form the most numerous and widespread group of all mycoses. Superficial mycoses are most often caused by dermatophytes and yeasts. Dermatophytes require keratin for growth and therefore cause diseases in body sites where keratin is present. These sites include the skin surface, hair, and nail. Treatment of *dermatophytosis* is often dependent on the clinical setting. Single cutaneous lesions can be adequately treated with a topical antifungal drug, but the treatment of some fungal infections on scalp and nail is often ineffective, and systemic therapy is usually needed to cure these disorders (Ghannoum et al. 2013; Zhang et al. 2007). Yeasts are part of human normal microflora, and invasive infections are observed due to disruption of the barrier or impairment of immune functions. Invasive superficial cutaneous infections include mostly the yeast infections with *Candida* as the absolutely dominating pathogen (Arendrup 2013; Zhang et al. 2007). *Candida albicans* (*C. albicans*) can invade virtually every part of the body including deep organs and tissues as well as superficial sites like skin, nails, and mucous membranes (Randhawa et al. 2015). Most infections due to yeasts respond well to topical treatment, whereas extensive lesions often require oral treatment (Kelly 2012).

Fungal infections of the oral and vaginal mucosa and eyes are also predominantly associated with *C. albicans* (Melkoumov et al. 2015). A number of effective antifungal agents administered either topically or systemically are available for the management of oral candidiasis. However, antifungal drug resistance and poor patient compliance can impair therapy and lead to chronic relapse of the infections (Munoz et al. 2010). Fungal diseases of vulva and vagina are also superficial infections predominantly caused by *C. albicans* (vulvovaginal candidiasis) (Kasper et al. 2015; Stock 2010). Most cases are accessible to the treatment with local and systemic antifungal agents, and different conventional vaginal formulations (creams, gels, suppositories, powder, ointment, etc.) are available. But these conventional formulations have limited efficacy due to self-cleansing action of vagina, resulting in diminishing residence time of drug on vaginal epithelium (Johal et al. 2014; Stock 2010). Fungal eye infections are less common as compared to infections caused by bacteria or viruses. Various regions of the eye that are attacked by the fungi are cornea and the interior segment of the eye (Kumar et al. 2014).

The main classes of systemic and topical antifungal treatment modalities include the allylamines, the azoles, the polyenes, and the echinocandins (Munoz et al. 2010; Zhang et al. 2007). A list of antifungal compounds which are approved or under investigation and administrated via oral, intravenous or topical route is given in Table 7.1. The conducted studies using nanocarrier approaches for effective delivery and targeting of these agents are also included.

7.2.1 Allylamine Antifungal Agents

Allylamine antifungals act through inhibiting of an essential enzyme (squalene epoxidase) in the ergosterol biosynthesis pathway of cell membrane formation of fungus. The increased cellular permeability and growth inhibition occurred due to alterations in fungal cellular membranes (Güngör et al. 2013; Kelly 2012).

Terbinafine has a broad-spectrum activity against yeast, fungi, molds, and dermatophytes and is indicated for both oral and topical treatment of mycoses. Topical formulations of terbinafine, such as dermal spray, cream, and film-forming solution, are also effective against tinea versicolor, an infection caused by *Malassezia* sp. Terbinafine is considered the second-line systemic therapy for tinea capitis in children, whereas griseofulvin is considered the first-line systemic therapy (Kelly 2012). Naftifine has been shown to be an effective

Table 7.1 Conventional commercial available dosage forms and nano-based delivery carriers of antifungal drugs investigated in literature

Antifungal compound	Conventional dosage form	Nano-based delivery technology	Reference
Azoles			
Imidazoles			
Clotrimazole	Cream	Micelle solution	Souto et al. (2004)
	Solution	Solid lipid nanoparticle	Bachhav et al. (2011)
	Lotion	Nanostructured lipid carrier	Hashem et al. (2011)
		Ethosome	Maheshwari et al. (2012)
		Transfersome	Basha et al. (2013)
		Microemulsion	Vanić and Škalko-Basnet (2013)
Ketoconazole	Cream	Liposome	Patel et al. (2011)
	Gel	Ethosome	Che et al. (2015)
	Foam	Transfersome	Guo et al. (2015)
	Shampoo	Microemulsion	
	Tablet		
Sertaconazole	Solution	Microemulsion	Sahoo et al. (2014a)
	Cream	Microemulsion-based hydrogel	Sahoo et al. (2014b)
Bifonazole	Solution	Microemulsion-based hydrogel	Sabale and Vora (2012)
	Cream		
	Powder		
Miconazole	Cream	Liposome	Elmoslemany et al. (2012)
	Ointment	Transfersome	Cerdeira et al. (2013)
	Lotion	Nanosuspension	Pandit et al. (2014)
	Solution		
	Powder		
Econazole	Cream	Solid lipid nanoparticle	Sanna et al. (2007)
		Nanostructured lipid carrier	Sharma and Pathak (2011)
		Nanosponge	Verma and Pathak (2012)
		Ethosome	Keshri and Pathak (2013)

(continued)

Table 7.1 (continued)

Antifungal compound	Conventional dosage form	Nano-based delivery technology	Reference
Triazoles			
Fluconazole	Tablet	Microemulsion	Hoeller et al. (2008)
	Oral suspension	Microemulsion-based hydrogel	Bachhav and Patravale (2009)
		Liposome	Bhalaria et al. (2009)
		Liposome gel	Patel et al. (2009)
		Niosome	El-Nesr et al. (2010)
		Ethosome	Gupta et al. (2010)
			Salerno et al. (2010)
			Gupta et al. (2011)
			Schwarz et al. (2011)
			Coneac et al. (2015)
			Oliveira Brito et al. (2015)
Itraconazole	Capsule	Microemulsion	Alomrani et al. (2014)
	Oral suspension	Microemulsion-based hydrogel	Barot et al. (2012a, b)
		Transfersome	Chudasama et al. (2011)
		Nanosuspension	da Fonseca Antunes et al. (2013)
			Kumar and Shishu (2015)
Voriconazole	Oral solution	Nanostructured lipid carrier	El-Hadidy et al. (2012)
	Tablet	Transethosome	Song et al. (2012)
	Intravenous solution	Microemulsion	Song et al. (2014)
Posaconazole	Oral suspension	Nanosizing	Merisko-Liversidge and Liversidge (2011)
Allylamines			
Terbinafine	Tablet	Solid lipid nanoparticle	Barot et al. (2012a, b)
	Solution	Microemulsion	Chen et al. (2012)
	Cream	Liposome	Zhang et al. (2012)
	Spray	Ethosome	Kumar et al. (2014)
	Gel	Transfersome	Tanrıverdi and Özer (2013)
			Koutsoulas et al. (2014)
			Sudhakar et al. (2014)
			Çelebi et al. (2015)
Naftifine	Solution	Niosome gel	Barakat et al. (2009)
	Cream		
	Gel		
Polyenes			
Amphotericin B	Intravenous solution	Liposome gel	Kang et al. (2010)
	Intravenous liposomes	Nanosizing	Randhawa et al. (2015)
Nystatin	Cream	Nanoemulsion	El-Ridy et al. (2011)
	Powder	Niosome	Fernandez-Campos et al. (2012)
	Ointment	Nanosizing	Melkoumov et al. (2015)
	Tablet		
Echinocandins			
Caspofungin	Intravenous lyophilized powder	–	–
Micafungin			
Anidulafungin			

(continued)

Table 7.1 (continued)

Antifungal compound	Conventional dosage form	Nano-based delivery technology	Reference
Other compounds			
Ciclopirox	Cream	Niosome	Shaikh et al. (2010)
	Gel	Liposome gel	Karimunnisa and Atmaram (2013)
	Suspension		
	Shampoo		
	Solution		
Griseofulvin	Tablet	Solid lipid nanoparticle	Aggarwal and Goindi (2012)
		Microemulsion	Aggarwal and Goindi (2013) and Aggarwal et al. (2013)
		Transfersome	

topical allylamine antifungal for the treatment of superficial dermatophytoses (Ghannoum et al. 2013). Cream, gel, and dermal spray formulations of naftifine are available in the market.

7.2.2 Azole Antifungal Agents

Azole antifungals act through selectively inhibiting the synthesis of fungal cell ergosterol, and they alter the permeability of cell membrane by binding with the phospholipids in the fungal cell membrane. The azole antifungal drugs used in the treatment composed of either two or three nitrogens in the azole ring and are thereby classified as imidazoles (e.g., ketoconazole and miconazole, clotrimazole) or triazoles (e.g., itraconazole and fluconazole), respectively (Thompson et al. 2009). Triazoles have a broad range of applications in the treatment of both superficial and systemic fungal infections, and they became standard therapy for invasive candidiasis in the late 1980s (Güngör et al. 2013; Lewis 2010; Sheehan et al. 1999).

The reader is referred to the table to see the conventional dosage formulations of azole antifungals that are currently in therapy. There are a number of licensed azole antifungal drugs; however, fluconazole, itraconazole, posaconazole, and voriconazole are the most frequently used ones in a clinical setting for treatment of systemic fungal infections (Brüggemann et al. 2009).

7.2.3 Polyene Antifungal Agents

The polyene antifungal agents exert their antifungal activity by binding irreversibly to cell membrane ergosterol of fungus. Therefore, the polyenes are fungicidal and have the broadest spectrum of antifungal activity of any of the clinically available agents. These antifungal agents are of amphipathic nature and chemically instable (Flückiger et al. 2006; Güngör et al. 2013; Lewis 2010).

Amphotericin B is the first polyene macrolide antifungal that has been successfully developed into a systemic antifungal agent for the treatment of invasive fungal infections (Adler-Moore and Proffitt 2004). Even though its clinical use is limited by its toxic side effects, amphotericin B remains an important treatment option due to the drug's broad spectrum and minimal cross-resistance with other antifungals. The therapy options already in use include intravenous infusion, lipid complex, colloidal dispersion, and liposomal dispersion of the drug. Parenteral liposomal amphotericin B (Ambiosome) is on the market since 1990, and it has been shown that this preparation demonstrates less nephrotoxicity when compared to amphotericin B lipid complex (Abelcet®) (Kumar et al. 2014). Amphotericin B has very limited aqueous solubility, thus the development of its novel formulations and administration strategies, which take advantage of the drug's unique pharmacokinetic and pharmacodynamic profile, will open new opportunities for safer and more effective uses of this polyene antifungal (Lewis 2010).

Nystatin is a polyene antifungal drug that has been used in the treatment of cutaneous, vaginal, and oral mucosal fungal infections since the 1950s (El-Ridy et al. 2011). Nystatin has exhibited broad antifungal activity in treating mucocutaneous fungal infections, and its conventional formulations include oral suspension, vaginal ovules, and ointment. The most common adverse effect reported with topical nystatin is allergic contact dermatitis.

7.2.4 Echinocandin Antifungal Agents

Echinocandins destabilize fungal cell walls due to inhibiting of the 1,3-β-glucan synthase complex. They are active against a broad range of *Candida* and *Aspergillus* species (Sable et al. 2008). The Infectious Diseases Society of America has published in 2009 a guideline recommending the use of three echinocandin compounds, namely, caspofungin, micafungin, and anidulafungin as initial therapy for the treatment of candidemia in adults, whereas in Europe only caspofungin and micafungin are indicated for candidemia in pediatric and adult patients (Munoz et al. 2010).

7.2.5 Others (Ciclopirox and Griseofulvin)

Ciclopirox is a synthetic hydroxypyridone derivative having antifungal, antibacterial, and anti-inflammatory activities. Ciclopirox inhibits essential enzymes interfering with fungal mitochondrial electron transport processes and energy production. It is active against many fungi including dermatophytes and yeast (Güngör et al. 2013). The conventional formulations of ciclopirox are dermal and vaginal creams and dermal solutions and sprays.

Griseofulvin is the first agent with antifungal activity and was isolated in 1939 (Sheehan et al. 1999). It is an orally administered antifungal drug and is active against all common dermatophytes and shows fungistatic effect in humans. Griseofulvin is accepted as a safe medication and its adverse effects are common, but mild, including nausea, headache, and urticaria (Kelly 2012).

7.3 Nanocarriers of Antifungal Agents

Numerous types of nano-drug carriers have been developed for the treatment of fungal infections. Overview of novel drug delivery approaches of antifungal agents is briefly shown in Fig. 7.1. The research studies focused on nano-sized carriers as a challenge for the effective treatment of fungal infections have been summarized below.

Vesicular carriers such as liposomes, niosomes, and transfersomes have also been intensively explored for delivering fungal agents in either topical or systemic treatment. Vesicular systems have the advantages of controlling

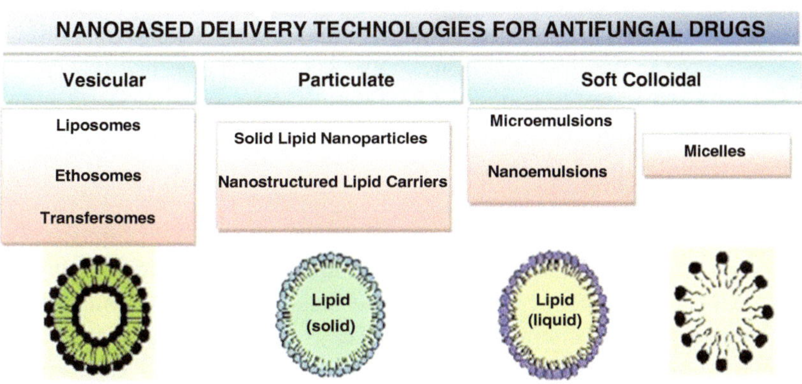

Fig. 7.1 Novel drug delivery technologies of antifungal agents

release rate of the active compound and providing localization of topical delivered drugs in dermal layers. Besides, vesicular systems help to carry drug molecules into systemic circulation to decrease toxicity of drugs (Cevc 1996; Akhtar 2014). At the end of the 1980s, the first liposome dermatological product of econazole nitrate (Pevaryl® Lipogel) had been approved by regulatory authorities, and it was introduced in various countries (Korting et al. 1997). Niosomes are liposomes prepared with nonionic surfactants. They have a potential to overcome the barrier characteristic of skin with the effect of high amount of surfactant in their content (Mahale et al. 2012; Marianecci et al. 2014). Transfersomes are defined as elastic vesicles that can be highly deformed. Classical liposomes have a diameter varying from 200 to 400 nm, which is too large to pass through the skin. On the other hand, transfersomes reach deeper dermal tissues and even the systemic circulation with their elastic and highly deformable characteristics (Benson 2009). Ethosomes contain phospholipids like classical liposomes; however, they contain high concentrations of alcohol in their structure. These carriers improve drug permeation due to the high alcohol content which acts as penetration enhancer as well as disruption of the structure of skin by the phospholipids in their content (Ainbinder et al. 2010; Mbah et al. 2014).

Microemulsions are one of the colloidal carriers studied intensively during the last decades by many scientists to enhance delivery of topical antifungal agents into targeted skin sites. They are thermodynamically stable, optically isotropic, colloidal nanocarriers with a dynamic microstructure that form spontaneously by combining oil, water, surfactant, and a cosurfactant. Microemulsions basically offer increased drug-solubilizing capacity and enhanced drug release and skin penetration. They can modify the diffusional barrier of the skin, which is resulted in enhancement of drug penetration into deeper skin layers compared to conventional formulations. The internal phase of microemulsions can act as a drug reservoir resulting in controlled and sustained drug release (Santos et al. 2008; Neubert 2011; Güngör et al. 2015a, b).

Solid lipid nanoparticle and nanostructured lipid carrier formulations have also been investigated intensively for the topical treatment of fungal skin infections due to their ability to enhance skin penetration of loaded drugs. While solid lipid nanoparticles are water-in-oil emulsions composed of solids as oil phase and are prepared from solid lipids or from blends of these lipids, nanostructured lipid carriers are new generation lipid particles, and they contain mixtures of different solid lipids blended with liquid oils. The principal advantage of these carriers is their low risk of toxicity (Iqbal et al. 2012; Scalia et al. 2015).

Micelles are colloidal carriers; they are composed of amphiphilic block polymers and characterized by core-shell morphology in nanoscale size. They have been investigated to facilitate the penetration of antifungal drugs for topical delivery via skin. Cutaneous delivery of drugs into micellar carriers, particularly for the treatment of skin diseases, would benefit in terms of improving aqueous solubility of hydrophobic drugs, targeting of drugs into skin layers (Güngör et al. 2015a, b).

The nano-sized carriers reported in current literature to improve delivery of antifungal drugs are summarized in Table 7.1.

7.3.1 Nanocarriers of Allylamine Antifungal Agents

Hydrogels and microemulsion-based gel formulations of terbinafine have been studied to evaluate the efficacy of these formulations for the treatment of fungal infections. In vitro examination of antifungal activity revealed that the Natrosol®-based hydrogel was a good candidate for the topical delivery of terbinafine (Çelebi et al. 2015). In another study, Barot et al. developed a microemulsion-based gel of terbinafine for the treatment of onychomycosis. They have indicated that the developed microemulsion-based gel formulation enhanced penetration and retention of terbinafine into the human cadaver skin as compared to its commercial cream. This gel formulation of terbinafine showed better activity against *C. albicans* and *Trichophyton rubrum* than the commercial cream (Barot et al. 2012a, b).

To resolve problems of long treatment durations and frequent administration in conventional therapy, solid lipid nanoparticles of terbinafine were developed. The skin of nude mice was used as a barrier membrane, and penetration levels of terbinafine of the formulations designed and its commercial product in the stratum corneum, viable epidermis, and dermis were measured. It was concluded that the application of terbinafine solid lipid particles for 12 h might have an efficacy comparable to that of the commercial topical product (Chen et al. 2012).

The skin permeation of ethosomes, binary ethosomes, and transfersomes of terbinafine was compared under nonocclusive conditions. The results indicated that the binary ethosomes most effectively permitted drug penetration through skin, whereas transfersomes made it easier for the drug to accumulate in the skin. Ethosomes had greater improvement in skin permeation of terbinafine than that of its skin deposition (Zhang et al. 2012).

Nails are skin appendages susceptible to fungal attack, and it is difficult to treat fungal infection of nail by conventional formulations because of the very low permeability of the nail plate (Kumar et al. 2014). Tanrıverdi and Özer prepared terbinafine hydrochloride-loaded liposome and ethosome formulations and performed nail characterization studies to examine the effect of formulations on nail surface. They concluded that vesicular carriers could be served as efficient formulations for ungual application of terbinafine hydrochloride (Tanrıverdi and Özer 2013).

In addition to these studies, liposomal gel formulations of terbinafine and an alcohol-free niosome gel containing naftifine hydrochloride were developed and characterized by various researchers (Barakat et al. 2009; Koutsoulas et al. 2014; Sudhakar et al. 2014).

7.3.2 Nanocarriers of Azole Antifungal Agents

Azole antifungal drugs show numerous drug-drug interactions (Brüggemann et al. 2009). The most significant common drug interactions are drug elevations of cyclosporine, tacrolimus and sirolimus due to most calcium channel blockers and benzodiazepines, many statins and steroids, and warfarin and rifabutin (Zonios and Bennett 2008). Furthermore, the increase in antifungal drug resistance, particularly to azoles, created a strong need for the identification of novel antifungal therapies and drug targets. Several novel vehicles, e.g., microemulsions, vesicular carriers, and lipid particles, have been proposed for the topical application of azole antifungal agents.

7.3.2.1 Imidazoles

Clotrimazole Clotrimazole was first synthesized in the late 1960s and was introduced to markets in 1973. This drug is accepted worldwide for the treatment of common fungal diseases such as vaginal yeast infections and athlete's foot. Microemulsions, ethosomes, and ultradeformable liposomes were formulated and evaluated by various researchers as clotrimazole nanocarriers for dermal, transdermal, vaginal, and ocular delivery (Basha et al. 2013; Maheshwari et al. 2012; Vanić and Škalko-Basnet 2013). The efficacy and tolerability of clotrimazole microemulsion in the treatment of various topical fungal infections were evaluated clinically (Hashem et al. 2011). Ultradeformable carriers showed prolonged ocular delivery of clotrimazole, and their antifungal activity against *C. albicans* was higher than niosomal formulation as well as the drug suspension (Basha et al. 2013). Clotrimazole-loaded solid lipid nanoparticles and nanostructured lipid carriers were prepared by the hot high-pressure homogenization technique to assess the physical stability of these particles, as well as the loading capacity and in vitro release pattern. The results obtained demonstrated the use of the lipid nanoparticles as modified release formulations for clotrimazole (Souto et al. 2004). Aqueous micelle solutions of several azole antifungals (clotrimazole, econazole nitrate, and fluconazole) were prepared using amphiphilic methoxy-poly (ethylene glycol)-hexyl substituted polylactide block copolymers. The significant increase in skin deposition demonstrated the ability of micelle-type carrier systems to improve cutaneous antifungal drug bioavailability (Bachhav et al. 2011).

Ketoconazole Ketoconazole is a broad-spectrum, systemic antifungal agent. Lipid vesicular systems including conventional liposomes, ethosomes, deformable liposomes, and ethanol-containing deformable liposomes were prepared as nanocarriers for ketoconazole, respectively. Characterization of the vesicles was based on particle size, zeta potential, entrapment efficiency, and transmission electron microscopy. The results demonstrated that ethanol-containing deformable liposomes improved both in vitro and in vivo skin deposition of ketoconazole (Guo et al. 2015). Microemulsion formulations of ketoconazole enhanced percutaneous absorption of the drug, and it has been shown that a microemulsion and cyclodextrin combination enhanced the antifungal activity of the drug in vitro against *Candida parapsilosis* test (Che et al. 2015; Patel et al. 2011).

Sertaconazole Sertaconazole is a broad-spectrum third-generation imidazole derivative that is effective and safe to cure superficial mycoses, such as tineas, candidiasis, and pityriasis versicolor. Microemulsion-based hydrogels for cutaneous delivery of sertaconazole were studied with an objective to increase the solubility and skin permeability of the drug. The results indicated that the microemulsion-based hydrogel studied provided higher antifungal activity against *C. albicans* when compared with commercial cream of sertaconazole (Sahoo et al. 2014a, b).

Bifonazole Bifonazole is a substituted imidazole antifungal agent possessing a broad spectrum of activity in vitro against dermatophytes, molds, and yeasts. It is an effective and well-tolerated treatment for superficial fungal infections of the skin. It has been reported that microemulsion-based hydrogel of bifonazole provided sustained drug release cutaneous application (Sabale and Vora 2012).

Miconazole Propylene glycol phospholipid vesicles and ultraflexible liposomes have been advocated as deformable lipid carriers for enhanced skin delivery of broad-spectrum imidazole antifungal agent miconazole. The results provided evidence of controlled drug delivery, higher rate of drug transfer across the skin, enhanced skin deposition of miconazole, and better antifungal activity as compared to traditional liposomes and plain drug solution (Elmoslemany et al. 2012; Pandit et al. 2014). Miconazole possesses only slight water solubility, which limits its bioavailability and antifungal efficacy. To increase the dissolution rate of miconazole and itraconazole, the particle size of these compounds was scaled down into nano-size by media milling, respectively. The nanosuspensions obtained had a mean particle size of 210 nm and showed enhanced in vitro performance (Cerdeira et al. 2013).

Econazole Solid lipid nanoparticles and nanostructured lipid carriers of econazole were designed for the treatment of deep-seated fungal infections. In vivo studies demonstrated that solid lipid nanoparticles promoted a rapid penetration of econazole through the stratum corneum and improved the diffusion of the drug in the deeper skin layers (Keshri and Pathak 2013; Sanna et al. 2007). Polymeric nanosponges were prepared as an alternative carrier for targeting econazole to the skin through topical hydrogel formulation (Sharma and Pathak 2011).

7.3.2.2 Triazoles

Fluconazole Fluconazole is a semisynthetic triazole, which is active against numerous yeasts. It remains one of mostly prescribed triazoles due to its good bioavailability and substantially fewer drug-drug interactions than other triazole compounds. However, systemic treatment with fluconazole can cause adverse effects, including hepatic enzyme elevation and drug interactions (Melkoumov et al. 2015). Microemulsions were investigated for topical delivery of fluconazole and some promising results have been achieved (Hoeller et al. 2008; Brito Oliveira et al. 2015; Patel et al. 2009). One of the major drawbacks of using microemulsions for topical delivery is their liquid nature and their consequently low residence time on the application site. Therefore microemulsion-based hydrogels have gained interest to improve the viscosity of different formulations. Microemulsion-based hydrogels were evaluated as fluconazole delivery systems for the treatment of cutaneous and vaginal mycoses. The small-scale clinical studies indicated that the microemulsion-based gel developed for the vaginal delivery of fluconazole showed faster onset of action than commercial preparation (Bachhav and Patravale 2009; Coneac et al. 2015; Salerno et al. 2010).

Fluconazole-entrapped liposomes and niosomes were prepared and evaluated in terms of in vitro skin permeation and skin retention. Cutaneous accumulation of fluconazole was affected by components and size of the vesicles. Physical stability studies showed superior potentials of lyophilized fluconazole liposomes after reconstitution in comparison with those of a solution product (El-Nesr et al. 2010; Gupta et al. 2011; Schwarz et al. 2011). To improve the stability and overall applicability of the formulations, fluconazole-loaded liposomes and niosomes have been further incorporated into polymer gels. The studies performed in vivo showed that liposomal gel could produce significantly higher drug, a localization in viable skin, compared with its plain gel. The antifungal study also confirmed the maximum therapeutic efficacy of liposomal gel showing its potential for topical delivery of fluconazole with increased accumulation of drug in the skin (Gupta et al. 2010). Bhalaria et al. prepared fluconazole-encapsulated ethosomes and incorporated in a dermatological base to assess the comparative clinical efficacy in the treatment of candidiasis patients. From the clinical evaluation, the novel delivery system developed demonstrated high antifungal activity in comparison to liposomal formulation, marketed formulation, and hydroethanolic solution of the drug (Bhalaria et al. 2009).

Itraconazole Itraconazole is a triazole antifungal agent which may be given orally or intravenously. A microemulsion-based gel of itraconazole was developed for overcoming the shortcomings and adverse effects of currently used therapies. The in vivo evaluation indicated the superiority of microemulsion hydrogel formulation for treating superficial fungal infections than those of its conventional approaches (Kumar and Shishu 2015). Microemulsion-based itraconazole gel was also evaluated for effective treatment of onychomycosis. Ex vivo permeation studies indicated better penetration and retention of itraconazole into nail plate and nail bed models (Barot et al. 2012a, b). Chudasama et al. determined in vitro itraconazole permeation from microemulsion gels across excised rat skin. In vitro antimycotic inhibitory activity of the gels was evaluated with agar cup method and *C. albicans* as a test organism. The results indicated that the microemulsion gel studied showed the widest zone of inhibition and therefore may be a promising vehicle for topical delivery of itraconazole (Chudasama et al. 2011). Deformable liposomes containing hydroxypropyl-β-cyclodextrin enhanced the amount of itraconazole in stratum corneum and deeper skin layers compared to conventional liposomes (Alomrani et al. 2014). Nanosuspensions formulated by ultrasound treatment improved the dissolution behavior of itraconazole (da Fonseca Antunes et al. 2013).

Voriconazole Voriconazole is a second-generation triazole available in both intravenous and oral formulations (Sable et al. 2008). It has revolutionized the treatment of aspergillosis in severely immunocompromised patients, but the use of voriconazole is compromised by complicated pharmacokinetics, notable drug interactions,

and relatively significant adverse events (Zonios and Bennett 2008). Microemulsions as vehicles for topical administration of voriconazole were investigated, and drug-loaded microemulsions were evaluated for their physicochemical characteristics and in vitro permeation studies using guinea pig skin. The presence of permeation enhancers in the formulation favored transdermal rather than dermal delivery of voriconazole (El-Hadidy et al. 2012).
Transethosomes composed of phospholipid, ethanol, water, and edge activator (surfactants) or penetration enhancer (oleic acid) were developed and evaluated for enhanced skin delivery of voriconazole. Transethosomes enhanced both in vitro and in vivo skin deposition of voriconazole in the dermis/epidermis region compared to conventional liposomes, ethosomes, and transfersomes (Song et al. 2012). In order to develop topical preparations of voriconazole for the treatment of mycotic infections of the skin, a nanostructured lipid carrier-based hydrogel formulation was developed and its physical characteristics, in vitro skin permeation, and retention profiles were examined. The in vitro skin permeation study showed that the nanostructured lipid carrier-based hydrogel was superior to conventional cream and microemulsion-based gel formulations of voriconazole. In addition, the formulation developed led to markedly greater accumulation of voriconazole in deeper skin layers as compared with the reference formulations (Song et al. 2014).

Posaconazole Posaconazole, a lipophilic second-generation antifungal triazole, has a significant role for the prophylaxis of invasive fungal infections in severely immunocompromised patients with extended antifungal spectrum, significant activity against the zygomycetes, and apparently, optimal safety profile. Posaconazole is only available as an oral formulation (Brüggemann et al. 2009). Multiple daily dosing, the need for fatty foods for absorption, and the absence of an intravenous formulation restrict its use to selected populations (Zonios and Bennett 2008). There has been consideration of a potential enhancement in solubility with decreasing of the particle size of drugs into nanometer size range. To enhance solubility and the rate of dissolution, posaconazole was nano-sized using wet media milling technology. The mean particle size of the processed crystals had a narrow distribution profile with a mean size of 185 nm (Merisko-Liversidge and Liversidge. 2011).

7.3.3 Nanocarriers of Polyene Antifungal Agents

The activity of nano-sized amphotericin B against the growth of *C. albicans* was investigated, and the conventional micro-sized amphotericin B was converted into nano-size (5–20 nm) using a ball milling machine. The results revealed that the antifungal activity of nano-sized drug was two to four times greater than the micro-counterpart. Nano-sized amphotericin B showed an enhanced surface activity by interacting with the walls of Candida yeasts (Randhawa et al. 2015). Kang et al. formulated amphotericin B-loaded cationic liposome gels for topical application and have found this thermosensitive gel to be more stable and less toxic than free amphotericin B for vaginal delivery (Kang et al. 2010).

Nystatin nanoemulsions have been evaluated for topical application. A faster pharmacokinetic release from the nanoemulsion formulation has been recorded than commercial nystatin ointment, and the amount of drug retained in the skin was sufficient to ensure an effective antifungal effect (Fernandez-Campos et al. 2012). In another study performing a comparative evaluation, a nanoparticulate nystatin formulation against a commercial nystatin preparation showed that nanonization of nystatin provides a novel approach to enhancing the efficacy of drug in the treatment of oral candidiasis (Melkoumov et al. 2015).

El-Ridy et al. have been encapsulated nystatin in niosomes to obtain a safe and effective formulation administered parenterally for neutropenic patients. Nystatin niosomes exerted less nephrotoxicity and hepatotoxicity in vivo, showed higher level of drug in significant organs, and improved efficacy in elimination of the fungal burden in experimental animals infected with *C. albicans* compared with those treated with free nystatin (El-Ridy et al. 2011).

7.3.4 Nanocarriers of Echinocandins

Caspofungin, micafungin, and anidulafungin exist in intravenous form (lyophilized powder for infusion) and are indicated in the management of invasive fungal infections such as esophageal candidiasis and candidemia (Sable et al. 2008). Currently there hasn't been any study published about nano-sized delivery systems of echinocandin antifungal agents. This fact might be due to the relatively big molecular weight of echinocandins (>1000 Da) when compared to conventional small-molecule drug compounds.

7.3.5 Nanocarriers of Ciclopirox and Griseofulvin

Aggarwal et al. developed a microemulsion formulation of griseofulvin intended for the treatment of dermatophytosis. The dermatopharmacokinetics and antifungal activity of the formulation were evaluated on guinea pig model. Treatment of guinea pigs with griseofulvin microemulsion resulted in a complete clinical and mycological cure in 7 days. The formulation was observed to be histopathologically safe and stable (Aggarwal et al. 2013).

Griseofulvin was encapsulated in the deformable membrane vesicles for dermal delivery. The optimized vesicles illustrated remarkably higher drug permeation and skin retention when compared with conventional liposomes. The results indicated that the topical deformable vesicles of griseofulvin could be considered as an alternative to reduce the obstacle of conventional oral formulations (Aggarwal and Goindi 2012). The same researchers optimized griseofulvin-loaded solid lipid nanoparticles by hot microemulsion technique. The cumulative amount of drug permeated through excised mice skin from solid lipid nanoparticles was more than fivefolds as compared to permeation from conventional cream base (Aggarwal and Goindi 2013).

Niosomal vesicular carriers were investigated for dermal delivery of ciclopirox olamine. Deposition of drug into rat skin from niosomal dispersion and niosomal gel formulation was significantly higher than that of solution formulation of ciclopirox olamine and its commercial product. Obtained niosomes possessed sufficient stability on storage (Shaikh et al. 2010). Karimunnisa and Atmaram investigated mucoadhesive liposomal ciclopiroxolamine gel for vaginal use. The gel exhibited good mucoadhesivity to sheep vaginal tissue, and furthermore liposome-entrapped drug displayed good antifungal activity (Karimunnisa and Atmaram 2013).

7.4 Conclusion

Delivery of antifungal agents used for the treatment of superficial and systemic infections has a great impact in terms of both therapeutic aspect and safety. In particular, treatment of deep-seated fungal infections could be inefficient due to poor penetration of drugs into targeted sites, resulting in unsatisfactory bioavailability. In addition, the controlling of drug release is also important to minimize systemic absorption and to decrease the toxicity of these agents. Therefore, novel drug delivery systems play a key role to overcome the limitations of conventional dosage forms to improve the efficacy and to ensure safety of the treatment. So far, optimization of different types of nanocarriers such as liposomes, niosomes, micelles, and transfersomes have been intensively investigated to improve efficiency and safety of antifungal agents. Some liposomal products of antifungal agents are already commercially available. At the end of the 1980s, the first liposome dermatological product (Pevaryl®Lipogel) has come into market. Liposome formulation of amphotericin B for parenteral administration has been first marketed under the name AmBisome in 1990. Still there is a great effort to optimize novel carriers of antifungal agents. In this context, a considerable amount of research focused on the development of nanocarriers of antifungal drugs, to some extent, has been published.

References

Adler-Moore JP, Proffitt RT (2004) Novel drug delivery systems for antifungal agents. In: Domer JE, Kobayashi GS (eds) The mycota XII human fungal pathogens. Springer, Heidelberg, pp 339–340

Aggarwal N, Goindi S (2012) Preparation and evaluation of antifungal efficacy of griseofulvin loaded deformable membrane vesicles in optimized guinea pig model of Microsporum canis dermatophytosis. Int J Pharm 437(1–2):277–287

Aggarwal N, Goindi S (2013) Preparation and in vivo evaluation of solid lipid nanoparticles of griseofulvin for dermal use. J Biomed Nanotechnol 9(4):564–576

Aggarwal N, Goindi S, Khurana R (2013) Formulation, characterization and evaluation of an optimized microemulsion formulation of griseofulvin for topical application. Colloids Surf B 105:158–166

Ainbinder D, Paolino D, Fresta M et al (2010) Drug delivery applications with ethosomes. J Biomed Nanotechnol 6:558–568

Akhtar N (2014) Vesicles: a recently developed novel carrier for enhanced topical drug delivery. Curr Drug Deliv 11:87–97

Alomrani AH, Shazly GA, Amara AA, Badran MM (2014) Itraconazole-hydroxypropyl-β-cyclodextrin loaded deformable liposomes: in vitro skin penetration studies and antifungal efficacy using Candida albicans as model. Colloids Surf B 121:74–81

Arendrup MC (2013) Candida and candidaemia. Susceptibility and epidemiology. Dan Med J 60(11):B4698

Bachhav YG, Patravale VB (2009) Microemulsion based vaginal gel of fluconazole: formulation, in vitro and in vivo evaluation. Int J Pharm 365(1–2):175–179

Bachhav YG, Mondon K, Kalia YN et al (2011) Novel micelle formulations to increase cutaneous bioavailability of azole antifungals. J Control Release 153 (2):126–132

Barakat HS, Darwish IA, El-Khordagui LK et al (2009) Development of naftifine hydrochloride alcohol-free niosome gel. Drug Dev Ind Pharm 35(5):631–637

Barot BS, Parejiya PB, Patel HK et al (2012a) Microemulsion-based gel of terbinafine for the treatment of onychomycosis: optimization of formulation using D-optimal design. AAPS PharmSciTech 13 (1):184–192

Barot BS, Parejiya PB, Patel HK et al (2012b) Microemulsion-based antifungal gel delivery to nail for the treatment of onychomycosis: formulation, optimization, and efficacy studies. Drug Deliv Transl Res 2(6):463–476

Basha M, Abd El-Alim SH, Shamma RN et al (2013) Design and optimization of surfactant-based nanovesicles for ocular delivery of Clotrimazole. J Liposome Res 23(3):203–210

Benson HA (2009) Elastic liposomes for topical and transdermal drug delivery. Curr Drug Deliv 6:217–226

Bhalaria MK, Naik S, Misra AN (2009) Ethosomes: a novel delivery system for antifungal drugs in the treatment of topical fungal diseases. Indian J Exp Biol 47 (5):368–375

Brito Oliveira M, Calixto G, Graminha M et al (2015) Development, characterization, and in vitro biological performance of fluconazole loaded microemulsions for the topical treatment of cutaneous leishmaniasis. Biomed Res Int. doi:10.1155/2015/396894

Brüggemann RJM, Alffenaar JWC, Blijlevens NMA et al (2009) Clinical relevance of the pharmacokinetic interactions of azole antifungal drugs with other coadministered agents. Clin Infect Dis 48:1441–1458

Carneiro G, Aguiar MG, Fernandes AP et al (2012) Drug delivery systems for the topical treatment of cutaneous leishmaniasis. Expert Opin Drug Deliv 9:1083–1097

Çelebi N, Ermiş S, Özkan S (2015) Development of topical hydrogels of terbinafine hydrochloride and evaluation of their antifungal activity. Drug Dev Ind Pharm 41(4):631–639

Cerdeira AM, Mazzotti M, Gander B (2013) Formulation and drying of miconazole and itraconazole nanosuspensions. Int J Pharm 443:209–220

Cevc G (1996) Transfersomes, liposomes and other lipid suspensions on the skin: permeation enhancement, vesicle penetration, and transdermal drug delivery. Crit Rev Ther Drug Carrier Syst 13:257–388

Che J, Wu Z, Shao W et al (2015) Synergetic skin targeting effect of hydroxypropyl-β-cyclodextrin combined with microemulsion for ketoconazole. Eur J Pharm Biopharm. doi:10.1016/j.ejpb.2015.03.028

Chen YC, Liu DZ, Liu JJ et al (2012) Development of terbinafine solid lipid nanoparticles as a topical delivery system. Int J Nanomedicine 7:4409–4418

Chudasama A, Patel V, Nivsarkar M et al (2011) Investigation of microemulsion system for transdermal delivery of itraconazole. J Adv Pharm Technol Res 2(1):30–38

Coneac G, Vlaia V, Olariu I et al (2015) Development and evaluation of new microemulsion-based hydrogel formulations for topical delivery of fluconazole. AAPS PharmSciTech. doi:10.1208/s12249-014-0275-8

da Fonseca Antunes AB, De Geest BG, Vervaet C et al (2013) Solvent-free drug crystal engineering for drug nano- and micro suspensions. Eur J Pharm Sci 48 (1–2):121–129

El-Hadidy GN, Ibrahim HK, Mohamed MI et al (2012) Microemulsions as vehicles for topical administration of voriconazole: formulation and in vitro evaluation. Drug Dev Ind Pharm 38(1):64–72

Elmoslemany RM, Abdallah OY, El-Khordagui LK et al (2012) Propylene glycol liposomes as a topical delivery system for miconazole nitrate: comparison with conventional liposomes. AAPS PharmSciTech 13(2):723–731

El-Nesr OH, Yahiya SA, El-Gazayerly ON (2010) Effect of formulation design and freeze-drying on properties of fluconazole multilamellar liposomes. Saudi Pharm J J18(4):217–224

El-Ridy MS, Abdelbary A, Essam T et al (2011) Niosomes as a potential drug delivery system for increasing the efficacy and safety of nystatin. Drug Dev Ind Pharm 37(12):1491–1508

Fernández Campos F, Calpena Campmany AC, Rodríguez Delgado G et al (2012) Development and characterization of a novel nystatin-loaded nanoemulsion for the buccal treatment of candidosis: ultrastructural effects and release studies. J Pharm Sci 101(10):3739–3752

Flückiger U, Marchetti O, Bille J, Eggimann P, Zimmerli S, Immof A, Garbino J, Ruef C, Pittet D, Tauber M, Glauser M, Calandra T (2006) Treatment options of invasive fungal infections in adults. Swiss Med Wkly 136:447–463

Ghannoum M, Isham N, Verma A et al (2013) In vitro antifungal activity of naftifine hydrochloride against dermatophytes. Antimicrob Agents Chemother 57 (9):4369–4372

Güngör S, Erdal MS, Aksu B (2013) New formulation strategies in topical antifungal therapy. J Cosm Dermatol Sci Appl 3:56–65

Güngör S, Erdal MS, Güngördük S (2015a) Colloidal carriers in topical treatment of dermatological diseases. In: Naik J (ed) Nano-based drug delivery. IAPC Open Book and Monograph Platform (OBP), Zagreb, pp 391–409

Güngör S, Kahraman E, Özsoy Y (2015b) Polymeric micelles for cutaneous drug delivery. In: Naik J (ed) - Nano-based drug delivery. IAPC Open Book and Monograph Platform (OBP), Zagreb, pp 369–387

Guo F, Wang J, Ma M et al (2015) Skin targeted lipid vesicles as novel nano-carrier of ketoconazole: characterization, in vitro and in vivo evaluation. J Mater Sci Mater Med. doi:10.1007/s10856-015-5487-2

Gupta M, Goyal AK, Paliwal SR et al (2010) Development and characterization of effective topical liposomal system for localized treatment of cutaneous candidiasis. J Liposome Res 20(4):341–350

Gupta M, Vaidya B, Mishra N et al (2011) Effect of surfactants on the characteristics of fluconazole niosomes for enhanced cutaneous delivery. Artif Cells Blood Substit Immobil Biotechnol 39(6):376–384

Hashem FM, Shaker DS, Ghorab MK et al (2011) Formulation, characterization, and clinical evaluation of microemulsion containing clotrimazole for topical delivery. AAPS PharmSciTech 12(3):879–886

Hoeller S, Klang V, Valenta C (2008) Skin-compatible lecithin drug delivery systems for fluconazole: effect of phosphatidylethanolamine and oleic acid on skin permeation. J Pharm Pharmacol 60(5):587–591

Iqbal MA, Md S, Sahni JK et al (2012) Nanostructured lipid carriers system: recent advances in drug delivery. J Drug Target 20:813–830

Johal HS, Garg T, Rath G et al (2014) Advanced topical drug delivery system for the management of vaginal candidiasis. Drug Deliv 24:1–14

Kang JW, Davaa E, Kim YT, Park JS (2010) A new vaginal delivery system of amphotericin B: a dispersion of cationic liposomes in a thermosensitive gel. J Drug Target 18(8):637–644

Karimunnisa S, Atmaram P (2013) Mucoadhesive nanoliposomal formulation for vaginal delivery of an antifungal. Drug Dev Ind Pharm 39(9):1328–1337

Kasper L, Miramón P, Jablonowski N et al (2015) Antifungal activity of clotrimazole against Candida albicans depends on carbon sources, growth phase, and morphology. J Med Microbiol. doi:10.1099/jmm.0.000082

Kaur P, Kakkar S (2010) Topical delivery of antifungal agents. Expert Opin Drug Deliv 7:1303–1327

Kelly BP (2012) Superficial fungal infections. Pediatr Rev 33:22–37

Keshri L, Pathak K (2013) Development of thermodynamically stable nanostructured lipid carrier system using central composite design for zero order permeation of econazole nitrate through epidermis. Pharm Dev Technol 18(3):634–644

Korting HC, Klovekornl W, Klovekorn G (1997) Comparative efficacy and tolerability of econazole liposomal gel 1%, branded econazole conventional cream 1% and generic clotrimazole cream 1% in tinea pedis. Clin Drug Invest 14:286–293

Koutsoulas C, Pippa N, Demetzos C et al (2014) Preparation of liposomal nanoparticles incorporating terbinafine in vitro drug release studies. J Nanosci Nanotechnol 14(6):4529–4533

Kumar N, Shishu (2015) D-optimal experimental approach for designing topical microemulsion of itraconazole: characterization and evaluation of antifungal efficacy against a standardized Tinea pedis infection model in Wistar rats. Eur J Pharm Sci 67:97–112

Kumar L, Verma S, Bhardwaj A, Vaidya S, Vaidya B (2014) Eradication of superficial fungal infections by conventional and novel approaches: a comprehensive review. Artif Cells Nanomed Biotechnol 42(1):32–46

Lewis RE (2010) Polyene antifungal agents. In: ComarúPasqualotto A (ed) Aspergillosis: from diagnosis to prevention. Springer, Heidelberg, pp 281–305

Madhu G, Agrawal U, Vyas S (2012) Nanocarrier-based topical drug delivery for the treatment of skin diseases. Expert Opin Drug Deliv 9:783–804

Mahale NB, Thakkar PD, Mali RG et al (2012) Niosomes: novel sustained release nonionic stable vesicular systems – an overview. Adv Colloid Interf Sci 183–184:46–54

Maheshwari RG, Tekade RK, Sharma PA et al (2012) Ethosomes and ultradeformable liposomes for transdermal delivery of clotrimazole: a comparative assessment. Saudi Pharm J 20(2):161–170

Marianecci C, Di Marzio L, Rinaldi F et al (2014) Niosomes from 80s to present: the state of the art. Adv Colloid Interf Sci 205:187–206

Mbah CC, Builders PF, Attama AA (2014) Nanovesicular carriers as alternative drug delivery systems: ethosomes in focus. Expert Opin Drug Deliv 11:45–59

Melkoumov A, Goupil M, Louhichi F et al (2015) Nystatin nanosizing enhances in vitro and in vivo antifungal activity against Candida albicans. J Antimicrob Chemother. doi:10.1093/jac/dkt137

Merisko-Liversidge E, Liversidge GG (2011) Nanosizing for oral and parenteral drug delivery: a perspective on formulating poorly-water soluble compounds using wet media milling technology. Adv Drug Deliv Rev 63:427–440

Munoz P, Guinea J, Rojas L et al (2010) New antifungal agents for the treatment of candidemia. Int J Antimicrob Agents 36:63–69

Neubert RH (2011) Potentials of new nanocarriers for dermal and transdermal drug delivery. Eur J Pharm Biopharm 77:1–2

Pandit J, Garg M, Jain NK (2014) Miconazole nitrate bearing ultraflexible liposomes for the treatment of fungal infection. J Liposome Res 24(2):163–169

Patel MR, Patel RB, Parikh JR et al (2009) Effect of formulation components on the in vitro permeation of microemulsion drug delivery system of fluconazole. AAPS PharmSciTech 10(3):917–923

Patel MR, Patel RB, Parikh JR et al (2011) Investigating effect of microemulsion components: in vitro permeation of ketoconazole. Pharm Dev Technol 16(3):250–258

Randhawa MA, Gondal MA, Al-Zahrani AH et al (2015) Synthesis, morphology and antifungal activity of nano-particulated amphotericin-B, ketoconazole and thymoquinone against Candida albicans yeasts and Candida biofilm. J Environ Sci Health Part A Toxic Hazard Subst Environ Eng 50(2):119–124

Sabale V, Vora S (2012) Formulation and evaluation of microemulsion-based hydrogel for topical delivery. Int J Pharm Invest 2(3):140–149

Sable CA, Strohmeier KM, Chodakewitz JA (2008) Advances in antifungal therapy. Annu Rev Med 59:361–379

Sahoo S, Pani NR, Sahoo SK (2014a) Effect of microemulsion in topical sertaconazole hydrogel: in vitro and in vivo study. Drug Deliv 20:1–8

Sahoo S, Pani NR, Sahoo SK (2014b) Microemulsion based topical hydrogel of sertaconazole: formulation, characterization and evaluation. Colloids Surf B 120:193–199

Salerno C, Carlucci AM, Bregni C (2010) Study of in vitro drug release and percutaneous absorption of fluconazole from topical dosage forms. AAPS PharmSciTech 11(2):986–993

Sanna V, Gavini E, Cossu M et al (2007) Solid lipid nanoparticles (SLN) as carriers for the topical delivery of econazole nitrate: in-vitro characterization, ex-vivo and in-vivo studies. J Pharm Pharmacol 59 (8):1057–1064

Santos P, Watkinson AC, Hadgraft J et al (2008) Application of microemulsions in dermal and transdermal drug delivery. Skin Pharmacol Physiol 21:246–259

Scalia S, Young PM, Traini D (2015) Solid lipid microparticles as an approach to drug delivery. Expert Opin Drug Deliv 12:583–599

Schwarz JC, Kählig H, Matsko NB et al (2011) Decrease of liposomal size and retarding effect on fluconazole skin permeation by lysine derivatives. J Pharm Sci 100 (7):2911–2919

Shaikh KS, Chellampillai B, Pawar AP (2010) Studies on nonionic surfactant bilayer vesicles of ciclopirox olamine. Drug Dev Ind Pharm 36(8):946–953

Sharma R, Pathak K (2011) Polymeric nanosponges as an alternative carrier for improved retention of econazole nitrate onto the skin through topical hydrogel formulation. Pharm Dev Technol 16(4):367–376

Sheehan DJ, Hitchcock CA, Sibley CM (1999) Current and emerging azole antifungal agents. Clin Microbiol Rev 12(1):40–79

Song CK, Balakrishnan P, Shim CK et al (2012) A novel vesicular carrier, transethosome, for enhanced skin delivery of voriconazole: characterization and in vitro/ in vivo evaluation. Colloids Surf B 92:299–304

Song SH, Lee KM, Kang JB et al (2014) Improved skin delivery of voriconazole with a nanostructured lipid carrier-based hydrogel formulation. Chem Pharm Bull 62(8):793–798

Souto EB, Wissing SA, Barbosa CM et al (2004) Development of a controlled release formulation based on SLN and NLC for topical clotrimazole delivery. Int J Pharm 278(1):71–77

Stock I (2010) Fungal diseases of vulva and vagina caused by Candida species. Med Monatsschr Pharm 33 (9):324–333

Sudhakar B, Varma JN, Murthy KV (2014) Formulation, characterization and ex vivo studies of terbinafine HCl liposomes for cutaneous delivery. Curr Drug Deliv 11 (4):521–530

Tanrıverdi ST, Özer Ö (2013) Novel topical formulations of Terbinafine-HCl for treatment of onychomycosis. Eur J Pharm Sci 48(4–5):628–636

Thompson GR, Cadena J, Patterson TF (2009) Overview of antifungal agents. Clin Chest Med 30:203–215

Torrado JJ, Espada R, Ballesteros MP et al (2008) Amphotericin B formulations and drug targeting. J Pharm Sci 97:2405–2425

Vanić Ž, Škalko-Basnet N (2013) Nanopharmaceuticals for improved topical vaginal therapy: can they deliver? Eur J Pharm Sci 50(1):29–41

Verma P, Pathak K (2012) Nanosized ethanolic vesicles loaded with econazole nitrate for the treatment of deep fungal infections through topical gel formulation. Nanomedicine 8:489–496

Zhang AY, Camp WL, Elewski BE (2007) Advances in topical and systemic antifungals. Dermatol Clin 25:165–183

Zhang JP, Wei YH, Zhou Y et al (2012) Ethosomes, binary ethosomes and transfersomes of terbinafine hydrochloride: a comparative study. Arch Pharm Res 35(1):109–117

Zonios DI, Bennett JE (2008) Update on azole antifungals. Semin Respir Crit Care Med 29(2):198–210

Synthetic Compounds for Antifungal Chemotherapy

8

Rupa Pegu, Rohan Borah, and Sanjay Pratihar

Abstract

Systemic fungal infections are major problems in phytopathology and infections caused by fungal species and to treat this various antimicrobial agents have been applied clinically, which is known as antifungal chemotherapy. Amongst various natural antifungal drugs, few are currently used in clinical practice to treat essentially systemic fungal infections because of their low spectrum of activity, toxicity and immunosuppressive nature. These limitation open up a market for new synthetic antifungal drugs or structural modifications of inactive molecules, which will provide a broad spectrum of activity with less toxicity. In this chapter, synthetic compounds for antifungal chemotherapy including various derivatives of azole, fluoropyrimidine, thiocarbamate, allyamines, morpholine, carabrol ester and carboline along with their mechanism, broad spectrum of activity, advantages and limitations have been discussed.

8.1 Introduction

We all are concerned about the coexistence of different microorganisms such as bacteria, viruses, and fungi with other living beings, i.e. humans, plants, and animals in the environment. Although some of these microorganisms are beneficial, many are pathogenic and are responsible for many life-threatening diseases. The therapeutics or therapy term deals with the treatment of these diseases, which may be done either by prevention or cure. The methods used in the therapy fall into three main categories: (i) physiotherapy, (ii) surgery, and (iii) pharmacological therapy (Loevenhart 1926). Pharmacology deals with the study of drugs and its action in the living systems to understand the properties of drugs and their actions, including the interactions of drug molecules with receptors and its effect. Chemotherapy (Northey 1943) is a phase of pharmacology and is broadly defined as the treatment of disease through administration of chemical agents or drugs that are highly toxic to the diseases but have low toxicity to the host. The 'chemotherapy' term was first introduced by Paul Ehrlich based on the affinity of a particular chemical to

R. Pegu • R. Borah • S. Pratihar (✉)
Department of Chemical Sciences, Tezpur University, Napaam, Assam 784028, India
e-mail: spratihar29@gmail.com; spratihar@tezu.ernet.in

A. Basak et al. (eds.), *Recent Trends in Antifungal Agents and Antifungal Therapy*,
DOI 10.1007/978-81-322-2782-3_8

certain living cells. Classification of chemotherapy is not easy as it has a broad field of application. However, on the basis of types of diseases or the nature of pathogen, they can be classified majorly in three categories: (i) anticancer, (ii) antimalarial, and (iii) antimicrobial chemotherapy. The antimicrobial chemotherapy is the clinical application of antimicrobial agents to treat the infectious disease caused by microscopic organisms and which may be categorised into three specific types like antibacterial, antiviral, and antifungal chemotherapy.

In this chapter, we will mainly focus on the available synthetic antifungal drugs for the treatment of various types of fungal infection. Various fungal species are widely distributed in soil, plant debris, and other organic substrates and causes a range of illnesses (mycoses) ranging from the chronic to the serious and can manifest themselves in a variety of ways. These infections can be *superficial*, that is, situated at or close to the surface of the skin, which can affect the body as a whole, instead of individual parts or organs. Diseases such as athlete's foot (tinea pedis), jock itch (tinea cruris), tinea manus (infection of the hand), thrush (oral and vaginal), and onychomycosis (affecting the nails) are some of the specific examples of superficial infections caused by the dermatophytes from the *Trichophyton*, *Microsporum*, *Candida*, and *Epidermophyton* species. The classifications of these diseases are difficult as infections may be caused by one or more of these organisms collectively and they can be developed in such a way that similar symptoms can be caused by different organisms. At the same time, detection of the disease from its symptom caused by a particular fungal agent is difficult because symptoms may vary depending upon the host immunity and physiological condition.

Systemic fungal infections are major problems in phytopathology, and infections caused by fungal species are most common in immune-compromised patient cases such as AIDS, cancer, old age, diabetes, cystic fibrosis, and organ transplants and other invasive surgical procedures. It is estimated that invasive fungal infections develop in 10–25 % of patients with acute leukaemia and those receiving bone marrow transplants (Sahin and Akova 2005). In the USA, *Candida* spp. is the fourth most common nosocomial pathogen with the highest crude mortality rate (40 %). In addition to *Aspergillus* spp., novel and emergent fungal pathogens such as zygomycetes, *Fusarium* spp., or *Scedosporium* spp. have become increasingly important pathogens. Their exposure to presently available antifungals may be limited, and resulting mortality rate is $\geq$70 % in patients with haematological malignancies. So, fungal infections are emerging as a major cause of morbidity and mortality in immune-compromised patients. Still, fungi are one of the most neglected pathogens, as demonstrated by the fact that the amphotericin B, a polyene antibiotic discovered as long ago as 1956, is still used as a 'gold standard' for antifungal therapy. The appearance of AIDS, the re-emergence of TB (including multidrug-resistant strains), and an overall increase in infectious disease mortality during the 1980s and early 1990s provide additional evidence that as long as microbes can evolve, new diseases will emerge. The advent of new diseases underscores the importance of disease prevention through antifungal chemotherapy and continual monitoring of underlying factors that may encourage the emergence or re-emergence of diseases (Spellberg et al. 2004). Following these problems, effort has been given in the hope of finding a substance that will be less toxic to the host and more potent in its particular type of pharmacological activity. In order to get success to treat a particular type of infection by using a synthetic compound, a collective effort of chemists, pharmacologists, and physicians are required.

8.2 Available Drugs in the Market

The growth of antifungal drugs was slow paced despite the knowledge of their existence well before that of bacteria. This is exemplified by the fact that the first antibacterial agent, penicillin, came into use in 1941, whereas the first antifungal agent, nystatin, was not discovered until 1949. The main reason believed for the slow development of broad-spectrum agents can be heaped on the nature of these fungal diseases; they are eukaryotic species which are biochemically similar to human hosts on contrary to prokaryotes like bacteria. This means that development of a drug to combat fungal infections leaving the human host unaffected is a very difficult job. The natural and synthetic drugs have been developed over the years to treat mycotic infections. Recognition of the rapid expansion of the fungal market has spurred efforts to discover alternative structural classes of antifungal agents. Despite extensive research on the development of new therapeutic strategies, there are few drugs available in the market to cure the invasive mycoses.

During the past two decades, a wide array of antifungal agents have been discovered, which have been or are currently being evaluated for use in treating invasive mycoses and are categorised by their site of action in fungal cells (Dismukes et al. 2000). Indeed, only four molecular classes, namely, (i) fluoropyrimidine analogues, (ii) polyenes, (iii) azoles, and (iv) echinocandins, are currently used in clinical practice to treat essentially systemic fungal infections, which are mainly targeted to three distinct fungal metabolic pathways. Several other classes of antifungal drugs, such as morpholines, allylamines, etc., are used only as topical agents due to either poor efficacy or severe adverse effects when administered systemically. These drugs acted upon fungal strains by three general mechanisms, which are inhibition of cell wall biosynthesis, disruption of the cell membrane, and inhibition of cell division. Inhibition with fungal cell wall biosynthesis has not been as successful and operative as penicillins or cephalosporins against bacteria. Till date, various chemicals have been discovered that interfere with numerous stages in fungal cell wall synthesis with excellent antifungal activity in vitro. Unfortunately, development of these agents into useful drugs has proven to be very difficult. Antifungal agents disrupt the cell membrane by targeting ergosterol, either by binding to the sterol with the formation of pores and effecting the membrane to become leaky (as with polyene antifungals) or inhibiting ergosterol biosynthesis (as seen with azole antifungal agents). Ergosterol is comparable to mammalian cholesterol; thus, agent-binding ergosterol may have a cytotoxic effect in the host tissue. A summary of various classes of antifungal drugs and their mechanism of action has been provided in Table 8.1.

Amongst various available classes of compounds, imidazoles such as clotrimazole, miconazole, sulconazole, and bifonazole work by binding to cytochrome P450, thus blocking ergosterol synthesis in the fungal cell membrane. The development of triazole antifungal agents that were less likely to cause hepatotoxicity possibly due to their diminutive effects on cytochrome P450-dependent enzymes. On the other hand, amongst various triazoles, terconazole was first developed, followed by itraconazole and finally fluconazole. Before the development of fluconazole, the two major antifungal agents existing were amphotericin B and flucytosine. The problems related with amphotericin B are that it has to be given intravenously and had harsh side effects, whereas flucytosine was orally active but had a limited range of action and resistance was easily developed. Later on, amphotericin B (AMB) was incorporated in three lipid formulations to reduce toxicity, while the first-generation triazoles (fluconazole and itraconazole) changed the epidemiology of *Candida* infections and presented a new treatment opportunity. A lot of these antifungals have proven to be less toxic and in some cases more effective than amphotericin B. However, most of the azole antifungals are highly lipophilic and are not effective orally because of their metabolic degradation. In this regard, Janssen Pharmaceutica had developed ketoconazole in mid-1978, which was proven to be effective orally, but it still degraded metabolically. Later on, Pfizer replaced the imidazole unit with 1,2,4-triazole derivative to get better activity. But they are not able to overcome the susceptibility to metabolism with those derivatives, and they found more activity in vivo but four times less potent in vitro. Various substituted

Table 8.1 Various marked antifungal drugs and their mechanism of action

Polyenes
Mechanism of action: binds to ergosterol in the fungal membrane causing the cell membrane to become leaky
Available drug: amphotericin B, nystatin, natamycin
Azoles (mainly imidazoles, triazoles, and thiazole)
Mechanism of action: inhibits ergosterol synthesis, thereby disrupting membrane permeability and also function of membrane-bound proteins
Available drug: ketoconazole, miconazole, clotrimazole, tioconazole, econazole, fenticonazole, sulconazole, sertaconazole, oxiconazole, tinidazole, enilconazole (Imazalil), parconazole, eberconazole, lanoconazole, bifonazole, lombazole, voriconazole, itraconazole, butoconazole, fluconazole, terconazole, posaconazole, abafungin
Allylamines
Mechanism of action: inhibits conversion of squalene to squalene-2,3-epoxide by squalene epoxidase. Squalene is toxic to the cell wall causing P^H imbalances and therefore membrane protein malfunction.
Available drug: amorolfin, butenafine, naftifine, terbinafine, tolnaftate
Echinocandins
Mechanism of action: inhibit the synthesis of glucan in the cell wall, via noncompetitive inhibition of the enzyme 1,3-β glucan synthase
Available drug: pneumocandins, echinocandin B, cilofungin, caspofungin, micafungin, anidulafungin
Thiocarbamate
Mechanism of action: inhibit squalene epoxidase, an important enzyme in the biosynthetic pathway of ergosterol
Available drug: tolnaftate, tolciclate
Other available drugs:
Flucytosine (inhibits cell division via incorporation into RNA leading to faulty RNA synthesis)
Griseofulvin (inhibits microtubule polymerisation, thus inhibiting the formation of mitotic spindle)
Tavaborole (inhibition of leucyl-tRNA synthetase)
Undecylenic acid (inhibition of fatty acid biosynthesis which disrupts germ tube (hyphae) formation)
Ciclopirox (disrupts various enzyme activities including cytochromes, catalase, peroxidise, Na-K ATPase, etc.)
Amorolfine (disrupts the ergosterol biosynthetic pathway by inhibiting δ14 reductase and D8-D7 isomerase enzymes)

bis-triazoles were synthesised and tried to get better activity. Finally, water-soluble fluconazole was discovered for intravenous administration and was found to be very effective for a wide variety of fungal infections and marketed in various countries.

8.3 Synthetic Compounds for Antifungal Chemotherapy

It has been found that the previously discovered natural antifungal drugs have very low spectrum of activity and their applications are limited due to their toxicity and immunosuppressive nature (e.g. amphotericin B). These limitation leads to the development of synthetic drugs. That is why people are giving constant effort to develop structurally new classes of antifungal agents with novel mechanisms of action as well as structural modifications to improve their spectrum of activity. Towards the search for a new antifungal agent, various groups of compounds have been tried. Amongst these various compounds, success has been achieved with some selected groups of compounds like imidazoles, triazoles and thiazole, echinocandins, fluoropyrimidine, morpholine, thiocarbamate, etc. In addition, a good number of synthetic drugs which are designed to bind independently to different biological targets are reported in literature. But due to lack of sufficient biological data, they remain ineffective for clinical trial. Herein, we will discuss various classes of synthetic compounds for antifungal chemotherapy along with their mechanism of action. In this regard, their broad spectrum of activity, advantage, and limitation will also be discussed.

8.3.1 Azoles

In the first place, azoles (mainly imidazoles, triazoles, and thiazole) are by far the most commonly used systemically acting synthetic antifungals in clinical practice. Although imidazoles, triazoles, and thiazole function under a similar mechanism, they behave differently with respect to their fungal spectrum, pharmacokinetics, and toxicities. These agents primarily inhibit the ergosterol biosynthesis by inhibiting the cytochrome P450-dependent enzyme 14-α-demethylase within the fungal cell (Bossche 1997; Andriole 1999; Sheehan et al. 1999). This inhibition occurs through the binding of the free nitrogen atom of the azole ring to the iron atom of the haem group in the enzyme. The subsequent accumulation and metabolism of 14-alpha-methylated sterol species causes the synthesis of toxic compounds, which are unable to successfully replace ergosterol (Carrillo-Munoz 2006). All azoles are fungistatic at their MIC, but some are fungicidal at higher concentrations. This class of compounds includes imidazoles (e.g. econazole, miconazole, bifonazole, butoconazole, tioconazole, ketoconazole, etc.), triazoles (e.g. terconazole, fluconazole, itraconazole, voriconazole, posaconazole, etc.), and thiazole (e.g. abafungin). In addition, many other azole-based antifungal drugs have been synthesised, of which a few are licensed for use in human therapy. Azoles possess very broad activity, with effects on dermatophytes, *Trichophyton*, *Microsporum*, and *Epidermophyton*; dimorphic fungi, *Histoplasma*, *Blastomyces*, and *Coccidioides*; filamentous fungi (moulds), *Aspergillus*; and yeasts, *Candida*, *Cryptococcus*, etc.

8.3.1.1 Imidazoles

Imidazole is an important member of the azole family. Till date, various imidazole compounds have been synthesised and tried for antifungal chemotherapy. Amongst various imidazole derivatives, some compounds fall in the first generation.

First generation: miconazole, econazole, isoconazole, bifonazole, eberconazole, lombazole, sertaconazole, tioconazole, lanoconazole, clotrimazole (Pelletier et al. 2000), fenticonazole, sulconazole, and tinidazole

These early imidazoles have very poor aqueous solubility because of the lipophilic nature and are generally ineffective for oral or systemic therapy. However, they have exceptional skin/mucous membrane compatibility and hence are beneficial for topical application. They are normally applied either as creams for dermatophyte infections or as 'nail varnish' formulations for onychomycosis. Amongst various first-generation imidazoles, miconazole provides an exception to the others because it is absorbed in the gut slightly better. Thus, a gel formulation of miconazole is available for oral and intestinal fungal infections. However, miconazole has strong binding affinity to lipoproteins (up to 95 % is protein bound and only 20–30 % of a dose is detectable in the blood). So, its 'effective' concentration is hence much decreased. The incidences of secondary resistance were uncommon with these drugs, but are now rising from 1990s onwards. Some of the common disadvantages associated with these drugs are (i) potential hepatotoxic effects in long-term therapy, (ii) possible teratogenic/embryotoxic effects, and (iii) that they may affect testosterone/corticosteroid synthesis (acting against other cytochrome P450 enzymes).

Janssen sought to improve upon miconazole and econazole, but keeps hold of the important advantages of the early imidazoles, namely, high antimycotic activity and broad spectrum of activity. Their main aim was to provide a drug that was bioavailable, i.e. that was more soluble in water and could be used for systemic infections either by injection or as a tablet. In 1979, ketoconazole (Bossche 1997), a *second-generation* drug containing imidazole group, was found to be suitable to treat fungal infections of the skin such as athlete's foot, jock itch, ringworm, and seborrhoea (dry, flaking skin) and is available as a cream, gel, and shampoo. Some of the drawbacks associated with the ketoconazole are (i) limited efficacy against *Aspergillus*, (*ii*) metabolically vulnerable (<5 % excreted unchanged), (iii) 95–99 %

of drug bound to lipoproteins in serum (activity drops 10–1,000-fold in the presence of serum with in vitro tests), (iv) poor penetration into CSF, and (v) best absorbed by the gut in acid conditions, so antacids should not be prescribed at the same time as ketoconazole. Some of the selected structure along with their MIC value against the fungal strain was described in Table 8.1.

8.3.1.2 Triazoles

The triazoles (Shalini et al. 2011) are also an important member of azole family, which are active against *Candida albicans*, non-*albicans* *Candida* spp., *Cryptococcus neoformans*, and the dimorphic fungi. Amongst non-*albicans* *Candida* spp., they are less active against *Candida glabrata* and inactive against *Candida krusei* with the exception of voriconazole and some investigational triazoles. Various *Aspergillus* spp. and dematiaceous patterns are only affected by itraconazole and voriconazole. The *first-generation* triazoles are fluconazole and itraconazole (Loevenhart 1926). Fluconazole, a fungistatic agent, is active against *Candida albicans*, *Candida tropicalis*, and *Candida glabrata*. Fluconazole is also used for the treatment of meningitis, cryptococcal meningitis, and systemic and mucosal candidiasis in both normal and immune-compromised patients. It is also used to treat coccidioidal meningitis, histoplasmosis, and infections due to chemotherapy or radiation therapy prior to a bone marrow transplant. Fluconazole circulates in plasma as the free form and is excreted unchanged through the kidneys. It is also orally bioavailable because it shows negligible hepatic metabolism. Almost 94 % fluconazole is absorbed and its oral bioavailability is not affected by food or gastric pH. On the other hand, itraconazole is largely metabolised in the liver by cytochrome P450 3A4, an active metabolite produced which is excreted in an inactive form via the liver and kidneys. It has strain-dependent fungicidal activity against filamentous fungi with the exception of some strains of *Cryptococcus neoformans*. It is taken orally as a capsule to treat fungal infections that start in the lungs and spread throughout the body. It can also be used to treat fungal infections of the nails and its oral solutions are useful to treat oral candidiasis. The most common side effects associated with itraconazole are constipation, heartburn, bleeding gums, headache, and dizziness. The more severe side effects with itraconazole can include excessive tiredness, nausea, vomiting, loss of appetite, tingling or numbness in the extremities, and difficulty in breathing or swallowing. The second-generation azole derivatives are voriconazole, posaconazole, and ravuconazole (Northey 1943). In 2002, voriconazole was first marketed and accepted as a first-line cure of oesophageal candidiasis; candidaemia; invasive aspergillosis; candidal infections of the skin; infections in the abdomen, kidney, and bladder wall and wounds; and infections caused by *Scedosporium apiospermum* and *Fusarium* spp. It is available for both oral and intravenous (i.v.) administration. Although it is orally bioavailable, it exhibits nonlinear pharmacokinetics. Voriconazole is metabolised in the liver, and, in case of renal failure, its excretion is not affected. Voriconazole interacts with drugs that are substrates of cytochrome P450 3A4 (terfenadine, cisapride, etc.), by increasing their serum levels. Its intake should be avoided in those patients who are taking cyclosporine, rifampicin, carbamazepine, ritonavir, and long-acting barbiturates.

Posaconazole is a hydroxylated analogue of itraconazole, which became available in Europe in 2005 and was approved by the Food and Drug Administration (FDA) in 2006. It has a wide range of antifungal activity against various fungal strains, viz. *Candida* spp. resistant to older azoles, *Cryptococcus neoformans*, *Aspergillus* spp., *Rhizopus* spp., *Bacillus dermatitidis*, *Candida immitis*, *Histoplasma capsulatum*, and other opportunistic filamentous and dimorphic fungi (Andriole 1999). Posaconazole is available only for oral administration, and it is 8–47 % bioavailable on an empty stomach, which increases by 400 % with the ingestion of a fatty meal. It is primarily metabolised by the liver, and approximately 77 % of the unaltered drug is excreted in the faeces and small amount in the urine. The posaconazole shows good efficiency for the treatment of zygomycosis, invasive fusariosis, cryptococcal meningitis, coccidioidomycosis, and other central nervous system fungal infections. A few common side effects associated with posaconazole are nausea, vomiting, headache, abdominal pain, and diarrhoea. Ravuconazole is another second-generation azole antifungal agent, which is highly active against a wide range of fungi, including *Candida* spp., *Candida neoformans*, and other yeast species, as well as isolates that are resistant to fluconazole (Gomez-Lopez et al. 2004). An in vitro study evaluating the activity of ravuconazole against 923 clinical isolates of non-dermatophyte filamentous fungi showed that the drug is highly active against *Aspergillus* spp. and has inhibitory activity against other species of hyaline filamentous fungi and zygomycetes and black moulds. The study further indicated that it is active against 56.2 % of *Mucorales* tested, whilst itraconazole was active only against one-third of the isolates, and voriconazole was inactive against almost the whole collection of *Mucorales*. Ravuconazole shows linear pharmacokinetics over the anticipated dosage range and undergoes hepatic metabolism, and it also has a significant potential for drug-drug interactions through the cytochrome P450 enzyme system (Table 8.2).

8.3.1.3 Other Imidazole Derivatives

In addition to the marketed imidazole antifungal drugs, several other imidazole derivatives have been synthesised so as to extend the spectrum of antifungal chemotherapy. Amongst various synthesised imidazoles, some selected compounds were found active against different fungal strains, which are summarised in Table 8.3. In this regard, Rauf et al. reported a series of two-substituted imidazole derivatives, which were found potent against fungal strains, *A. niger*, *C. albicans*, *Penicillium* spp., *T. viride*, and *Helminthosporium oryzae*, with MIC of 75–150 μg/ml when evaluated via disc diffusion method. All compounds are good antimycotics against yeasts and dermatophytes and are more active than bifonazole. The unsubstituted derivative of **1b** has MIC value ranging from 0.4 to 33 μg/ml which appeared to be most active in this series. On the other hand, Ogata et al. reported 1-[1-[2-[(3-chlorobenzyl)oxy]phenyl]vinyl]-1H-imidazole hydrochloride (**1e**) that was found to be active against typical dermatophyte species with MIC 0.16–1.25 μg/ml and *Aspergillus* spp. and *Penicillium* spp. with MIC 0.63–5 μg/ml. Mamolo et al. reported 1-(3,5-diaryl-4,5-dihydro-1H-pyrazol-4-yl)-1H-imidazole derivatives (**1h**) which show better in vitro antifungal activity than miconazole. Karaburan et al. reported aryl[3-(imidazol-1-ylmethyl)benzofuran-2yl] ketoximes (**1l**) with moderate antifungal potency of MIC 5–25 μg/ml. On the other hand, Na et al. reported a series of 1-benzyl-3-(imidazol-1-ylmethyl)indole derivatives (**1m**), and all the compounds were tested against two human fungal pathogens, *C. albicans* and *A. fumigatus* with good activity. Mullen et al. also reported 3,5-diphenyl-3-(1H-imidazol-l-ylmethyl)-2-alkylisoxazolidine derivatives (**1o**) which were found to be the most potent compounds in vitro with MIC values ranging between 0.2 and 7.0 μg/ml (Rani et al. 2013).

8.3.1.4 Fluoropyrimidines

Fluoropyrimidines (Vandeputte et al. 2011) are a class of compounds, of which only 5-fluorocytosine (5-FC) and 5-fluorouracil (5-FU) are used in human drugs and are synthetic structural analogues of the DNA nucleotide cytosine of candidiasis. It inhibits pyrimidine metabolism via intrusion with RNA and protein synthesis in the fungal cell (Spellberg et al. 2004). In 1964, 5-FC was shown to be active in vivo (mice) against experimental cryptococcal and candidal infections. Its spectrum of activity is narrow and 8–10 % of *Candida* strains are resistant, which is a major drawback for this compound and is limited to yeast infections. Thus, it is used for cryptococcal meningitis and in case of dissemination in combination with

Table 8.2 The structure and activity against particular fungal strain of various azoles

Compound	Fungal strains	MIC (μg/ml)
Miconazole	*T. mentagrophytes*	0.312–80
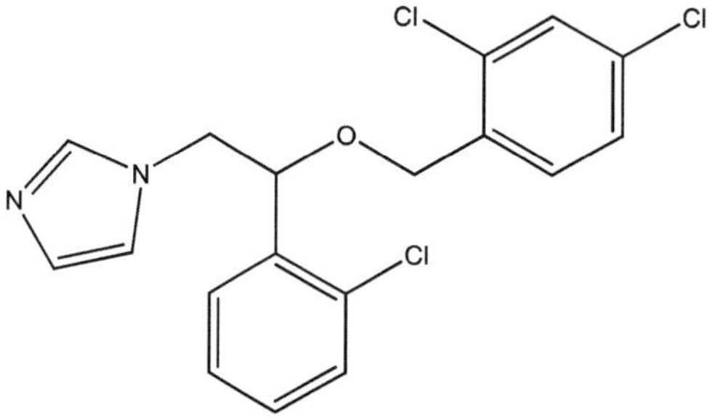	*T. rubrum*	0.312–0.55
	T. tonsurans	10
	T. verrucosum	8
	M. audouinii	9–10
	M. gypseum	10–12
	Epidermophyton floccosum	0.4–0.625
	C. albicans	0.25–32
	C. parapsilosis	0.12–0.5
	C. tropicalis	0.5–16
Meerpoel et al. (2010)	*C. krusei*	2–4
	C. guilliermondii	0.25–1
	C. glabrata	<0.05–3.125
	Cryptococcus neoformans	0.625–1.95
	Geotrichum candidum	40
	Torulopsis glabrata	1.25
	Malassezia furfur	0.88–6.25
	A. fumigatus	33.1–160
	A. niger	22–40
	A. flavus	12
Econazole	*T. mentagrophytes*	0.60–0.80
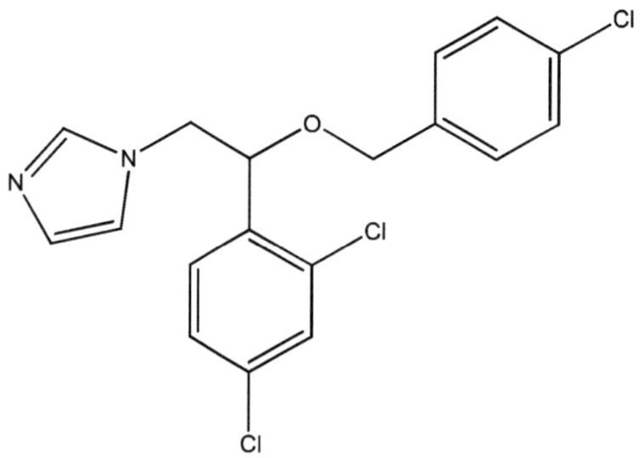	*T. rubrum*	0.58
	T. tonsurans	4
	T. verrucosum	10
	M. audouinii	10
	M. gypseum	1.3
	Epidermophyton floccosum	0.25–32
	C. albicans	0.25–4
	C. parapsilosis	0.5–16
	C. tropicalis	4–8
	C. krusei	0.5–4
	C. guilliermondii	<0.05–12.5
Meerpoel et al. (2010)	*C. glabrata*	2.25–2.62
	Cryptococcus neoformans	40
	Geotrichum candidum	0.31
	Torulopsis glabrata	0.78–6.25
	Malassezia furfur	28
	A. fumigatus	18
Isoconazole	*C. albicans*	0.12–4
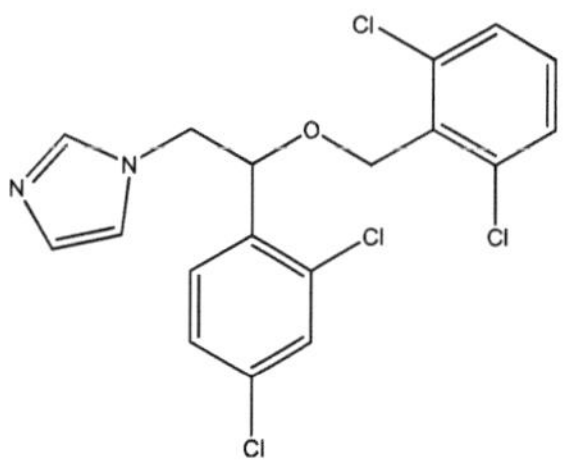	*C. parapsilosis*	0.12–0.25
	C. tropicalis	0.12–2
	C. krusei	1–2
	C. guilliermondii	0.12–1
Meerpoel et al. (2010)		

(continued)

Table 8.2 (continued)

Compound	Fungal strains	MIC (μg/ml)
Bifonazole 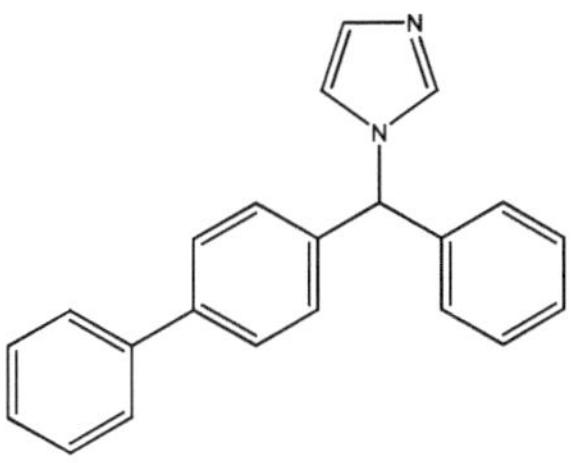 Baziard et al. (2011)	*A. niger*	4
	C. albicans	128
	C. krusei	128
	T. rubrum	4
	C. tropicalis	–
	C. glabrata	–
	Candida spp.	–
	Aspergillus spp., *Trichophyton* spp.	–
Eberconazole 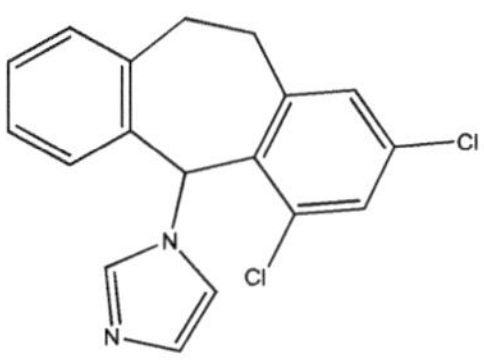 Guarro et al. (2003)	*E. floccosum*	0.03–1
	M. audouinii	0.03–2
	M. canis	0.01–4
	M. canis	0.03–0.5
	M. ferrugineum	0.03–2
	M. fulvum	0.03–1
	M. gallinae	0.03–8
	M. gypseum	>16
	M. racemosum	0.03–2
	T. balcaneum	0.03–4
	T. rubrum	0.03–4
	T. tonsurans	0.125–>16
	T. simii	0.03–8
Lombazole 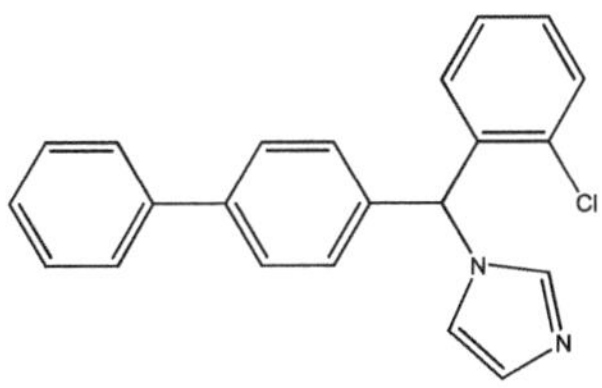	*C. albicans*	–
Sertaconazole 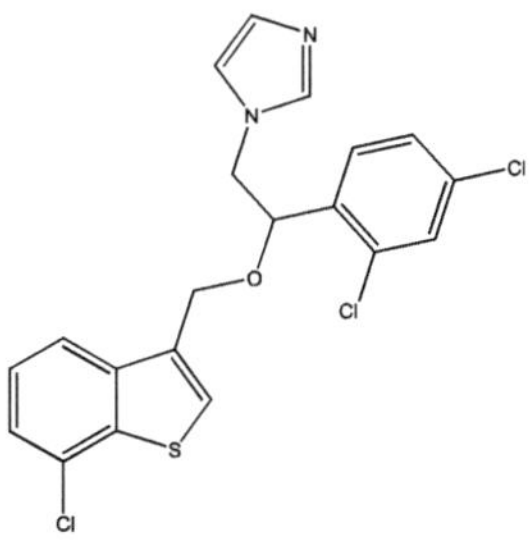 Meerpoel et al. (2010)	*C. albicans*	0.03–1
	C. parapsilosis	0.06–0.5
	C. tropicalis	0.125–8
	C. krusei	0.06–4
	C. glabrata	0.125–0.5

(continued)

Table 8.2 (continued)

Compound	Fungal strains	MIC (μg/ml)
Tioconazole 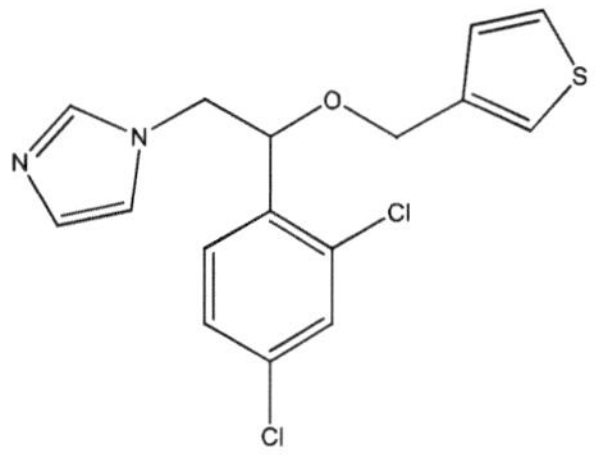 Jevons et al. (1979)	*C. albicans*	1.6–12.5
	Candida spp.	<0.01–6.3
	Cryptococcus neoformans	0.01–0.4
	Torulopsis glabrata	0.05–0.07
	Aspergillus fumigatus	3.1–9.4
	Aspergillus fumigatus	3.1–6.2
	Aspergillus fumigatus	0.08–0.18
	T. mentagrophytes	0.5
	T. rubrum	0.3–0.5
Lanoconazole 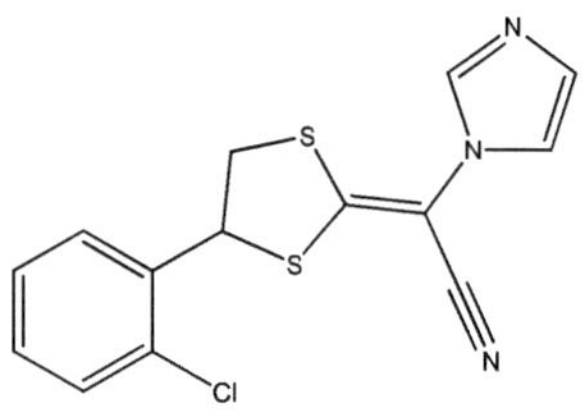 Koga et al. (2006) Niwano et al. (1995)	*C. albicans*	0.63–5
	Trichophyton rubrum	0.00024–0.016
	T. mentagrophytes	0.0005–0.004
	Epidermophyton floccosum	0.001
Clotrimazole 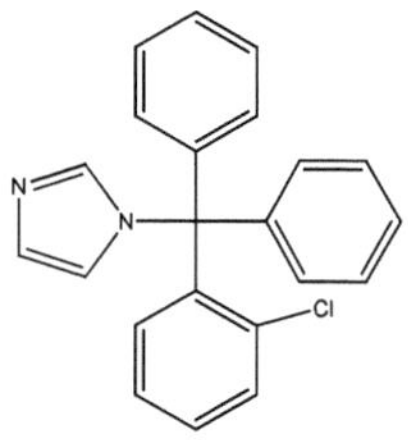 Borelli et al. (2007)	*C. albicans*	2–32
	C. glabrata	1–32
	C. guilliermondii	1–>64
	C. krusei	1–2
	C. tropicalis	2–8
	C. parapsilosis	1–>64
	T. rubrum	0.06
	T. mentagrophytes	0.06–1.0
	A. fumigatus	1–2
	A. niger	1–4
	M. gypseum	0.06
Fenticonazole 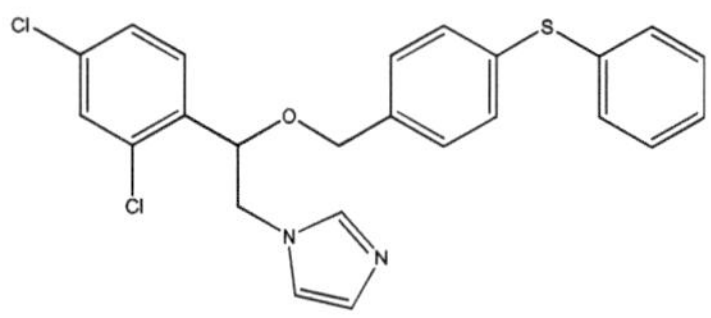	*T. mentagrophytes*	0.28–20
	T. rubrum	0.33–0.625
	T. verrucosum	1.25
	T. tonsurans	0.25
	M. canis	5–20
	M. audouinii	8–9.2
	M. gypseum	5–8.8
	Epidermophyton floccosum	0.625–1.1
	C. albicans	0.25–32
	C. parapsilosis	0.25–1
	C. tropicalis	0.5–16
	C. krusei	4–16
	C. guilliermondii	2–8
	C. glabrata	0.03–2
	Cryptococcus neoformans	0.312–0.55
	Geotrichum candidum	40
	Torulopsis glabrata	40
	A. fumigatus	22.3–160
	A. niger	10–37.5
	A. flavus	10

(continued)

Table 8.2 (continued)

Compound	Fungal strains	MIC (μg/ml)
Sulconazole 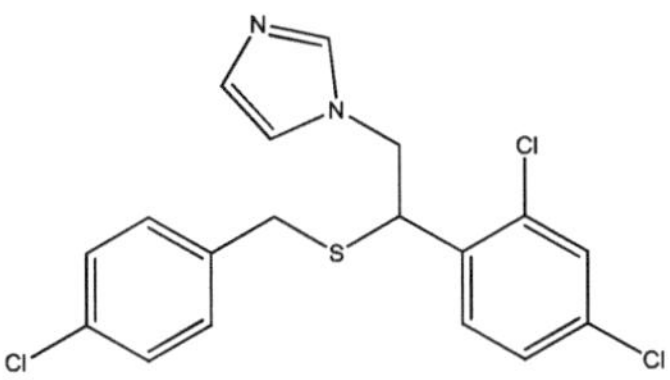 Meerpoel et al. (2010)	*C. albicans*	0.12–8
	C. parapsilosis	0.12–8
	C. tropicalis	0.25–8
	C. krusei	4–8
	C. guilliermondii	0.5–2
Tinidazole X 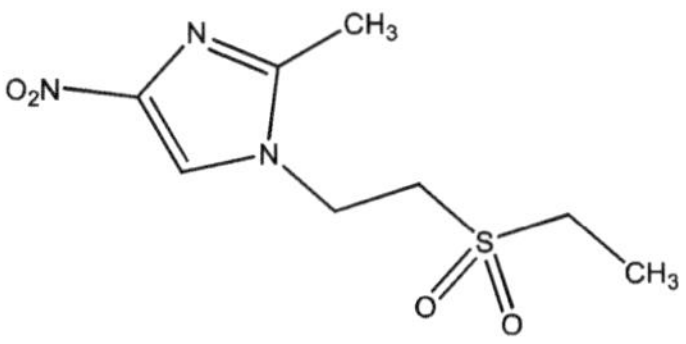	*B. fragilis* subsp. *fragilis*	0.4–1.6
Ketoconazole 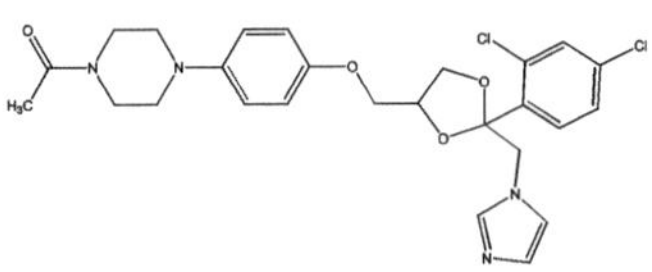 Meerpoel et al. (2010)	*C. albicans*	0.03–0.25
	C. parapsilosis	0.03–0.06
	C. tropicalis	0.03–4
	C. krusei	0.03–
	C. guilliermondii	0.5–4
	C. glabrata	0.03–2
Terconazole 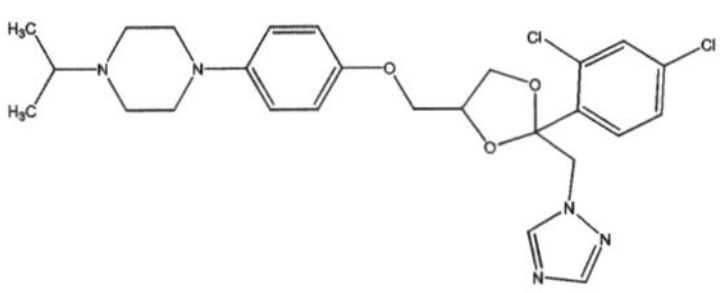	*M. canis*	
	T. mentagrophytes	
	T. rubrum	
	Cryptococcus neoformans	
	C. albicans	
	C. tropicalis	
	Phialophora verrucosa	
	Sporothrix schenckii	
	A fumigatus	
	Mucor spp.	
	Saprolegnia spp.	
Fluconazole 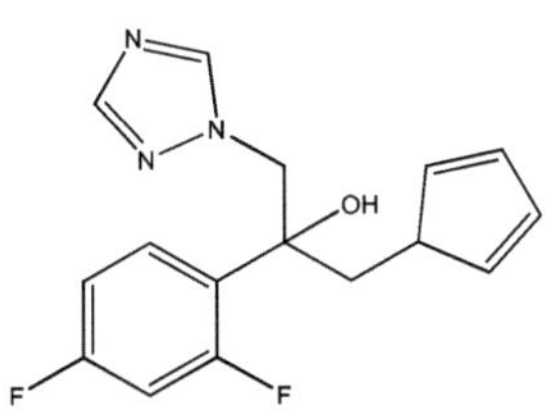 Sanclemente et al. (2009)	*C. parapsilosis*	0.125–8
	C. albicans	0.125–4
	C. guilliermondii	0.125–64
	C. tropicalis	0.125–1
	T. rubrum	1–64
	T. mentagrophytes	16–64
	M. gypseum	64
Posaconazole 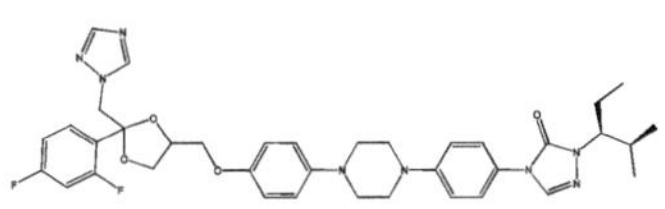 Pfaller et al. (2011)	*C. albicans*	0.007–2
	C. glabrata	0.007–8
	C. parapsilosis	0.007–1
	C. tropicalis	0.007–2
	C. krusei	0.015–1
	C. lusitaniae	0.007–1
	C. guilliermondii	0.015–2
	C. kefyr	0.007–25

(continued)

Table 8.2 (continued)

Compound	Fungal strains	MIC (μg/ml)
Itraconazole	*C. parapsilosis*	0.031–0.5
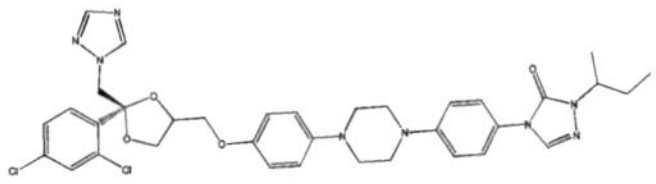	*C. albicans*	0.031–0.5
	C. guilliermondii	0.031–16
	C. tropicalis	0.031–0.125
Sanclemente et al. (2009)	*T. rubrum*	0.06–1
	T. mentagrophytes	0.03–0.5
	M. gypseum	0.25–1.0
Voriconazole	*C. albicans*	0.031–0.5
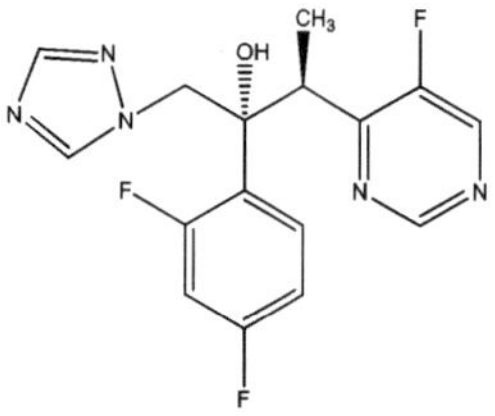	*C. glabrata*	
	C. parapsilosis	
	C. tropicalis	
	C. Krusei	
	C. lusitaniae	
Cuenca-Estrella et al. (2004)	*C. guilliermondii*	
Ravuconazole	*C. albicans*	0.031–16
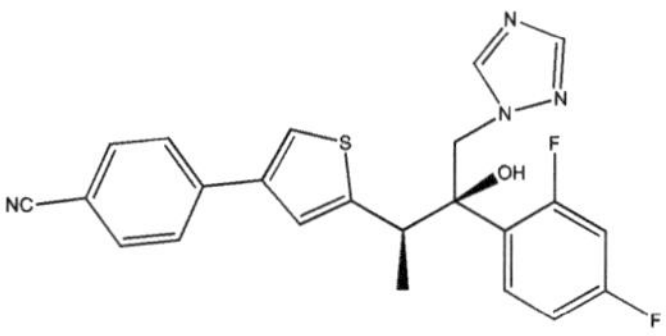	*C. glabrata*	
	C. parapsilosis	
	C. tropicalis	
	C. krusei	
	C. guilliermondii	
Estrella et al. (2004)	*C. lusitaniae*	
	C. haemulonii	
	C. dubliniensis	
	C. rugosa	
	C. inconspicua	
	C. zeylanoides	
	D. hansenii	
	S. cerevisiae	
	Y. lipolytica	
	Arxiozyma telluris	
	T. delbrueckii	
Abafungin	*T. mentagrophytes*	0.031–0.125
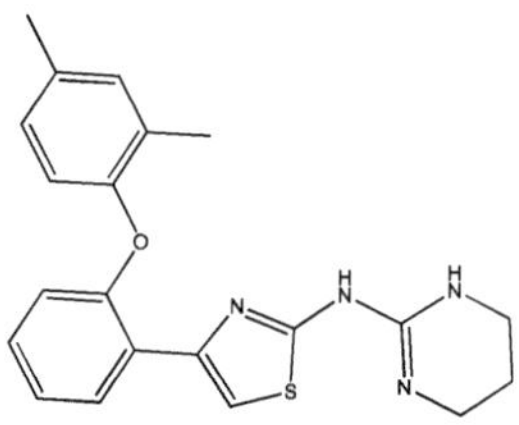	*T. rubrum*	
	T. equinum	
	C. albicans	
	C. glabrata	
	C. krusei	
	C. tropicalis	
	C. parapsilosis	
C. Borelli et al. (2008)	*A. fumigatus*	
	A. niger	
	A. amstelodami	
	M. mycetomatis	

Table 8.3 Structures and activities against specific fungal strains of some imidazole derivatives

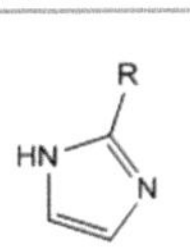

1a

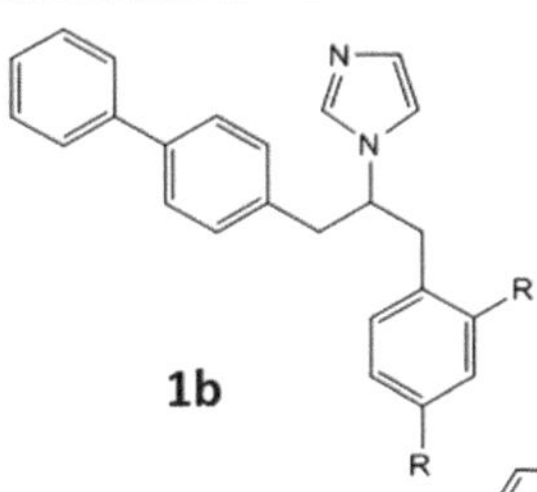

1b

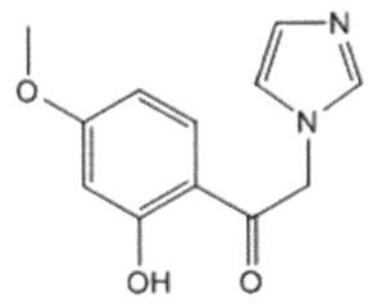

1c

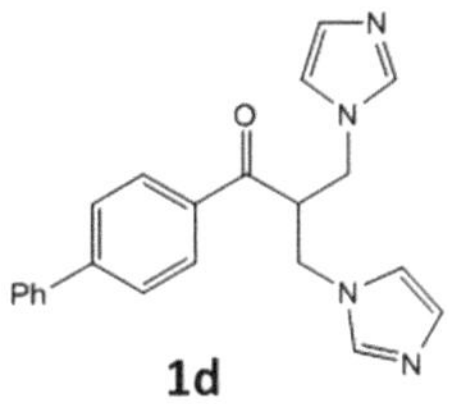

1d

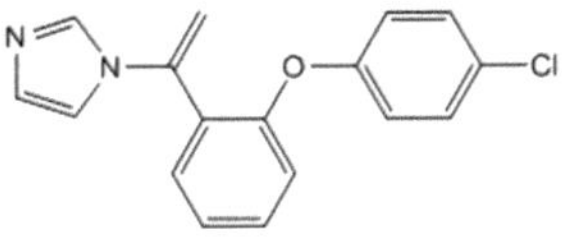

1e

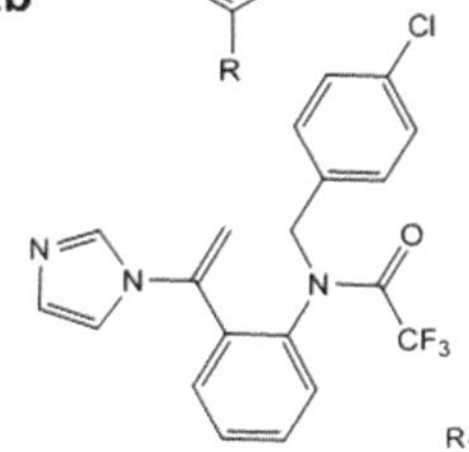

1f

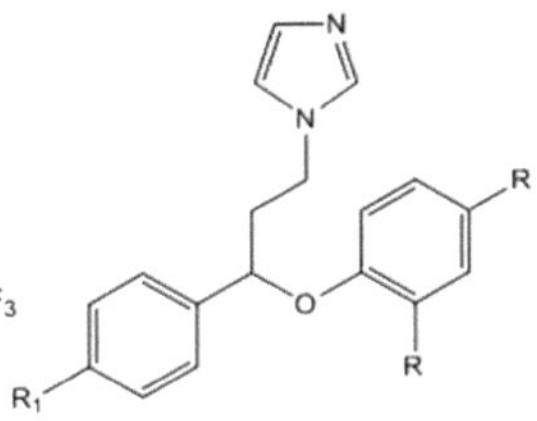

1g

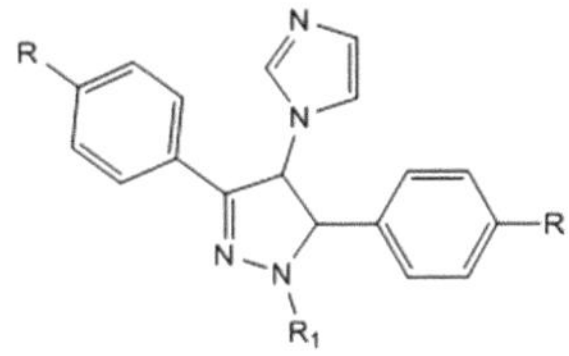

1h

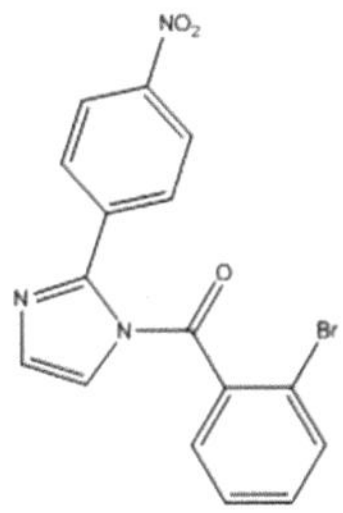

1i

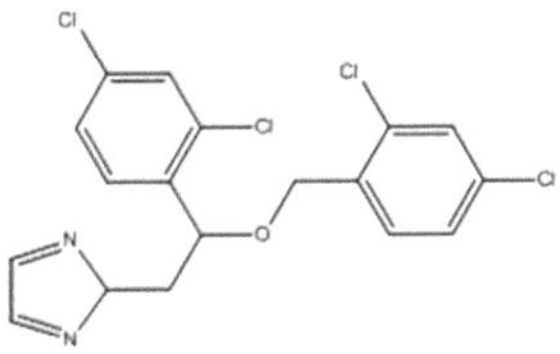

1j

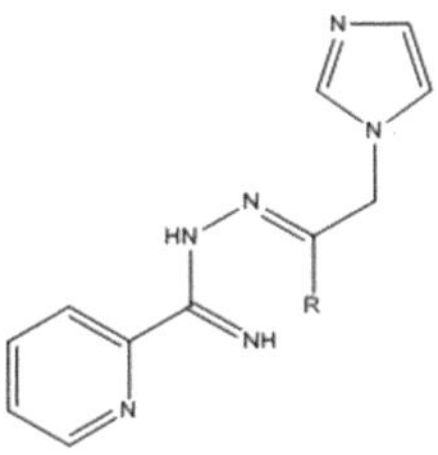

1k

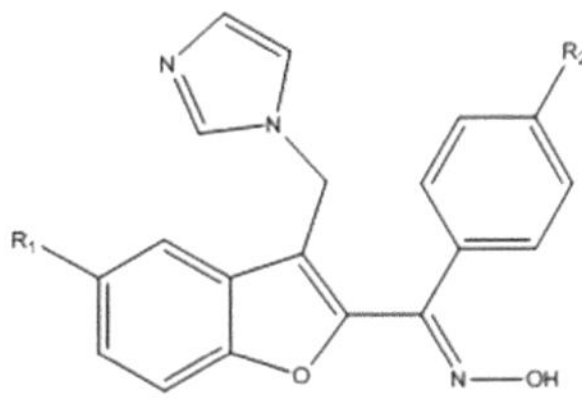

1l

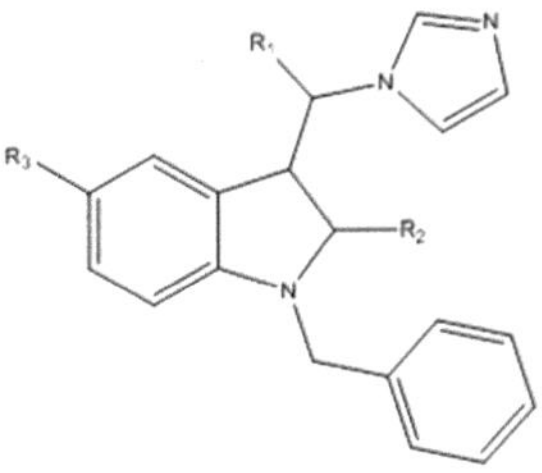

1m

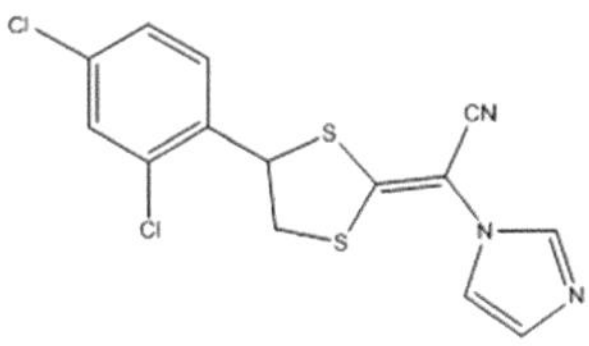

1n

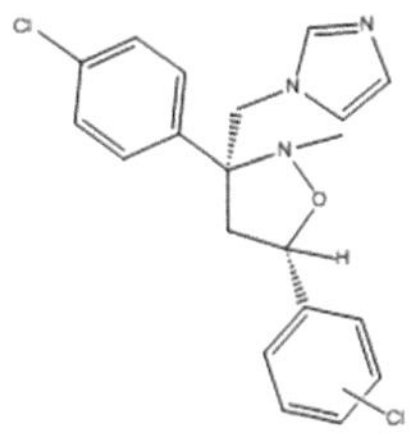

1o

Compounds	Fungal strain	MIC (μg/ml)
1a	*A. niger*	0.06–1
	C. albicans	
	Penicillium spp.	
	T. viride	
	Helminthosporium oryzae	
1b	*C. albicans*	0.03–0.5
	C. parapsilosis	
	C. neoformans	
	T. verrucosum	
	T. rubrum	
	M. gypseum	
	A. fumigates	

(continued)

Table 8.3 (continued)

Compounds	Fungal strain	MIC (μg/ml)
1c	*C. albicans*	0.25–1.0
	C. parapsilosis	
	C. neoformans	
	T. verrucosum	
	T. rubrum	
	M. gypseum	
1d	*C. albicans*	2–4
1e	*Dermatophyte* spp.	0.16–1.25
	Aspergillus spp. and *Penicillium* spp.	0.63–5.0
	Candida spp. and other yeasts	10–80
1f	*C. albicans*	1.6–50
	A. fumigates	
	T. asteroids	
1g	*C. albicans*	0.125–128
1h	*C. albicans*	0.06
1i	*A. niger*	4
	C. albicans	16
1j	*E. floccosum*	25–100
	P. notatum	
	A. niger	
	A. fumigates	
	C. albicans	
	Torulopsis glabrata	
1k	*C. albicans*	0.125–2
1l	*Candida*	5–25
1m	*C. albicans*	1–6
	A. fumigatus	
1n	*T. mentagrophytes*	0.0093
	T. rubrum	0.0015
1o	*C. parapsilosis*	0.2–7.0
	C. neoformans	
	C. albicans	
	T. verrucosum	
	T. rubrum	
	M. gypseum	

amphotericin B. The structure along with the MIC value against fungal strain has been described in Table 8.4. In spite of its numerous pharmacological advantages, the use of 5-FC in clinical practice is shrinking because of the frequent incidence of innate or acquired resistance to this drug in fungal pathogens. The common side effect associated with 5-FC are mainly due to the long courses of therapy that include gastrointestinal (GI) disorders, nausea, and diarrhoea and in serious cases ulceration and enterocolitis. On the other hand, its excretion occurs through the kidneys, so renal function is sometimes impaired with extended dosing.

8.3.1.5 Thiocarbamates

The thiocarbamates were found to have strong and selective in vitro antifungal properties against *Trichophyton* spp. They are mostly used for the topical treatment of skin mycoses like athlete's foot, but are not very effective against nail and scalp mycoses. However, they are very

Table 8.4 The structure and activity against particular fungal strain of 5-FC

Compound	Sensitive fungal strains with species	MIC (μg/ml)
Flucytosine	*C. albicans*	0.12–128.0
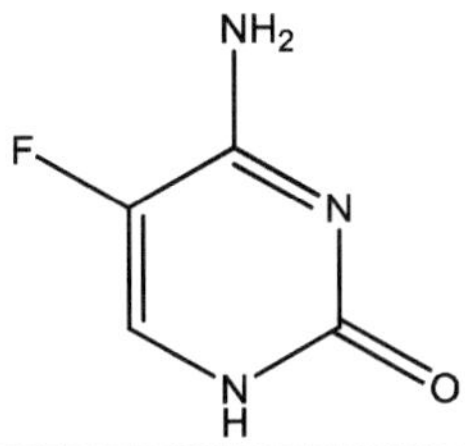	FLZ-R	
Estrella et al. (2004)	FLZ-S	
	Candida parapsilosis	0.12–1.0
	FLZ-R	
	Candida tropicalis	0.12–16.0
	FLZ-R	
	Candida glabrata	0.06–4.0
	FLZ-R	
	Candida krusei	0.50–8.0
	FLZ-R	
	Candida lusitaniae	0.25–1.0
	FLZ-R	
	Candida dubliniensis	0.12
	FLZ-R	0.12
	FLZ-S	
	Candida rugosa	1.0
	FLZ-R	0.25–1.0
	FLZ-S	
	Candida inconspicua	1.0
	FLZ-R	0.25
	FLZ-S	
	Candida zeylanoides	0.25–1.0
	FLZ-R	
	Debaryomyces hansenii	0.12–0.25
	FLZ-S	
	Saccharomyces cerevisiae	0.25–4.0
	FLZ-R	0.12–16.0
	FLZ-S	
	Torulaspora delbrueckii	0.25
	FLZ-R	0.12–0.25
	FLZ-S	
	Pichia membranifaciens	0.50–4.0
	FLZ-R	
	Pichia norvegensis	1.0–16.0
	FLZ-R	
	Rhodotorula glutinis	0.25–1.0
	FLZ-R	

active against actively growing cells only. The thiocarbamates inhibit the activity of squalene epoxidase, a key enzyme in the biosynthetic pathway of sterol production (Ryder et al. 1986). For topical application, they are commonly administered as a 1 % solution or

Table 8.5 In vitro MICs of thiocarbamates against various fungal strains

Compound	Fungal strains	MIC (μg/ml)
Tolnaftate or napthiomate T 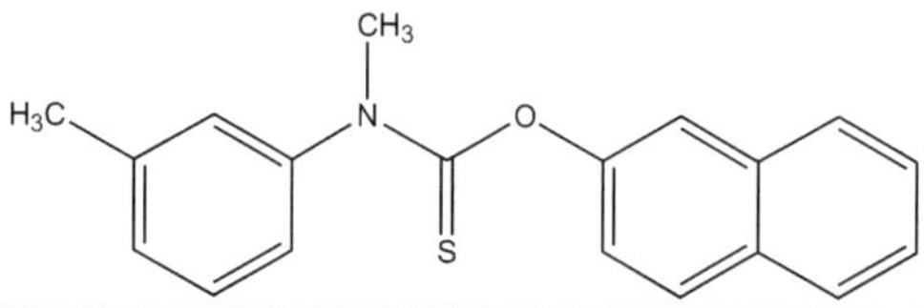 Boa et al. (course synopsis)	Dermatophytes	0.1–1
Napthiomate N 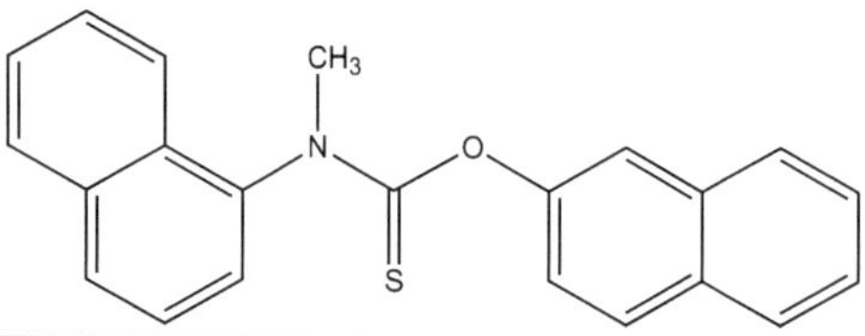 Boa et al. (course synopsis)	Dermatophytes	–
Tolciclate 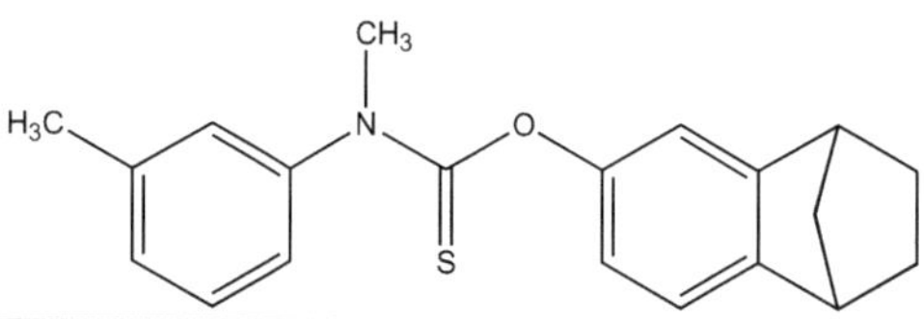 Ryder et. al. (1986)	*C. albicans* *C. parapsilosis* *T. mentagrophytes*	–

cream in polyethylene glycol. The structures of various thiocarbamate derivatives as well as their in vitro MIC values are provided in Table 8.5.

8.3.1.6 Allylamine Derivatives

In this category, the general structural motif for all active compounds is a tertiary allylamine.

In this family, naftifine is topically active against dermatophytes and shows some activity against yeasts (Georgopoulos et al. 1981). On the other hand, terbinafine is active against a wider range of fungi including *Aspergillus*, *Histoplasma*, and *Candida* species. The *tert-butyl* side chain of terbinafine is essential for activity against these particular fungi, and it is most active against dermatophytes (Petranyi et al. 1987). Like thiocarbamates, allylamines also inhibit the squalene epoxidase (Ryder and Dupont 1985; Ryder 1989), a key enzyme in the biosynthetic pathway of sterol production, and activity appears to be due to the intolerance of the fungal cell to increased squalene levels. However, the fungicidal *Candida albicans* can tolerate higher levels of squalene and show only fungistatic susceptibility to allylamines, whereas terbinafine is 100-fold more active compare to naftifine and is used to treat for ringworm and nail infections where oral administration is considered appropriate. Terbinafine may also be administered topically as a 0.5–1.0 % cream. Moreover, groupings of azoles and terbinafine are predicted to exhibit positive result in the treatment of mycosis (Garcia-Effron et al. 2004; Andriole 1998). Terbinafine has been reported to cause gastrointestinal disturbance and nausea when used orally, and in rare cases serious skin reactions can be triggered (Table 8.6).

8.3.1.7 Morpholine Derivatives

The morpholines were discovered in the late 1960s as a sequel from research into plant growth regulators and were initially used as agricultural fungicides. Tridemorph and dodemorph are the examples of the agricultural fungicides. Tridemorph has been used to treat yellow rust

Table 8.6 In vitro MICs of various allylamines against various fungal strains

Compound	Fungal strains	MIC (μg/ml)
Naftifine	*Trichophyton* spp.	0.1–0.2
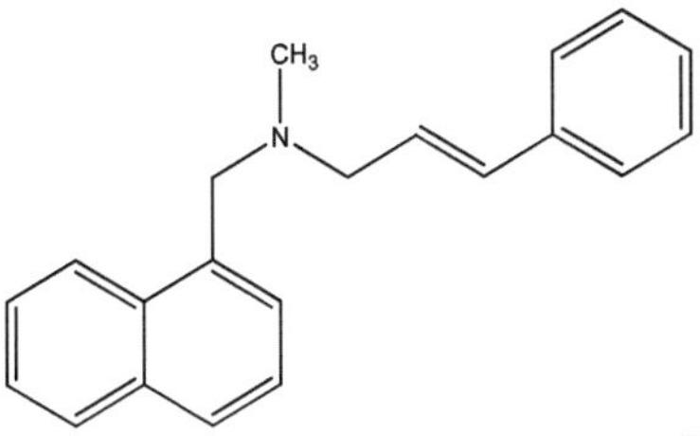	*E. floccosum*	0.1–0.2
Georgopoulos et al. (1981)	*Microsporum* spp.	0.1–0.2
	Aspergillus spp.	0.8–12.5
	Sporothrix schenckii	0.8–1.5
	Candida spp.	1.5–≥100
Terbinafine	*T. rubrum*	0.007–0.06
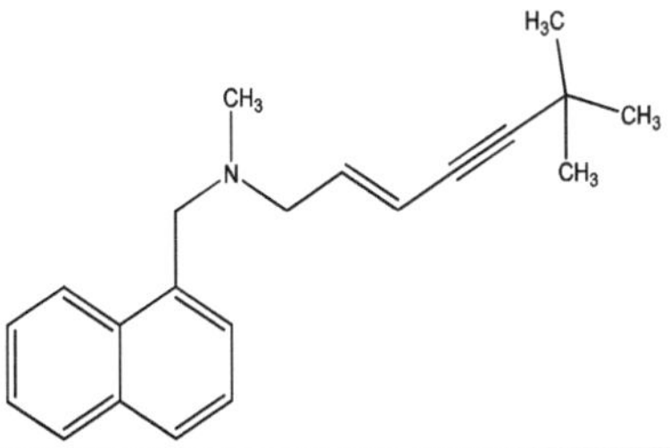	*T. mentagrophytes*	0.007–0.06
Sanclemente et al. (2009)	*M. gypseum*	0.015–0.125
	C. albicans	8–>64
Borelli et al. (2008)	*C. glabrata*	32–>64
	C. krusei	>64
	C. guilliermondii	>64
	C. tropicalis	>64
	C. parapsilosis	>64
Butenafine	Dermatophytes	0.0015–0.05
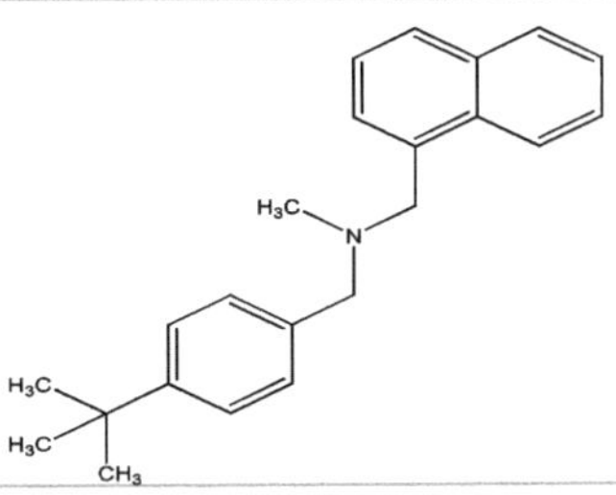	*Aspergillus*	0.025–0.78
	Cryptococcus neoformans	0.025–0.78
	Candida	0.78–1.56

in cereals, Dutch elm disease, and the banana fungal disease sigatoka. On the other hand, dodemorph was used to treat powdery mildews. The early morpholines were 4-alkyl-cis-2,6-dimethylmorpholines and were at first incorrectly believed to act in a similar fashion to the polyenes as the 4-alkyl group was most often a long or large-ring lipophilic alkyl group. These plant fungicides led the foundation for search of morpholines with applications in the field of human mycoses via fenpropimorph and culminating in the discovery of amorolfine in 1981. Amorolfine and fenpropimorph are strong inhibitors of fungal pathogens in both plants and humans and are highly active against dermatophytes. Depending upon the substituent,

Table 8.7 In vitro MICs of various morpholines against various fungal strains

Compound	Fungal strains	MIC (μg/ml)
Amorolfine	Dermatophytes	0.06
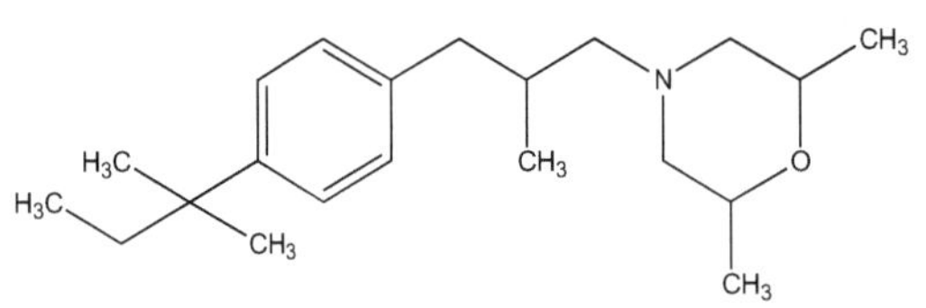	Dimorphic fungi	0.1
Marcireau et al. (1990)	Dematiaceous fungi	0.07
	Yeasts	0.5
	Alternaria	0.3
	Scopulariopsis	0.3
Fenpropimorph	*S. cerevisiae*	0.1 for cholestenol
		0.1–0.3, lanosterol
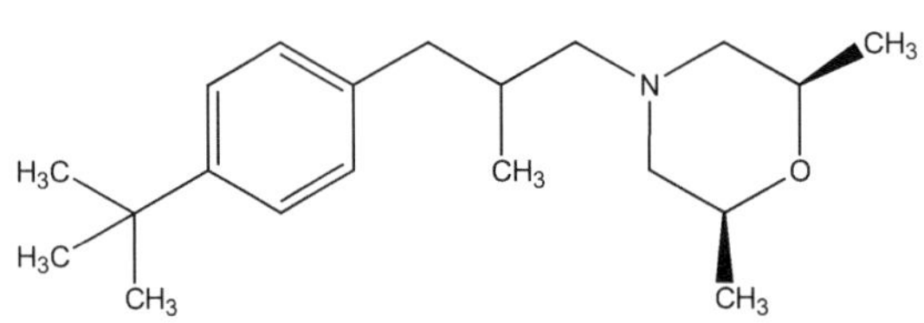		
Marcireau et al. (1990)		>60, ergosterol

morpholines act at multiple sites in the ergosterol biosynthetic pathway each to a different degree. The structure of various morpholines along with their in vitro MICs is provided in Table 8.7. Fenpropimorph and amorolfine are about 100-fold more active than tridemorph against the δ14 reductase, perhaps indicating this enzyme's significance to treat human mycoses with this class of drugs. Topical use of amorolfine can lead to temporary burning sensations, erythema, and pruritus.

8.3.1.8 Carabrol Ester Derivatives

In an important contribution, Zhang et al. synthesised and characterised 38 new ester derivatives of carabrol (Feng et al. 2012). The antifungal activities of all the synthesised compounds were evaluated using a spore germination assay against the fungal pathogen *Colletotrichum lagenarium*. They observed that out of those 38 ester derivatives, 16 have higher antifungal activity than that of carabrol and 7 compounds have higher antifungal activity than that of carabrone. The study revealed that the C-4 position of carabrol was a key position for possessing its antifungal activity, which exhibited the deviation of 50 % inhibition concentration (IC50) from 2.70 to 52.33 μg/ml. The electron-withdrawing groups at the phenyl ring of carabrol ester derivative showed higher activity compared to electron-donating groups. Two ester derivatives, carabryl 4-cynaobenzoate (**2p**, IC50, 2.70 μg/ml) and carabryl 4-isopropylbenzoate (**2q**, IC50, 2.82 μg/ml), exhibited only somewhat lower antifungal activity than that of the positive control chlorothalonil (IC50, 0.87 μg/ml) and have been recognised as prospective leads for progress of new environmentally friendly fungicides (Table 8.8).

8.3.1.9 Carboline Derivatives

β-Carboline alkaloids (Cao et al. 2007) are a large collection of natural and synthetic indole alkaloids and are widely distributed in nature,

Table 8.8 The IC50 of some selected carabrol ester derivatives against spore germination of *C. lagenarium*

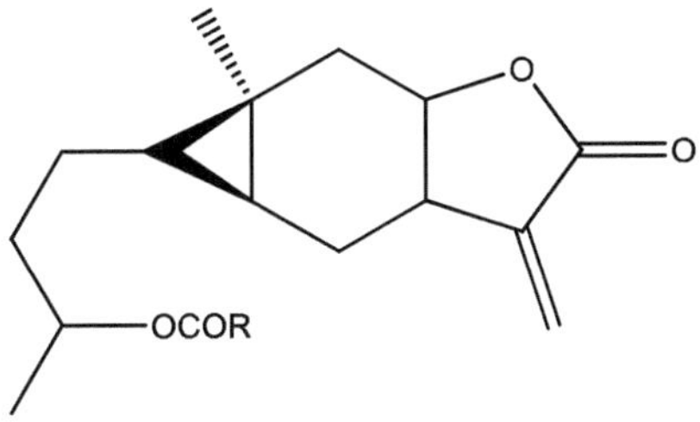

R= CH_3 (**2a**), CH_2Cl (**2b**), CH_2CH_3 (**2c**), $CH_2CH_2CH_3$ (**2d**), $CH_2(CH2)_2CH_3$ (**2e**)

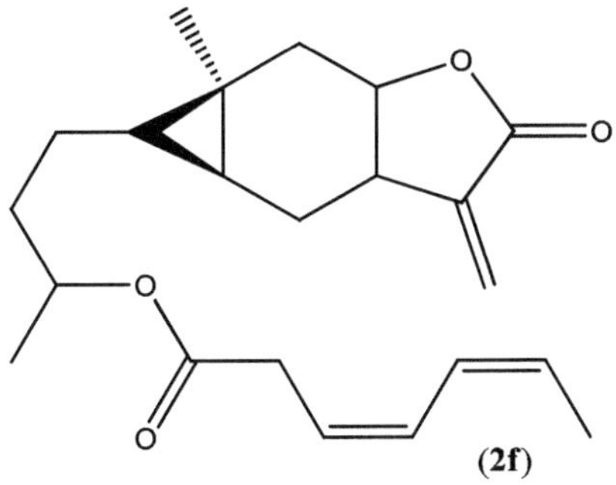

(**2f**)

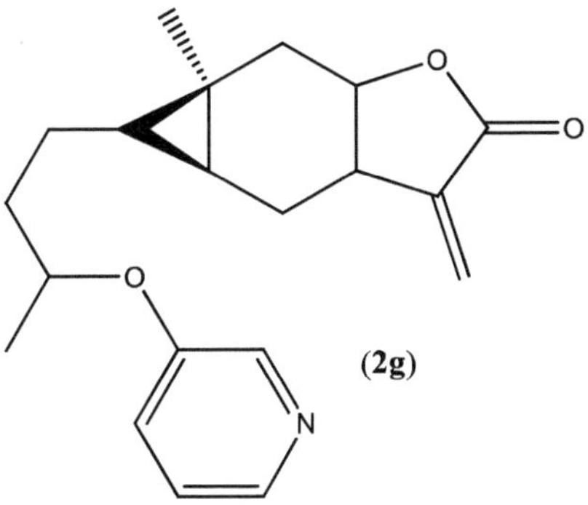

(**2g**)

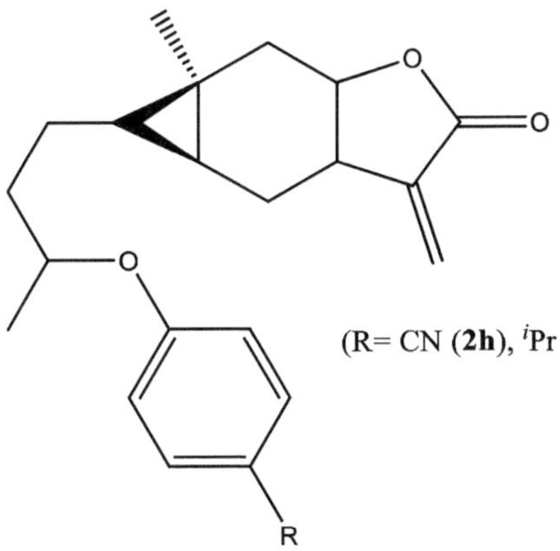

(R= CN (**2h**), ^{i}Pr (**2i**)

Fungal strain	Compounds	IC_{50}(μg/ml)[a]
C. lagenarium	**2a**	10.77 ± 0.09
	2b	4.51 ± 0.06
	2c	23.28 ± 0.57
	2d	28.84 ± 0.41
	2e	31.19 ± 0.59
	2f	8.41 ± 0.53
	2g	8.26 ± 0.25
	2h	2.70 ± 0.13
	2i	2.82 ± 0.12

[a]All values are presented as means ± SD ($n = 3$). And chlorothalonil was used as the positive control

comprising various foodstuffs, plants, marine creatures, insects, mammalians as well as human tissues and body fluids. Most of these compounds are of great interest due to their diverse biological activities. In this regard, Zhang et al. have designed and synthesised a series of novel antifungal carboline derivatives which showed broad-spectrum antifungal activity. Particularly, compound **4d** showed comparable in vitro antifungal activity to fluconazole and showed almost negligible toxicity to human embryonic lung cells. It also showed good fungicidal activity against both fluconazole- sensitive and fluconazole-resistant *Candida albicans* cells. Moreover, **4d** exhibited good synergistic antifungal activity in combination with fluconazole against fluconazole-resistant *Candida* species. Preliminary mechanistic study revealed that **4d** might act by inhibiting the synthesis of fungal cell wall (Wang et al. 2014) (Table 8.9).

Table 8.9 Structure and activity against specific strains of some carboline derivatives

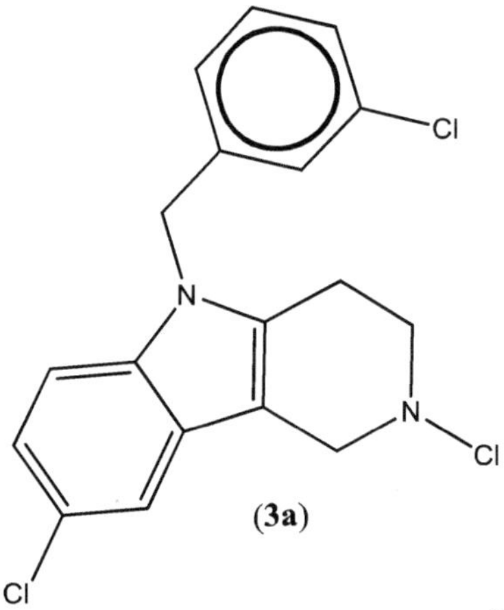

(**3a**)

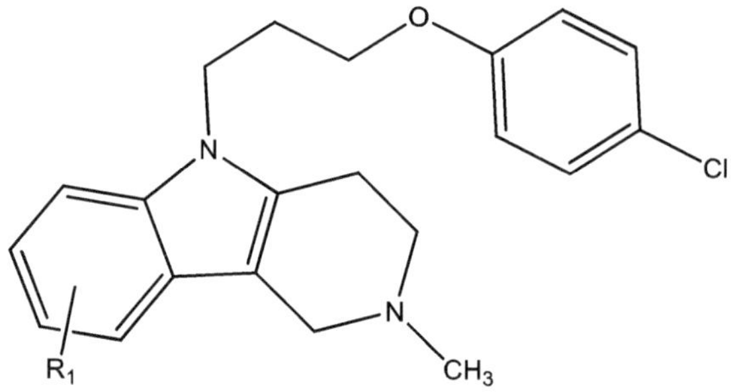

R_1 = 2-CH_3 (**3b**); R_1= 2-OCH_3 (**3c**)

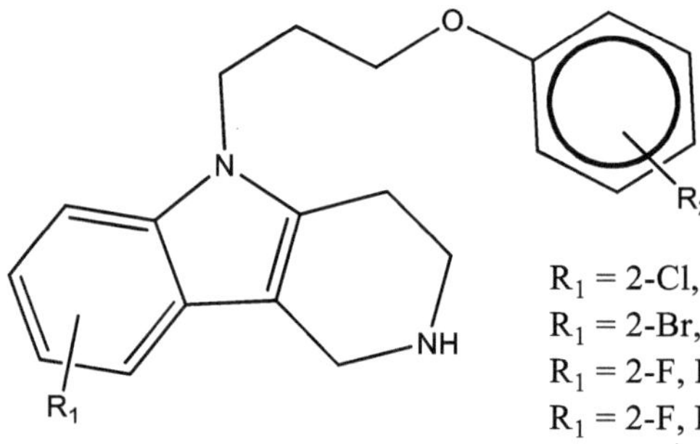

R_1 = 2-Cl, R_2 = 4-F (**4a**)
R_1 = 2-Br, R_2 = 4-F (**4b**)
R_1 = 2-F, R_2 = 4-^{t}Bu (**4c**)
R_1 = 2-F, R_2 = 4-Cl (**4d**)

Compounds	Fungal strain	MIC_{80} (μg/ml)
3a	*C albicans*	64
	C. parapsilosis	32
	C. neoformans	16
	C. glabrata	16
	A. fumigatus	32
	T. rubrum	8
	M. gypseum	16
3b	*C albicans*	32
	C. parapsilosis	16
	C. neoformans	8
	C. glabrata	16
	A. fumigatus	16
	T. rubrum	8
	M. gypseum	8
3c	*C albicans*	32
	C. parapsilosis	32
	C. neoformans	8
	C. glabrata	>64
	A. fumigatus	32
	T. rubrum	8
	M. gypseum	8

(continued)

Table 8.9 (continued)

Compounds	Fungal strain	MIC_{80} (μg/ml)
4a	*C albicans*	4
	C. parapsilosis	4
	C. neoformans	4
	C. glabrata	4
	A. fumigatus	4
	T. rubrum	8
	M. gypseum	4
4b	*C albicans*	4
	C. parapsilosis	4
	C. neoformans	4
	C. glabrata	16
	A. fumigatus	4
	T. rubrum	8
	M. gypseum	4
4c	*C albicans*	8
	C. parapsilosis	8
	C. neoformans	4
	C. glabrata	4
	A. fumigatus	16
	T. rubrum	4
	M. gypseum	4
4d	*C albicans*	1
	C. parapsilosis	2
	C. neoformans	4
	C. glabrata	4
	A. fumigatus	4
	T. rubrum	2
	M. gypseum	1

8.3.1.10 Miscellaneous

8.3.1.10.1 5-Amino-6-arylamino-1*H*-pyrazolo[3,4-*d*]pyrimidin-4(5*H*)-one Derivatives

Wang et al. synthesised some new 5-amino-6-arylamino-1*H*-pyrazolo[3,4-*d*]pyrimidin-4(5*H*)-one derivatives (Wang et al. 2008) via tandem aza-Wittig and annulation reactions of iminophosphorane and aromatic isocyanates. From bioassay analysis, they found that both the derivatives (5a-5c and 6a-6d) possessed high antifungal activity against *Botrytis cinerea* Pers. and *Sclerotinia sclerotiorum*. Especially, compounds **6a**, **6b**, and **6c** shows 100 % inhibition against *Sclerotinia* at the concentration in 50 mg/L. At the same time, it shows almost 80–90 % inhibition activity against *Sclerotinia* at 10 mg/L. Both the derivatives were found to be active against *Botrytis* at 50 mg/L. For example, inhibitory rates of compounds **5a**, **5b**, **5c**, **6a**, **6b**, **6c**, and **6d** were 93 %, 87 %, 99 %, 91 %, 99 %, 99 %, and 97 %, respectively.

8.3.1.10.2 β-Trifluoroalkyl Aminovinyl Ketone Derivatives

Gellerman et al. synthesised ten β-trifluoroalkyl aminovinyl ketone derivatives and their inhibitory effect on several phytopathogenic fungi, an oomycete and plants were tested. The fungal strains used in their study were *Sclerotinia sclerotiorum*, *Alternaria alternata*, *Botrytis cinerea*, *Rhizoctonia solani*, *Penicillium digitatum*, *Pythium aphanidermatum*, and *Neurospora crassa*, and they were cultured on potato dextrose agar. Almost all the synthesised compounds exhibited some degree of toxicity towards the filamentous fungi in different extent. Most of the compounds were found to be fungitoxic at the dose of 10–100 μg/ml. Amongst all the tested compounds, (Z)-3-amino-4,4,4-trifluoro-1-(4-chlorophenyl)but-2-en-1-one (**7a**) showed the highest inhibitory effect, whilst **7b** and **7c** were the less significantly toxic to the most of tested pathogens. From the study, they found that *Neurospora crassa* is the most sensitive fungi, whereas *Alternaria alternate* and *Penicillium digitatum* are the least susceptible fungi to most of the tested compounds (Gellerman et al. 2009).

8.3.1.10.3 Dithiocarbamate and Rhodanine Derivatives

Chauhan et al. reported a series of 27 dithiocarbamate and the rhodanine derivatives containing pyridine moiety to check their antifungal activity. All the synthesised compounds (**8a–f** and **9**) were tested in vitro (Table 8.10) against the various fungal strains such as *Candida parapsilosis*, *Cryptococcus neoformans*, *Sporothrix schenckii*, *Trichophyton mentagrophytes*, *Aspergillus fumigatus*, and *C. albicans*. Of all the compounds **8d**, **8e**, and **9** showed better antifungal potency than fluconazole against *Aspergillus fumigatus* and *Trichophyton mentagrophytes* against *Aspergillus fumigatus*. Compounds **8b** and **8d** showed almost 12-fold better activity compared to fluconazole and clotrimazole against *Sporothrix schenckii*. Further, these active compounds were tested for in vitro anti-candidal activity and amphotericin B-resistant strain of *Candida*.

Particularly compounds **8a–f** and **9** were found to be more effective (MIC = 0.39–3.12 μg/ml) against *Candida albicans* MTCC183. Further, these compounds except **8e** and **8f** showed significantly better activity against amphotericin B-resistant strain of *Candida albicans* in comparison to fluconazole (Chauhana et al. 2012).

8.4 Future Prospects

Invasive fungal infections are a principal problem in immunocompromised patients. They are found difficult to treat because their clinical manifestations are not specific unlike other infective diseases. The complete eradications of this invasive infection are not possible as the fungi are common in the environment. However, prevention can be achieved by detection and identification of the causal agent and using proper diagnosis techniques and therapeutic drug monitoring. Antifungal susceptibility testing is the one novel approach to understand the clinical significance of antifungal agents and has become an important medium for the recognition of microbiologically resistant organisms and for optimal selection of antimicrobial therapy. Many standardised methods for testing the in vitro susceptibility of fungi are now available, and concentration-effect relationships are increasingly explored. The availability of drugs with different molecular targets has opened new avenues for exploring combination therapies of two or even more drugs. The apparent aims of combination therapies are to extend the antifungal spectrum, to reduce the selection of resilient organisms, and to develop overall antifungal ability without conceding patient safety. However, combination therapy is not the universal approach for all established or alleged fungal infections but will be earmarked for selected subgroups. Of late, the increase of antifungal drug research has occurred because there is a crucial necessity for new antifungal agents to treat these fatal invasive infections. Although we have innovative, less toxic, antifungal agents that are present for clinical use, their clinical effectiveness in certain invasive fungal infections, such as aspergillosis and fusariosis, is not ideal. Many antifungal drugs targeting different molecules in the fungal cell have been developed. But route for identification of novel target and development of entirely new

Table 8.10 Structure and active strains of some miscellaneous antifungal agents

Structure	Fungal strain	Reference
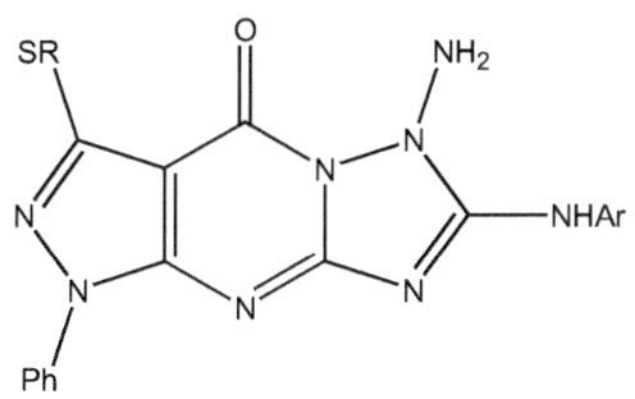	*Botrytis cinerea* Pers.	Wang et al. (2008)
5a, R=Me Ph= o-FC6H4 **5b**, R=Me Ph=m-FC6H4 **5c**, R= Me Ph=m-ClC6H4	*Sclerotinia sclerotiorum*	
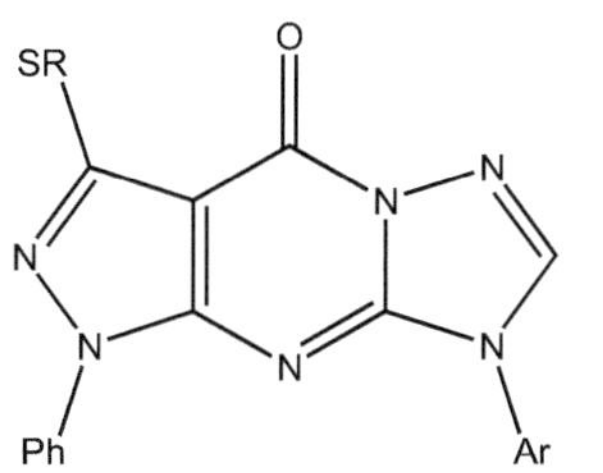	*Botrytis cinerea* Pers.	Wang et al. (2008)
6a, R=Me Ph=m-FC6H4 6b, R=Me Ph=C6H5 6c, R= Me Ph=m-ClC6H4 6d, R=Me Ph=m-MeC6H4	*Sclerotinia sclerotiorum*	
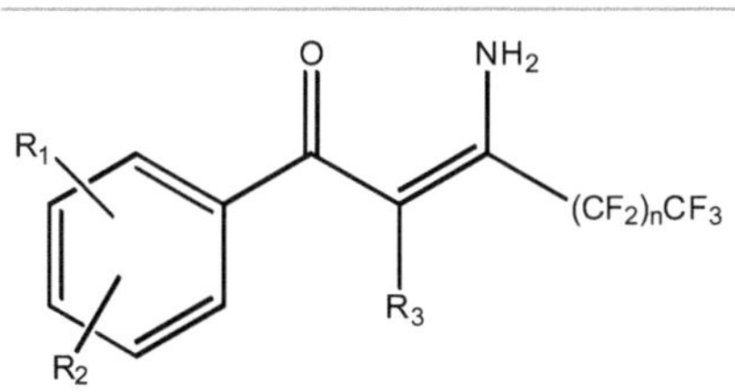	*S. sclerotiorum*	Gellerman et al. (2009)
	B. cinerea	
7a, R_1= *p*-Cl, R_2= H, R_3= H, n= 0	*Penicillium digitatum*	
7b, R_1= *p*-Cl, R_2= H, R_3= Et, n= 0		
7c, R_1= *p*-NHBoc, R_2= H, R_3= H, n= 0	*Pythium aphanidermatum*, *Alternaria alternate*, *Rhizoctonia solani* and *Neurospora crassa*	
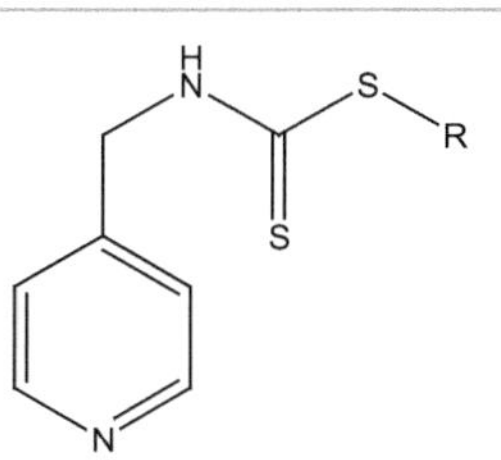	*C. albicans*	Chauhan et al. (2012)
8a-f, R = ethyl, n-propyl, n-butyl, allyl, benzyl, 4-trifluoromethylbenzyl	*C. parapsilosis*	
	S. schenckii	
	T. mentagrophytes	
	A. fumigatus	
	C. neoformans	
	C. albicans	

(continued)

Table 8.10 (continued)

Structure	Fungal strain	Reference
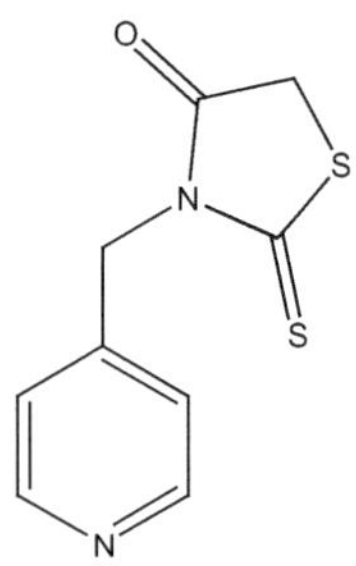	*C. parapsilosis*	
9	*S. schenckii*	
	T. mentagrophytes	
	A. fumigatus	
	C. neoformans	

therapeutics is difficult because of the eukaryotic nature of fungal cells. Thus, concerted effort in antifungal drug discovery is still required to develop more promising and effective antifungal agents for use in the clinical arena. Furthermore, developments of new methodology for antifungal susceptibility testing as well as laboratory and well-planned concerted clinical trials are essential in order to augment utility in clinical guidance of antifungal therapy.

References

Andriole VT (1998) Current and future therapy of invasive fungal infections. Curr Clin Top Infect Dis 18:19–36

Andriole VT (1999) Current and future antifungal therapy: new targets for antifungal agents. J Antimicrob Chemother 44:151–162

Borelli C, Klövekorn G, Ernst T-M, Bödeker R-H, Korting HC, Neumeister C et al (2007) Comparative study of 2% sertaconazole solution and cream formulations in patients with tinea corporis, tinea pedis interdigitalis, or a corresponding candidosis. Am J Clin Dermatol 8(6):371–378

Bossche HV (1997) Mechanisms of antifungal resistance. Rev Iberoam Micol 14:44–49

Bueno JG, Martinez C, Zapata B, Sanclemente G, Gallego M, Mesa AC et al (2009) In vitro activity of fluconazole, itraconazole, voriconazole and terbinafine against fungi causing onychomycosis. Clin Exp Dermatol 35(6):658–663

Cao R, Peng W, Wang Z, Xu A et al (2007) β-Carboline alkaloids: biochemical and pharmacological functions. Curr Med Chem 14:479–500

Carrillo-Munoz AJ (2006) Antifungal agents: mode of action in yeast cells. Rev Esp Quimioter 19 (2):130–139

Chauhana K, Sharmaa M, Singhb P, Kumarc V, Shuklab PK, Siddiqic MI, Chauhan PMS et al (2012) Discovery of a new class of dithiocarbamates and rhodanine scaffolds as potent antifungal agents: synthesis, biology and molecular docking. Med ChemComm 3 (9):1104–1110

Cuenca-Estrella M, Gomez-Lopez A, Mellado E, Garcia-Effron G, Rodriguez-Tudela JL et al (2004) In vitro activities of ravuconazole and four other antifungal agents against fluconazole-resistant or -susceptible clinical yeast isolates. Antimicrob Agents Chemother 48(8):3107–3111

Dismukes WE et al (2000) Introduction to antifungal drugs. Clin Infec Dis 30:653–657

El Hage S, Lajoie B, Feuillolay C, Roques C, Baziard G et al (2011) Arch Pharm Chem Life Sci 11:402–410

Feng JT, Wang H, Ren SX, He J, Liu Y, Zhang X et al (2012) Synthesis and antifungal activities of carabrol ester derivatives. J Agric Food Chem 60:3817–3823

Garcia-Effron G, Gomez-Lopez A, Mellado E, Monzon A, Rodriguez-Tudela JL, Cuenca-Estrella M et al (2004) In vitro activity of terbinafine against medically important non-dermatophyte species of filamentous fungi. J Antimicrob Chemother 53:1086–1089

Gellerman G, Pariente N, Paz, Shnaiderman A, Yarden O (2009) Synthesis and antifungal activity of β-trifluoroalkyl aminovinyl ketone derivatives. J Agric Food Chem 57:8303–8307

Georgopoulos A, Petranyi G, Mieth H, Drews J (1981) In vitro activity of naftifine, a new antifungal agent. Antimicrob Agents Chemother 19:386–389

Gomez-Lopez A, Mellado E, Garcia-Effron G, Rodriguez-Tudela JL, Cuenca-Estrella M et al (2004) In vitro activities of ravuconazole and four other antifungal agents against fluconazole-resistant or -susceptible clinical yeast isolates. Antimicrob Agents Chemother 8:3107–3111

Guarro J, Alves SH, Gene J, Grazziotin NA, Mazzuco R, Dalmagro C, Capilla J, Zaror L, Mayayo E et al (2003) Two cases of subcutaneous infection due to Phaeoacremonium spp. J Clin Microbiol 41 (3):1332–1336

Jevons S, Gymer GE, Brammer KW, Cox DA, Leeming MRG et al (1979) Antifungal activity of tioconazole (UK-20,349), a new imidazole derivative. Antimicrob Agents Chemother 15(4):597–602

Koga H, Nanjoh Y, Kaneda H, Yamaguchi H, Tsuboic R et al (2006) Short-term therapy with luliconazole, a novel topical antifungal imidazole, in guinea pig models of tinea corporis and tinea pedis. Antimicrob Agents Chemother 56(6):3138–3143

Loevenhart AS (1926) Future chemotherapy. Ind Eng Chem 18(12):1268–1272

Marcireau C, Guilloton M, Karst F (1990) In vivo effects of fenpropimorph on the yeast Saccharomyces cerevisiae and determination of the molecular basis of the antifungal property. Antimicrob Agents Chemother 34(6):989–993

Meerpoel L, Heeres J, Back LJJ, Van der Veken LJE, Hendrickx R, Corens D, Groot AD, Leurs S, Vander Eycken L, Weerts J, Luyts P, Van Gerven F, Woestenborghs FAA, Van Breda A, Oris M, van Dorsselaer P, Willemsens GHM, Bellens D, Marichal PJ MG, Vanden Bossche HF, OddsPro FC et al (2010) Synthesis and in vitro and in vivo antifungal activity of the hydroxy metabolites of saperconazole. Chem Med Chem 5(5):757–769

Niwano Y, Tabuchi T, Kanai K, Hamaguchi H, Uchida K, Yamaguchi H et al (1995) Short-term topical therapy of experimental tinea pedis in guinea pigs with lanoconazole, a new imidazole antimycotic agent. Antimicrob Agents Chemother 39(10):2353–2355

Northey EH (1943) Chemical side of chemotherapy. Ind Eng Chem 35(8):829–836

Pelletier R, Peter J, Antin C, Gonzalez C, Wood L, Walsh TJ et al (2000) Emergence of resistance of candida albicans to clotrimazole in human immunodeficiency virus-infected children: in vitro and clinical correlations. J Clin Microbiol 38(4):1563–1568

Petranyi G, Meingassner JG, Mieth H et al (1987) Antifungal activity of the allylamine derivative terbinafine in vitro. Antimicrob Agents Chemother 31:1365–1368

Pfaller MA, Boyken L, Hollis RJ, Kroeger J, Messer SA, Tendolkar S, Diekema DJ et al (2011) Wild-type MIC distributions and epidemiological cutoff values for posaconazole and voriconazole and Candida spp. as determined by 24-hour CLSI broth microdilution. J Clin Microbiol 49(2):630–637

Rani N, Sharma A, Gupta GK, Singh R et al (2013) Imidazoles as potential antifungal agents: a review. Mini-Reviews in Med Chem 13:1626–1655

Ryder NS (1989) Effect of allylamine antimycotic agents on fungal sterol biosynthesis measured by sterol side-chain methylation. J Gen Microbiol 131(1):595–1602

Ryder NS, Dupont MC (1985) Inhibition of squalene epoxidase by allylamine antimycotic compounds. Biochem 230:765–770

Ryder NS, Frank I, Dupont MC et al (1986) Ergosterol biosynthesis inhibition by the thiocarbamate antifungal agents tolnaftate and tolciclate. Antimicrob Agents Chemother 29:858–860

Sahin GO, Akova M et al (2005) Emerging and life-threatening fungal pathogens in the immunocompromised host. Uropeanoncology Rev:1–5

Shalini K, Kumar N, Drabu S, Sharma PK et al (2011) Advances in synthetic approach to and antifungal activity of triazoles. Beilstein J Org Chem 7:668–677

Sheehan DJ, Hitchcock CA, Sibley CM (1999) Current and emerging azole antifungal agents. Clin Microbiol Rev 12:40–79

Spellberg B, Powers JH, Brass EP, Miller LG, Edwards J et al (2004) Trends in antimicrobial drug development: implications for the future. Antimicrob Res Dev 38:1279–1286

Vandeputte P, Ferrari S, Coste AT et al (2011) Antifungal resistance and new strategies to control fungal infections. Int J Microbiol 2012:1–26

Wang HQ, Zhou WP, Wang YY, Lin CR (2008) Synthesis and antifungal activities of novel 5-Amino-6-arylamino-1*H*-pyrazolo[3,4-*d*]pyrimidin-4(5*H*)-one derivatives. J Agric Food Chem 56:7321–7325

Wang S, Wang Y, Liu W, Liu N, Zhang Y, Dong G, Liu Y, Li Y, He Z, Miao Z, Yao J, Li J, Zhang W, Sheng C et al (2014) Novel carboline derivatives as potent antifungal lead compounds: design, synthesis, and biological evaluation. ACS Med Chem Lett 5:506–511

Antifungal Therapy in Eye Infections: New Drugs, New Trends 9

Joveeta Joseph and Savitri Sharma

9.1 Introduction

The treatment of ocular fungal infections is a challenge for any ophthalmologist, and although ocular fungal infections are relatively uncommon in developed countries compared with bacterial infections (O'Brien 1999), in tropical countries it is being increasingly recognized as important causes of morbidity and blindness, accounting for nearly 50 % of all cases of microbial keratitis (Srinivasan et al. 1997). The most common causative pathogens for these infections include *Fusarium* spp. and *Aspergillus* spp., these vary geographically among areas. The two most common sites for fungal infections of the eye are the cornea (Fig. 9.1) and retina or vitreous (Fig. 9.2) but they can also involve other periocular tissues, including the lacrimal apparatus, conjunctiva, sclera (Fig. 9.3), eyelids, and bony orbit.

Successful medical treatment of ophthalmic mycotic infections depends on the spectrum of antifungal activity of a drug, the pharmacokinetic properties, the concentration achieved in the target tissue, the route by which it has been administered, the duration of contact with the target ocular tissue, and the bioavailability of the compound in the eye (Behrens-Baumann 1999; Manzouri et al. 2001). Compounds with a molecular mass exceeding 500 Da, such as amphotericin B (924.10 Da), natamycin (665.75 Da), and ketoconazole (531.44 Da), barely penetrate an intact corneal epithelium because the force of friction reduces diffusion (Manzouri et al. 2001). The ocular penetration of molecules with intermediate molecular masses, such as miconazole (416.12 Da) and fluconazole (306.30 Da), is determined by the solubility in lipid-rich tissues and the force of friction. Lipophilic compounds, such as itraconazole, easily cross the lipid-rich epithelial and endothelial cell membranes and the blood-aqueous barrier. Comparatively, hydrophilic compounds more easily cross the corneal stroma and biphasic compounds penetrate all corneal layers (Friedberg et al. 1991). However, the exact relevance of these considerations in therapy especially in mycotic keratitis, where the integrity of the corneal epithelium and stroma is usually breached by the disease process itself, remains unanswered. Selection of the most appropriate antifungal remains a challenge, hindered by the diverse clinical presentation, delay in clinical and laboratory diagnosis, lack of ocular pharmacological profiles of common antifungals, and more recently the emergence of resistance. This chapter reviews the main antifungals currently used in the therapy of ocular fungal infections.

J. Joseph (✉) • S. Sharma
Jhaveri Microbiology Centre, L V Prasad Eye Institute, Kallam Anji Reddy Campus, L.V. Prasad, Marg Banjara Hills, Hyderabad 500034, India
e-mail: joveeta@lvpei.org; savitri@lvpei.org

A. Basak et al. (eds.), *Recent Trends in Antifungal Agents and Antifungal Therapy*,
DOI 10.1007/978-81-322-2782-3_9

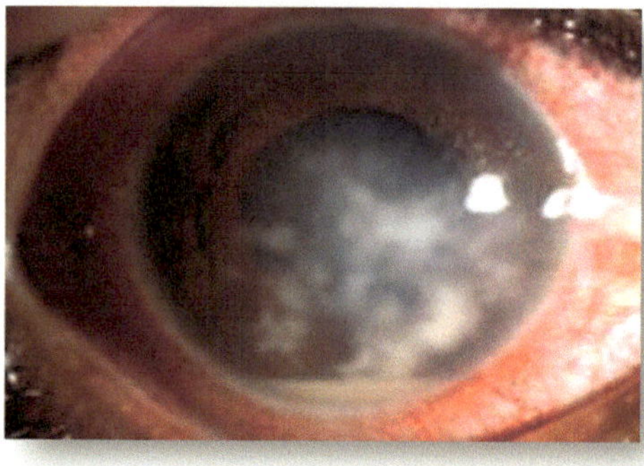

Fig. 9.1 Slit lamp biomicroscopic picture of cornea showing multiple stromal infiltrates with hyphate margins suggestive of fungal keratitis

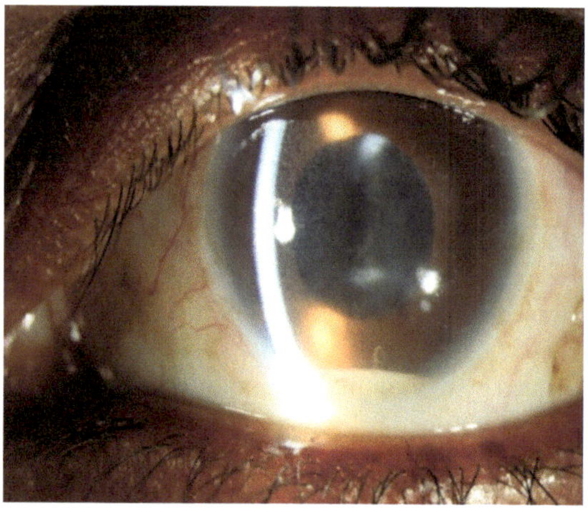

Fig. 9.2 Slit lamp biomicroscopic picture of the eye showing white fluffy exudates in the pupillary region and hypopyon in the anterior chamber, suggestive of fungal endophthalmitis

9.2 Antifungal Agents for Eye Infections

9.2.1 History

Some of the earlier reports in the treatment of mycotic keratitis involved the use of 2.3 % povidone iodine to successfully treat keratitis by *Candida albicans* and *Acremonium strictum* (Muller et al. 2013). Another study reported repeated cauterization by carbolic acid (Phenol) and povidone Iodine for treating mycotic corneal

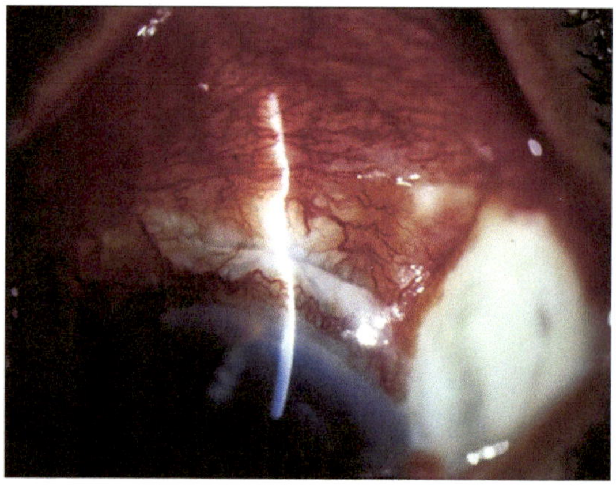

Fig. 9.3 Clinical picture of a patient diagnosed to have fungal scleritis showing severe hyperemia of the conjunctiva along with scleral necrosis

ulcer (Katz et al. 2004). Similarly, Fiscella et al. (1997) in an experimental study showed that polyhexamethylene biguanide (PHBM) 0.02 % is effective in significantly reducing fungal growth in a New Zealand white rabbit induced model of *F. solani* keratitis. However, there are no comparative studies between PHMB and other antifungal agents. Attempts have also been made to evaluate the efficacy of 0.2 % chlorhexidine gluconate (Rehman et al. 1998) compared to 2.5 % natamycin therapy in patients with suspected mycotic keratitis, and it was found that though none of the severe ulcers was fully healed at 21 days, three of those treated with chlorhexidine eventually healed within 60 days. These results could be misleading, since an effective formulation of natamycin was not used. Attempts to use chlorhexidine in the treatment of mycotic keratitis in two locations in Africa have also not had encouraging results (Johnson 1998). In another study, the efficacy of a 1 % silver sulfadiazine ointment was compared with that of 1 % miconazole in therapy of clinical mycotic keratitis in a prospective, controlled, randomized, double-blind clinical study (Mohan et al. 1988). Overall, a higher success rate was achieved with silver sulfadiazine (80 %) than with miconazole (55 %), although the response of *Aspergillus* keratitis was comparable in the two groups. Lack of experimental studies using other fungal species and controlled clinical trials of all these compounds proved them ineffective over time and are now not being recommended in the treatment of ophthalmic mycosis.

9.2.2 Classification

The drugs that are now being used in the treatment of ophthalmic mycosis can be broadly classified as first, the naturally occurring antibiotics like polyenes and echinocandins and second, the synthetic drugs including azoles and fluorinated pyrimidines.

9.2.2.1 Polyenes

The polyenes are the oldest group of antifungal agents and continue to be an important component of the ocular antifungal armamentarium. They work by directly bonding to ergosterol, a sterol unique to fungal cell membranes and the integrity of these membranes is disrupted, resulting in the leakage of essential intracellular constituents. The agents commonly used in this group for the treatment of ocular mycoses are amphotericin-B and nystatin and natamycin. Nystatin has not been used to treat eye infections for several decades due to its low tissue penetration, toxicity, and reports of resistance (Reddy et al. 1982; Oliveira et al. 2001). However, amphotericin-B and natamycin remain as the primary drugs in current use for the treatment of ophthalmic mycoses (Bernauer et al. 1998).

9.2.2.1.1 Natamycin

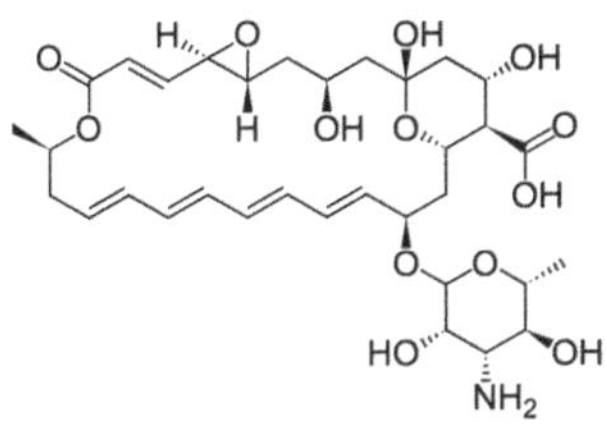

Natamycin, also known as pimaricin, is a naturally occurring antifungal agent produced during fermentation by the bacterium *Streptomyces natalensis*, commonly found in soil (Sweetman 2004). Natamycin was the first antifungal specifically developed for topical ophthalmic use and is currently the only topical ophthalmic antifungal compound approved by the Food and Drug Administration of the United States for the treatment of fungal keratitis (O'Brien 1999). It is also used as a pesticide and as a preservative in the food industry (Jones 2004). Natamycin has a long molecule with low water solubility. Presented as a suspension, it needs to be shaken before administration. It is stable in a 5 % suspension and, in this form, adheres well to the cornea for clinically useful periods. Natamycin is a broad-spectrum agent, especially against filamentous fungi, including species of *Fusarium*, *Aspergillus*, *Acremonium*, *Penicillium*, *Lasiodiplodia*, and *Candida* (O'Brien 1999). Due to its high molecular weight, natamycin has low corneal penetration and is only indicated as a monotherapy in the treatment of superficial infections (Tu et al. 2007; Kaur et al. 2008). In deep infections or those involving intraocular structures, natamycin should be associated with other antifungal agents using a different route of administration (Alfonso and Miller 2011; Jones et al. 1970; Rosa et al. 1994). Epithelial debridement is recommended as an adjuvant therapy for greater adherence of the drug to the de-epithelized surface so that higher concentrations can be achieved in the corneal stroma (Kaur et al. 2008). The dosing interval is every 4 h to maintain therapeutic concentrations in the cornea with good long-term tolerability (Prajna et al. 2003).

Toxicity Topical natamycin 5 % suspension is well tolerated in the cornea. Corneal toxicity does occur, however, with prolonged use, usually in the form of a punctate keratitis. Subconjunctival administration is discouraged due to serious complications, such as scleritis and conjunctival necrosis (O'Day 1987; Dursun et al. 2003). There are no reports of administration of natamycin through other routes (intracameral, intravitreal, intrastromal, or systemic). Though several studies have highlighted the superiority of natamycin in the treatment of infections by *Fusarium* spp. (Li et al. 2008; Xuguang et al. 2007; Pearse et al. 2009), some authors have shown that about one-third of *Fusarium* infections do not respond to natamycin (Jones et al. 1972; Polack et al. 1971). In such cases, natamycin should be replaced by or associated with another drug. Lalitha et al. (2007), in a comparative study on the minimum inhibitory concentrations (MIC) of different antifungal agents, reported that natamycin has a lower relative MIC than amp-B both against *Fusarium* and *Aspergillus* species. Kalavathy et al. (2005) compared the efficacy of natamycin and itraconazole and found better results in the group treated with natamycin, although the difference was not significant. Additionally natamycin remains the drug of choice with other treatment modalities for therapy of mycotic scleritis (Fig. 9.3) (Sullivan et al. 1994), conjunctivitis, and endophthalmitis (O'Brien 1999); however, controlled clinical trials are needed to confirm the efficacy of natamycin for these indications.

9.2.2.1.2 Amphotericin-B

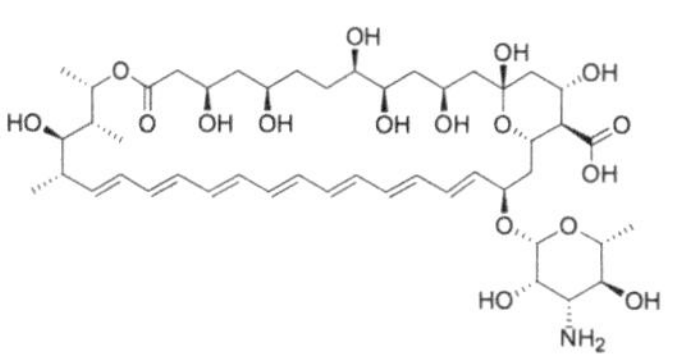

Amphotericin-B (Amp-B) belongs to the family of polyene macrolide antibiotics and was the first broad-spectrum antifungal agent to be discovered. Isolated in the 1950s, amp-B is produced by the actinomycete *Streptomyces nodosus*. It became popular after approval by the FDA in the 1960s due to its great efficiency in controlling disseminated fungal infections (O'Brien 1999). In ophthalmology, it is still considered the gold standard of antifungal agents to which other drugs are compared. Amp-B acts by increasing cell permeability through the formation of pores or channels in the fungal cell membrane upon binding to ergosterol and by promoting oxidative action on cells, thus altering their metabolic functions. It also binds to cholesterol in human cells, which is the main reason for its side effects (Gallis et al. 1990). The drug's name is derived from its amphoteric properties (soluble in extreme pHs, both acidic and basic). It has low water solubility and needs to be diluted in deoxycholate for administration. Amp-B acts on both yeasts and filamentous fungi. It has an excellent spectrum, being effective against *Candida* spp., *Aspergillus* spp., *Penicillium marneffei*, *Cryptococus* spp., and the causative agents of mucormycosis. It is also effective, to a lesser extent, against the *Fusarium* spp. (O'Brien 1999). Its action is primarily fungistatic, with fungicidal action depending on the concentration reached in the target tissue (Khoo et al. 1994). Systemic administration of Amp-B produces a little penetration into ocular tissues and does not reach therapeutic levels in the cornea, aqueous or vitreous humour (Green et al. 1965; Goldblum et al. 2000; Kaur et al. 2008; O'Day et al. 1985). Furthermore, its side effects discourage systemic administration (Mora-Duarte et al. 2002; Khoo et al. 1994). Amp-B also promotes immunopotentiation by binding to cholesterol on the cell membrane of lymphocytes. Suppressor T lymphocytes have higher concentrations of cholesterol in their cell membrane than B and T helper lymphocytes, therefore Amp-B leads to a reduction in suppressor cells with a relative increase in pro-inflammatory cells (Shirley and Little 1979a, b).

Direct *in situ* administration is therefore the main form of treatment. It is one of the few drugs described in the literature as being used through the subconjunctival, intrastromal, intracameral, and intravitreal routes, as well as topically. Topical administration in concentrations of 1.5–5 mg/ml at hourly intervals at the beginning of treatment, and then every 4 h after the therapeutic response is observed is commonly the first choice in the treatment of fungal keratitis. Periodic debridement of the corneal epithelium is also recommended for better penetration of the drug into the cornea (Qu et al. 2010; O'Day et al. 1984, 1986). The drug showed good tolerability and efficiency when used both as eye drops and ointment (Hirose et al. 1997; Wood and Williford 1976). Intracorneal administration provides better results than subconjunctival administration in patients with low adherence to treatment, (O'Day et al. 1987, 1991) due to fewer complications and more sustained corneal concentrations than topical or intracameral administration. Several cases of keratitis unresponsive to topical treatment are successfully resolved after intrastromal administration at a concentration of 5–10 μg/0.1 ml (Qu et al. 2010, Garcia- Valenzuela E, Song 2005), but further controlled studies are still needed. Intracameral injection is indicated in deep infections that penetrate Descemet's membrane and affect the anterior chamber and the lens. There are, however, reports of side effects like cataract after administration and transient increase in chamber reaction within 24 h due to the immunopotentiating effect of amp-B, along with iritis and corneal oedema, but they are reversible (Kermani and Aggarwal 2000; Yoon et al. 2007; Kaushik et al. 2001; Kuriakose et al. 2002).

For the treatment of fungal endophthalmitis, intravitreal injection of Amp-B is the therapy of choice. Amphotericin B is effective against a wide range of fungal pathogens, but its utility in treating fungal endophthalmitis is limited by poor ocular penetration and the potential for intraocular toxicity. Intravitreal doses of 5–10 μg of amphotericin B have been used increasingly since Axelrod et al. (1973) found that these doses did not cause clinical, histopathologic, or electroretinographic changes in rabbit eyes. Although these doses are generally well tolerated, another study (Souri and Green 1974) showed that retinal damage can occur with doses as low as 1 μg. The two studies agreed that intravitreal doses greater than 25 μg were extremely toxic and may cause profound intraocular inflammation, retinal necrosis, and cataract formation (Serracarbassa et al. 1998). In cases of fungal endogenous endophthalmitis, treatment depends on the extent of ocular involvement. Vitrectomy along with intravitreal antifungal injections is the preferred mode of treatment, for which usually intravenous amphotericin B is the drug of choice (Riddell et al. 2011). Intravenous amphotericin B continues to be the treatment of choice for invasive fungal infections of the orbit (Levin et al. 1996; Seiff et al.1999).

Pharmacokinetics It is not absorbed by oral route. Administered IV infusion as colloidal suspension made with deoxycholate (DOC). It is distributed widely in the body but penetration in the CSF is poor. Its $t_{1/2}$ is 15 days. Excretion occurs slowly in urine and bile. It takes more than 2 months for complete clearance of the drug.

Adverse Effects

1. *Toxicity* is high, acute infusion related side effects like aches, fever, chills, and rigors.
2. Long-term toxicity:
    1. *Nephrotoxicity* – It is most important toxicity. It manifests as hypokalemia, renal tubular acidosis, and inability to concentrate urine. Nephrotoxicity can be minimized by mannitol administration, sodium loading and alternate day administration. However, caution may be taken while liberalizing salt intake in patients with CHF, renal failure, or cirrhosis with ascitis. The lipid-based amphoterecin B products (e.g. Abelcet, Amphotec, Ambisome) have all shown a lower incidence of nephrotoxicity. In addition, none the lipid-based products demonstrate superior efficacy when compared with amphotericin B against invasive candidiasis and cryptococcal meningitis.
    2. *Thrombophlebitis* can be minimized by using dilute solutions (<0.1 mg/ml) and addition of heparin 500–1000 U per liter of solutions.
    3. *Anaemia*
    4. *CNS toxicity*: On intrathecal injection it results in headache and vomiting.

9.2.2.2 Azoles

Introduced into medical practice in the 1970s, azoles represented an important advance in antifungal therapy. Compared to amp-B, they have a broader spectrum of action and cause fewer adverse effects. Azoles bind to a cytochrome P-450 fungal enzyme involved in the 14α demethylation of either lanosterol or 25 methylenedihydro lanosterol, resulting in a decrease in ergosterol synthesis and an accumulation of 14α-methylated sterols; this leads to increased permeability of the fungal-cell membrane, alteration of membrane enzymes, inhibition of growth, and ultimate death of the fungal cell (Yamaguchi et al. 1993). All azoles, except fluconazole, appear to decrease the function of immune system cells, especially lymphocytes; this may lessen the degree of tissue damage occurring with the inflammatory reaction thereby affecting the efficacy of the azoles *in vivo* and are thus considered as fungistatics when used in ocular fungal infections (Jones and O'Day 1988). The imidazoles used more often in ophthalmology include miconazole, econazole, and ketoconazole. Among the first-generation triazoles, the most used are itraconazole and fluconazole. Second-generation triazoles were introduced into clinical practice in the past decade and include voriconazole and posaconazole.

9.2.2.2.1 Miconazole

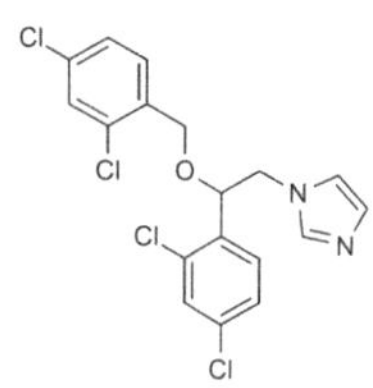

Miconazole was initially developed for use as a topical cream in the treatment of superficial mycoses. (Sawyer et al. 1975). Systemic administration produces good results but is in disuse due to its cardiovascular and hepatotoxic side effects (Coley and Crain 1997; Barasch and Griffin 2008). Miconazole not only acts on the synthesis of ergosterol, similar to other azoles, but also promotes the inhibition of peroxidases, resulting in an accumulation of free radicals in the fungal cytoplasm, which leads to cell death (Fothergill 2006; Kobayashi et al. 2002; Thevissen et al. 2007; Sud et al. 1981). Topical use at a concentration of 10 mg/ml miconazole is available as a solution for intravenous administration; it can also be used for topical administration as a 1 % (10-mg/ml) solution and has good penetration after epithelial scraping is done (Foster and Stefanyszyn 1979; Mohan et al. 1989), or for subconjunctival administration (5–10 mg) in the treatment of infections caused by *Candida*, *Fusarium*, *Curvularia*, and *Aspergillus* (Foster 1981; Foster and Stefanyszyn 1979; Ishibashi and Kaufman 1985). Fitzsimons and Peters (1986) have shown that in a study of 20 patients with fungal keratitis, treatment with topical and subconjunctival miconazole along with ketoconazole resulted in healing in 13 cases. *In vitro*, it was reported to be more effective than ketoconazole and itraconazole against *Aspergillus* spp., *Candida albicans*, and non-*albicans* (Foster and Stefanyszyn 1979; Uchida et al. 2006). Further comparative controlled studies are needed to demonstrate the real benefits of this drug.

Toxicity Topical miconazole 1 % exhibits minimal toxicity that is characterized by conjunctival injection and punctate corneal epithelial erosions. Animal studies suggest that the rate of healing of epithelial defects is not retarded by topical administration of the drug (Foster et al. 1981). Subconjunctival administration is generally well tolerated. Systemic administration of miconazole has been successful in a few reported cases, but this form of therapy has been superseded by the triazoles, which have greatly improved pharmacokinetics.

9.2.2.2.2 Econazole

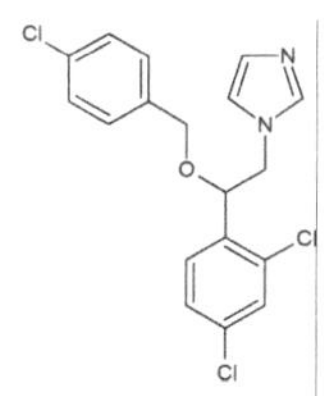

Econazole, an imidazole with a similar molecular structure to miconazole, is used primarily in the treatment of superficial mycoses, with some studies involving systemic use (Heel et al. 1978). Mahashabde et al. (1987) suggest the use of econazole ointment 1 % as prophylactic treatment after ocular trauma with risk of fungal infection. In a study by Prajna et al. (2004), it was concluded that even concurrent use of 5 % natamycin and 2 % econazole did not appear to offer additional benefits over monotherapy with 5 % natamycin for the management of fungal keratitis. Hence this drug is not widely used nowadays.

9.2.2.2.3 Ketoconazole

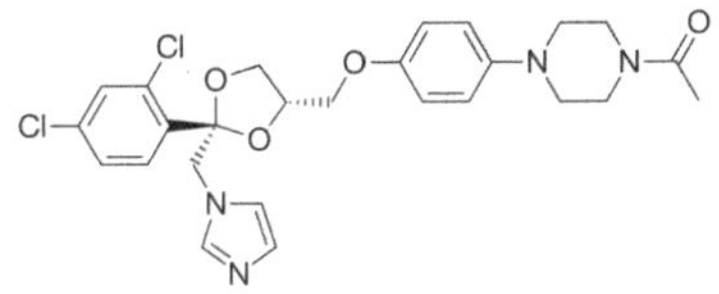

Ketoconazole was the first systemic imidazole to be used successfully, but its use is limited in internal medicine and has now been replaced by itraconazole due to the latter's milder influence on the metabolism of glucocorticoids and extended antifungal spectrum (Kaur et al. 2008). It is used at a dose of 100–400 mg every 12 h; its oral absorption depends on gastric pH (below 3), therefore it should be taken without food or gastric acid-suppressive agents (Kaur et al. 2008). Although its penetration into the cerebrospinal fluid and urine is low, its penetration into ocular tissues is significant when used systemically. There are numerous reports of therapeutic success with oral ketoconazole with or without topical natamycin or Amp-B in the treatment of fungal keratitis. Some authors suggest its routine use in all cases of fungal keratitis (Ishibashi et al. 1983, 1986; Thomas et al. 1987) but this is not supported by controlled studies. There are case reports treated exclusively with topical ketoconazole (10–50 mg/ml) (Torres et al. 1985); however, in comparative studies by Komadina et al. (1985) and Singh et al. (1989) wherein topical and oral ketoconazole was compared with natamycin, ketoconazole was found to have better response with oral administration, combined with topical natamycin. *In vitro* studies with strains of *Aspergillus* spp., *Fusarium* spp., and *Candida* spp., exhibited a lower susceptibility of these organisms to ketoconazole compared to natamycin and voriconazole (Pearse et al. 2009; Marangon et al. 2004). Currently, systemic ketoconazole is indicated only for the adjuvant treatment of deep fungal keratitis.

Toxicity Ketoconazole is much less toxic than amphotericin B. It is less effective and more toxic than newer azoles. Ketoconazole interferes with the biosynthesis of adrenal and gonadal steroid hormones producing significant endocrine effects, such as gynaecomastia, infertility, and menstrual irregularities. This effect is less with flucanozole, itraconazole, and voriconazole. Gastric upset is seen with all orally used azoles. Hepatic impairment occurs with ketoconazole. There is asymptomatic increase in serum transaminases (Foster et al. 1981).

9.2.2.2.4 Itraconazole

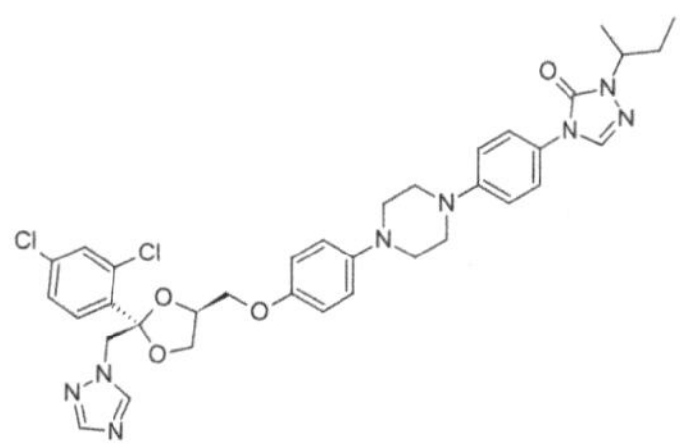

Itraconazole is more frequently used in general practice than ketoconazole and has fewer side effects when administered systemically. However, when administered orally it exhibits lower bioavailability, solubility, and penetration into ocular tissues than other azoles (Jones 2004; Kaur et al. 2008; Rajasekaran et al. 1987; Abad and Foster 1996). Studies in rats showed that Itraconazole has a lower teratogenic risk than ketoconazole (Kaur et al. 2008). Systemic administration at 400 mg/day was effective in the treatment of infections by *Candida* spp. (Klotz et al. 1996). However, in infections by *Fusarium* spp., topical use of itraconazole (10 mg/ml) has been reported ineffective compared to natamycin 5 % (Kalavathy et al. 2005). *In vitro* studies found that itraconazole had a higher MIC than amp-B and natamycin (Li et al. 2008; Rajasekaran et al. 1987), and even found some drug resistance among all analyzed strains (Lalitha et al. 2007). Itraconazole was effective against *Aspergillus* spp., but not as effective as ketoconazole (Marangon et al. 2004). Systemic use should be limited only to the adjuvant treatment of eye infections by yeasts.

9.2.2.2.5 Fluconazole

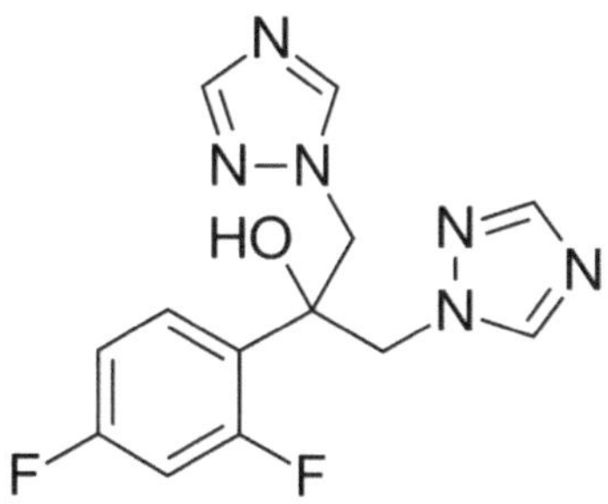

Unlike itraconazole and ketoconazole, flucoconazole shows excellent absorption from the gastrointestinal tract and is not influenced by gastric acidity. Its plasma concentrations with oral use reach almost the same levels as with intravenous administration. Penetration into ocular tissues is effective, reaching aqueous concentrations similar to those in the plasma (O'Day et al.1990). Oral use at 200–400 mg/day was effective in the treatment of eye infections, with or without topical natamycin (Thakar 1994; Urbak and Degn 1994). When used subconjunctivally in association with topical amp-B, a broader antifungal spectrum was observed with less toxicity than isolated amp-B (Mahdy et al. 2010a, b). Yilmaz and Maden (2005) managed to treat 60 % of cases of fungal keratitis with subconjunctival injections (2 mg/1 ml) administered daily for 10 days of flucoconazole alone. Flucoconazole eye drops (2 mg/ml) achieved good intracorneal therapeutic levels against strains of *Aspergillus fumigatus* in rabbits (Yee et al. 1997; Avunduk et al. 2003). Flucoconazole is less effective than other drugs in the treatment of fungal endophthalmitis. Despite its good vitreal penetration when administered orally, its ineffectiveness against filamentous fungi (*Aspergillus* and *Fusarium* spp.), which exhibited marked resistance (Hariprasad et al. 2008; Li et al. 2008; Marangon et al. 2004), discourages its use as an adjuvant. However, there are reports of successful treatment of endogenous endophthalmitis by *Candida* spp. with flucoconazole (Muller et al. 2013; Christmas and Smiddy 1996; Wellington and Gigliotti 2001).

9.2.2.2.6 Voriconazole

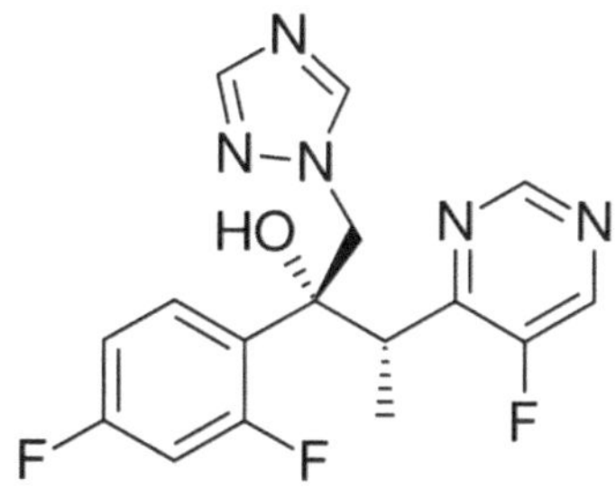

Voriconazole was developed from the fluconazole molecule and presents better efficacy at lower MICs than the first triazoles, is more effective in blocking the synthesis of ergosterol, thereby increasing its effectiveness against filamentous fungi (Kaur et al. 2008). Because of its great efficacy in treating disseminated fungal infections, with lower toxicity compared to amp-B, voriconazole is currently the drug of choice in the treatment of invasive aspergillosis (Herbrecht et al. 2002). It is metabolized by the liver, therefore liver enzymes should be monitored during therapy. Among its side effects are visual disorders, which are present in about 30 % of patients using the drug and are usually reversible. Similar to flucoconazole, it presents good gastric absorption and bioavailability (Hariprasad et al. 2008). Administered orally at a dose of 200 mg every 12 h, voriconazole reaches peak plasma concentrations after 2–3 h. The drug has been extensively studied in the treatment of keratitis and endophthalmitis due to its good concentrations in several ocular tissues (Freda 2006). Alfonso et al. (2011)

suggest voriconazole as the drug of choice for oral use in the treatment of deep keratitis, scleritis, and endophthalmitis and as prophylaxis after penetrating keratoplasty. Hariprasad et al. (2008) also suggest oral voriconazole as prophylaxis in cases of ocular trauma with plant material. Administered topically at a concentration of 1 mg/ml, it was effective in the treatment of keratitis by *Candida*, *Aspergillus*, *Fusarium*, *Scedosporium*, and *Paecilomyces*, among others (Anderson et al. 2004; Bunya et al. 2007; Lee et al. 2009; Nulens et al. 2003). Its advantages compared to polyenes include its greater stability to light and temperature, remaining effective for up to 30 days (Al-Badriyeh et al. 2010; Dupuis et al. 2009). Studies in horses showed drug penetration even with epithelial integrity (Clode et al. 2006). Some reports support the use of intracorneal voriconazole in cases of deep keratitis unresponsive to topical and/or oral administration. Prakash et al. (2008) report success in three cases of keratitis unresponsive to topical natamycin using intrastromal injection of voriconazole 50 μg/0.1 ml. Recently, Siatiri et al. (2011) described three cases of *Fusarium* keratitis unresponsive to topical treatment that resolved after intracorneal voriconazole. Sharma et al. (2011) in a series with 13 patients also suggest the use of intrastromal voriconazole in refractory keratitis. However, there are a few studies of treatment failure with voriconazole. Giaconi et al. (2006) reported two cases, a keratitis by *Fusarium oxysporum* and another by *Colletotrichum dematium*, which were unresponsive to topical therapy with voriconazole. In vitro studies have demonstrated the superiority of voriconazole to amp-B against *Aspergillus* spp. (Diekema et al. 2003; Johnson and Kauffman 2003) Against *Fusarium* spp., the absolute MIC of voriconazole, natamycin, and amp-B were similar, with voriconazole having a lower relative MIC than the polyenes (Lalitha et al. 2007). In another study, the minimum inhibitory concentration of voriconazole for *Fusarium* spp. was found to be higher than that for *Candida* and *Aspergillus* spp. (Marangon et al. 2004).

There is increasing literature on intravitreal injection of voriconazole as well. An in vitro study using human retinal pigment epithelium cells exposed to voriconazole at concentrations from 25 up to 10,000 μg/ml showed that in concentrations <250 μg/ml, no toxic effects was observed at any dosage level, and intravitreal administration was shown to be safe in experimental rat model as well with no changes in electroretinography (Gao et al. 2003). Thus, it is suggested that voriconazole concentrations of up to 25 μg/ml in the vitreous are safe. Injection of voriconazole is suggested to be safer than amp-B because it immediately achieves high levels of the drug in the vitreous, whereas serum levels from systemic administration gradually reach a steady state. Clinical efficacy of systemic voriconazole in patients with *Candida* endophthalmitis as well as endophthalmitis with filamentous fungi (Hariprasad et al. 2008; Jang et al. 2005) has been reported in having a rapid response with good outcome. One added advantage of voriconazole over fluconazole is that it has activity against *Aspergillus* spp. and fluconazole resistant *Candida* spp., such as *Candida glabrata* and *Candida krusei* (Riddell et al. 2011; Motukupally et al. 2015).

9.2.2.2.7 Posaconazole

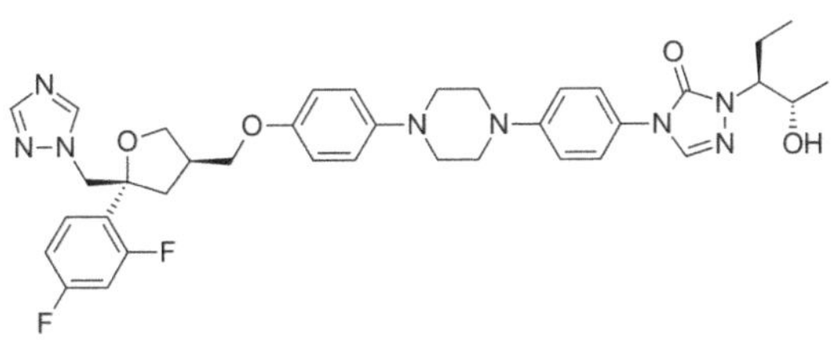

Similar to voriconazole, posaconazole is a second-generation triazole recently introduced into medical practice. It results from an improvement in the molecule of itraconazole and is primarily indicated for the treatment of invasive fungal infections in onco-hematological patients. It is only available as an oral solution and should be administered at a dose of 200 mg four times daily or 400 mg twice daily. A parenteral presentation is currently being developed. Gastrointestinal complaints are the only adverse effects reported to date (Ullmann et al. 2006). In vitro and in vivo studies show that posaconazole has a broad spectrum against *Candida* spp., *Cryptococcus neoformans*, *Aspergillus* spp., and *Fusarium* spp., among others. Posaconazole was effective against most agents resistant to itraconazole and fluconazole (Cuenca-Estrella et al. 2006; Torres et al. 2005) and, together with voriconazole, had the lowest MIC against multiple agents (Lalitha et al. 2007). Experience with its use in ocular infections is still limited, but initial results are encouraging. Another report demonstrated resolution of *Fusarium* endophthalmitis in two patients in whom voriconazole treatment had failed (Tu et al. 2007). Sponsel et al. (2002) also described a case of keratitis by *Fusarium solani* resistant to amp-B and natamycin but successfully treated with oral posaconazole 200 mg four times daily associated with topical use (100 mg/ml prepared from an oral solution). However, comparative controlled studies with main antifungal agents are still lacking.

9.2.2.3 Pyrimidines

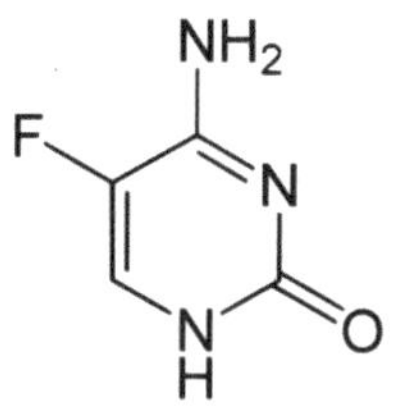

Pyrimidines are represented by 5-fluorocytosine (5-FC) or flucytosine, which is the only antifungal agent with intracellular action. After being absorbed by the fungus it is converted into 5-fluorouracil, a powerful antimetabolic which acts by inhibiting the synthesis of DNA (Vermes et al. 2000). Topical (1 %) solution well tolerated and can be given orally (150 mg/kg/day) or intravenously. Active against *Candida*, *Cryptococcus*, and related fungi, it has varied action against some strains of *Aspergillus*. It is mostly useful as adjunctive therapy for yeast keratitis and cannot be administered alone in treating *Candida* or *Cryptococcus* keratitis due to rapid emergence of resistance. It has limited spectrum of activity against *Fusarium* species (Riddell et al. 2011).

Flucytosine is an adjunctive agent that can be used in combination with amp-B for the treatment of *Candida* endophthalmitis (Pappas et al. 2009). It is synergistic with amp-B in killing *Candida* and achieves high levels in all intraocular compartments in rabbits and humans (Louie et al. 1999). Its use in eye infections is restricted due to its narrow antifungal spectrum and low penetration into ocular tissues (O'Day et al. 1985).

Adverse Effects Bone marrow depression, loss of hair

9.2.2.4 Echinocandins

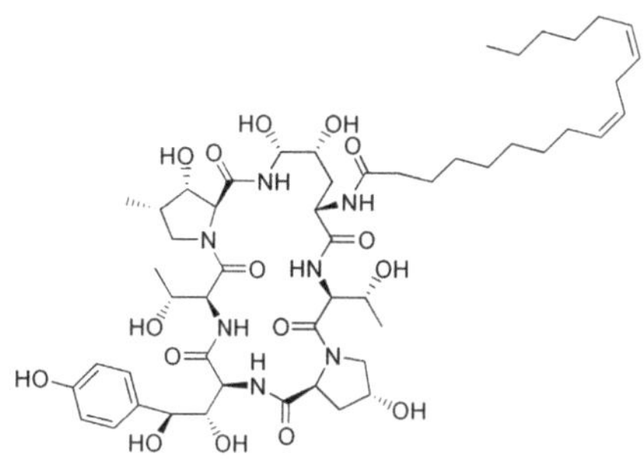

Echinocandins are semisynthetic lipopeptides that inhibit the synthesis of glucan in the fungal cell wall through non-competitive inhibition of the enzyme 1,3-β-glucan synthase, causing osmotic imbalance and cell lysis (Morris and Villmann 2006a, b). This class of drugs includes caspofungin, micafungin, and anidulafungin. Used in yeast infections, echinocandins have rapid fungicidal action against most *Candida* spp., including strains resistant to fluconazole, but not against *Cryptococcus*, *Rhodotorula*, and *Trichosporon* (Bachmann et al. 2002). Echinocandins have fungistatic action against some filamentous fungi such as *Aspergillus*, but not against *Fusarium* and *Rhizopus* (Wagner et al. 2006). Topical caspofungin at a concentration 1.5–5 mg/ml was as effective as amp-B in the treatment of corneal ulcer by *Candida albicans* in an animal model (Goldblum et al. 2005). Two other studies involving 1 mg/ml topical micafungin found an efficacy comparable or superior to fluconazole in the treatment of keratitis by *Candida albicans* and *Candida parapsilosis* (Matsumoto et al. 2005, 2011).

In a double-blind multicenter trial of 239 patients, caspofungin was found to be equally effective as amphotericin B in the treatment of candidemia (Mora-Duarte et al. 2002). Gauthier et al. (2005), however, reported a case of *Candida* endophthalmitis that failed treatment with oral caspofungin (50 mg/day) because of its poor penetration into the vitreous cavity. In other studies in experimental animals, it has been shown that concentrations of micafungin in the vitreous after systemic administration in non inflamed eyes are very low, ranging from undetectable to 0.034 μg/ml (Groll et al. 2001a). Similarly, Anidulafungin levels in the vitreous have ranged from undetectable to 0.184 μg/ml when very high dosages were used systemically (Groll et al. 2001b). Given the existing literature and its limited vitreal penetration, the role of echinocandins remains to be determined in fungal endophthalmitis.

Toxicity Systemic adverse effects includes fever, headache, abdominal pain, nausea, vomiting, diarrhoea, rash, itching, redness, and pain around injection site may occur. Altered liver enzyme level and blood abnormalities may also occur (Foster et al. 1981).

9.2.3 Clinical Trials in Antifungal Therapy of Eye Infections

9.2.3.1 Keratitis

The gold standard for treatment of fungal keratitis remains elusive. A recent Cochrane Database systematic review (FlorCruz and Evans 2015) of medical interventions for mycotic keratitis analyzed 12 trials in this review that included 981 people; the evidence is current up to March 2015. Ten trials were conducted in India, one in Bangladesh, and one in Egypt (Table 9.1). Participants were randomized to the following comparisons: topical 5 % natamycin compared to topical 1 % voriconazole; topical 5 % natamycin compared to topical 2 % econazole; topical 5 % natamycin compared to topical chlorhexidine gluconate (0.05 %, 0.1 % and 0.2 %); topical 1 % voriconazole compared to intrastromal voriconazole 50 g/0.1 ml (both treatments combined with topical 5 % natamycin); topical 1 % voriconazole combined with oral voriconazole compared to both oral voriconazole and oral itraconazole (both combined with topical 5%natamycin); topical 1 % itraconazole compared to topical 1 % itraconazole combined with oral itraconazole; topical amphotericin B compared to topical amphotericin B combined with subconjunctival injection of fluconazole; intracameral injection of amphotericin B with conventional treatment compared to conventional treatment alone; topical 0.5 % and 1 % silver sulphadiazine compared to topical 1 % miconazole. Overall the results were inconclusive because for most comparisons only one small trial was available. The exception was the comparison of topical natamycin and topical voriconazole for which three trials were available (Prajna et al. 2010; Arora et al. 2011; Sharma et al. 2013). In one of these trials (Arora et al. 2011), clinical cure (healed ulcer) was reported in all 15 people allocated to natamycin and in 14/15 people allocated to voriconazole (risk ratio (RR) 1.07; 95 % confidence interval (CI) 0.89–1.28, low quality evidence). In one trial (Sun et al. 2010), people randomized to natamycin were more likely to have a microbiological cure at 6 days (RR 1.64; 95 % CI 1.38–1.94, 299 participants). On average, people randomized to natamycin had better spectacle-corrected visual acuity at 2–3 months compared to people randomized to voriconazole but the estimate was uncertain and the 95 % confidence intervals included 0 (no difference) (mean difference −0.12 logMAR, 95 % CI −0.31–0.06, 434 participants; 3 studies, low quality evidence) and a decreased risk of corneal perforation or therapeutic penetrating keratoplasty, or both

(RR 0.61; 95 % CI 0.40–0.94, 434 participants, high quality evidence). There was inconclusive evidence on time to clinical cure (FlorCruz and Evans 2015). In our analysis, we have included another randomized controlled study by Sharma et al. (2015a, b), which randomized 118 patients with fungal keratitis to an identical dosage schedule with 5 % natamycin or 1 % voriconazole eye drops (Table 9.1). This study concluded that natamycin was more effective in the treatment of fungal keratitis, especially *Fusarium* keratitis leaving the choice of voriconazole as the second best.

The trials included in this review were of variable quality and were generally underpowered. There is evidence that natamycin is more effective than voriconazole in the treatment of fungal ulcers. Future research should evaluate treatment effects according to fungus species.

9.2.3.2 Endophthalmitis

Unlike keratitis, prospective, randomized control trials are lacking to evaluate the treatment of fungal endophthalmitis and most data are based on retrospective, interventional case series. In a recent paper by Mittal et al. (2015) wherein 12 eyes of 12 consecutive cases of filamentous fungal endophthalmitis were treated with a combination of intravitreal amphotericin-B and intravitreal voriconazole (AmB-Vo Regime) along with pars plana vitrectomy at a single center, the visual acuity at final follow-up was 20/400 or better in 7/12 eyes (58.33 %) and 20/60 in 2/12 eyes (range 20/60 to LP). Globe salvage was achieved in all cases. The authors concluded that combining intravitreal amphotericin B and voriconazole could be a novel treatment strategy in the management of endophthalmitis caused by filamentous fungus.

In a double-blind multicenter trial of 239 patients, caspofungin was found to be equally effective as amphotericin B in the treatment of candidemia. Patients were infused with either amphotericin B (0.6–1 mg/kg/day) or caspofungin as a single loading dose of 70 mg with a daily maintenance dose of 50 mg. In this cohort of patients, seven patients with endophthalmitis were included and the endophthalmitis resolved in all seven patients. Unfortunately, the authors failed to mention whether or not these seven patients received caspofungin or amphotericin B (Mora-Duarte et al. 2002). It is also not clear whether the vitreous sample of these patients was investigated for fungus. Review of the five randomized controlled treatment trials that studied echinocandins for candidemia and invasive candidiasis revealed that 21 of the 1028 patients who received an echinocandin were noted to have endophthalmitis (Pappas et al. 2009; Kuse et al. 2007; Reboli et al. 2007; Betts et al. 2009). It is difficult to draw any conclusions about the efficacy of systemic echinocandins for the treatment of fungal endophthalmitis.

9.2.4 In Vitro Antifungal Susceptibility Testing of Ocular Fungal Isolates

The clinical relevance of antifungal susceptibility testing is thought to lie in guiding the clinician in the selection of an appropriate antifungal compound. Such tests are reported to help in the selection of the appropriate antifungal in different ophthalmic mycoses. Antifungal susceptibility studies with ocular fungal isolates, however, are limited, likely due to the absence of established minimum inhibitory concentration (MIC) clinical breakpoints, which classify isolates as susceptible, intermediate, or resistant to an antimicrobial agent. Determination of susceptibility of a fungal isolate to different antifungal agents may aid rational specific antifungal therapy, but most laboratories do not routinely perform such tests because the methods in use are diverse. Disc diffusion method similar to bacterial susceptibility testing can be adopted for yeasts (Clinical and Laboratory Standard Institute 2009) and some of the non-dermatophyte filamentous fungi (Clinical and Laboratory Standard Institute 2010). For determination of minimum inhibitory concentration (MIC) of antifungal drugs, broth or agar dilution methods are used. Drugs commonly tested include 5-fluorocytosine (5-FC or flucytosine), ketoconazole, miconazole, fluconazole, itraconazole, and amphotericin B. In a study, the

Table 9.1 Summary of findings of randomized controlled trials for the treatment of Fungal keratitis

S. No	Author (year) reported	Total no. of participants in trial	Interventions	Outcome measures
1.	Mohan et al. (1987)	30	Topical 0.5 % and 1 % silver sulphadiazine versus topical 1 % miconazole	In silver sulphadiazine group, 15/20 (75 %) people had a healed ulcer at 2–4 weeks compared to 6/10 (60 %) of the 1 % miconazole group
2.	Rahman et al. (1997)	60	Topical chlorhexidine gluconate 0.05 % (N = 8) Topical chlorhexidine gluconate 0.1 % (N = 17) Topical chlorhexidine gluconate 0.2 % (N = 17) Natamycin 5 % (N = 18)	Compared with the response to natamycin as the referent, the relative efficacy was 1.17 with chlorhexidine 0.05 %, 1.43 with 0.1 % and 2.00 with 0.2 %. The superiority of 0.2 % chlorhexidine over natamycin was statistically significant (relative efficacy 2.20, p = 0.043) in patients not having had prior antifungal treatment
3.	Rahman et al. (1998)	70	Topical chlorhexidine gluconate 0.2 % versus topical natamycin 2.5 %	None of the severe ulcers healed at 21 days, but three allocated to chlorhexidine eventually healed in 60 days. Of the non-severe ulcers, 66.7 % were healed at 21 days with chlorhexidine and 36.0 % with natamycin
4.	Agarwal et al. (2001)	54	Topical (what %) itraconazole versus topical and oral itraconazole	Forty-two out of 54 (78 %) participants "responded favorably" to treatment but the comparison between topical and topical and oral groups was not clearly presented making it difficult to draw conclusions
5.	Prajna et al. (2003)	116	Topical 5 % natamycin versus topical 2 % econazole	There was no significant difference (log rank 0.52, P = 0.47) between the two arms for success which was defined as a healed or healing ulcer at 4 weeks
6.	Prajna et al. (2013)	299	Topical 5 % natamycin versus topical 1 % voriconazole	Did not report on clinical cure but did report on microbiological cure at six days. In participants randomised to natamycin 132/155 (85.2 %) were culture negative compared to 75/144 (52.1 %) people in the voriconazole group
7.	Prajna et al. (2010)	120	Topical 5 % natamycin versus topical 1 % voriconazole	No significant differences in visual acuity, scar size, and perforations between voriconazole- and natamycin-treated patients
8.	Mahdy et al. (2010a, b)	48	Topical amphotericin B (what %) versus topical amphotericin B and subconjunctival injection (how much?) of fluconazole	Higher proportion of ulcers healed with combination treatment (amphotericin B and fluconazole) 20/24 (83 %) compared to amphotericin alone 16/24 (67 %)

(continued)

Table 9.1 (continued)

S. No	Author (year) reported	Total no. of participants in trial	Interventions	Outcome measures
9.	Arora et al. (2011)	30	Topical 5 % natamycin versus topical 1 % voriconazole	No added advantage of using topical 1 % voriconazole over topical natamycin as primary treatment
10.	Parchand et al. (2012)	45	Oral and topical 1 % voriconazole versus oral (how much?) voriconazole and topical 5 % natamycin versus oral (how much?) itraconazole and topical 5 % natamycin	At 3 months treatment success was observed in 10/15 (66.7 %) participants allocated to topical and oral voriconazole compared to 11/15 (73.3 %) people receiving oral voriconazole and topical natamycin and 10/15 (66.7 %) in the itraconazole and natamycin group
11.	Sharma et al. (2013)	40	Topical 1 % voriconazole versus intrastromal voriconazole 50 g/ 0.1 mL (both treatments combined with topical 5 % natamycin)	Treatment was successful in 19/20 (95 %) people receiving topical voriconazole compared to 16/20 (80 %) people receiving intrastromal voriconazole
12.	Sharma S et al. (2015)	118	1 % topical voriconazole vs 5 % topical natamycin	Natamycin was more effective in the treatment of fungal keratitis, especially *Fusarium* keratitis
13.	Sharma N et al. 2015	45	Intracameral injection of amphotericin B Vs (topical + oral antifungal + drainage of hypopyon)	Intracameral Amp-B does not offer any benefit over topical antifungal therapy when performed alone or associated with drainage of hypopyon

[a]oral fluconazole 150–200 mg twice a day for 3 weeks + topical natamycin 5 % every hour + topical amp- B 0.15 % every hour

susceptibility testing of 90 filamentous fungal isolates (*Aspergillus*-41, *Fusarium*-38, others-11) from patients with keratitis showed lowest MIC of triazoles and caspofungin against *Aspergillus* spp. and lowest MIC of voriconazole, amphotericin B, and posaconazole against *Fusarium* spp. Isolates of *Fusarium* spp. were not inhibited by itraconazole and caspofungin (Lalitha et al. 2007). In a study that determined the MICs of natamycin and voriconazole on isolates from fungal keratitis, a higher MIC was significantly associated with an increased likelihood of perforation; however, there was no significant association between MIC and 3-week or 3-month visual acuity or between MIC and 3-week or 3-month infiltrate/scar size (Lalitha et al. 2012). It should be however added that susceptibility breakpoints for natamycin have not been described so far in CLSI guidelines, and an MIC of 16 μg/ml or less is considered to indicate susceptibility of a fungal isolate based on a notable publication by Lalitha et al. (2008), who convincingly showed that eye drop preparations are an alternative to pharmaceutical grade natamycin for testing antifungal susceptibility. Since then other studies have used the same criteria to evaluate the MIC of natamycin against fungi from mycotic keratitis (Pradhan et al. 2011). Reports of application of E-test, a convenient method for determination of MIC of several antifungal antibiotics against yeast and certain filamentous fungi, is also now available in the literature (Motukupally et al. 2015, Qiu et al. 2005).

The real value of in vitro antifungal susceptibility testing in mycotic keratitis lies, perhaps, not in predicting responses in individual cases but in providing important baseline data on the spectrum of activity of antifungal compounds against ocular fungal isolates. Interestingly, when the in vitro antifungal susceptibilities of nine isolates of filamentous fungi were determined by the NCCLS method in 11 different laboratories and compared to antifungal treatment outcomes in animal infection models, only a limited association between MIC and treatment outcome was seen, due to drawbacks in the models used (Odds et al. 1998). In systemic bacterial infections, in vitro susceptibility is thought to predict the response of bacterial infections according the "90–60 rule", which states that susceptible organisms respond 90 % of the time, and resistant organisms respond 60 % of the time (Rex and Pfaller 2002; Weinstein et al. 1983). More recently, a similar predictive utility has been suggested for ocular fungal infections, and susceptibility testing demonstrated that a lower MIC was significantly associated with a good outcome (Lalitha et al. 2012). Many factors play a role in the healing of a corneal ulcer, in addition to the effectiveness of the antimicrobial. Above all, the relationship between in vitro susceptibility data and clinical response to topical antifungal medication needs to be clarified; and a larger sample size may be needed to demonstrate whether or not there is a true association between visual acuity and infiltrate/scar size and MIC.

9.3 Conclusion

There are many options of antifungal agents and routes of administration, and the choice depends on both the etiologic agent and the location and extent of the infection. Standard therapy with polyenes remains effective. Despite the numerous reports of infections that do not respond to first-line drugs and improve after the introduction of other agents, particularly second-generation triazoles, comparative studies demonstrating the superiority of the latter are lacking. Until the real benefit of the new generation of antifungal agents is demonstrated, we believe those drugs should only be used as an alternative to standard therapy. A few significant advances have been made in ophthalmic antifungal research; problems remain to be dealt with and many challenges await answers in this area.

References

Abad JC, Foster CS (1996) Fungal keratitis. Int Ophthalmol Clin 36(3):1–15

Agarwal PK, Roy P, Das A, Banerjee A, Maity PK, Banerjee AR (2001) Efficacy of topical and systemic itraconazole as a broadspectrum antifungal agent in mycotic corneal ulcer. A preliminary study. Indian J Ophthalmol 49(3):173–176

Al-Badriyeh D, Neoh CF, Stewart K, Kong DC (2010) Clinical utility of voriconazole eye drops in ophthalmic fungal keratitis. Clin Ophthalmol 4:391–405

Alfonso ECGA, Miller D (2011) Fungal keratitis. In: Krachmer JH, Mannis MJ, Holland EJ (eds) Cornea: fundamentals, diagnosis and management, 3rd edn. Mosby Elsevier, New York

Anderson KL, Mitra S, Salouti R, Pham TA, Taylor HR (2004) Fungal keratitis caused by *Paecilomyces lilacinus* associated with a retained intracorneal hair. Cornea 23(5):516–521

Arora R, Gupta D, Goyal J, Kaul R (2011) Voriconazole versus natamycin as primary treatment in fungal corneal ulcers. Clin Exp Ophthalmol 39:434–440

Avunduk AM, Beuerman RW, Warnel ED, Kaufman HE, Greer D (2003) Comparison of efficacy of topical and oral fluconazole treatment in experimental *Aspergillus* keratitis. Curr Eye Res 26(2):113–117

Axelrod AJ, Peyman GA, Apple DJ (1973) Toxicity of intravitreal injection of amphotericin B. Am J Ophthalmol 76(4):578–583

Bachmann SP, VandeWalle K, Ramage G, Patterson TF, Wickes BL, Graybill JR, López-Ribot JL (2002) In vitro activity of caspofungin against *Candida albicans* biofilms. Antimicrob Agents Chemother 46 (11):3591–3596

Barasch A, Griffin AV (2008) Miconazole revisited: new evidence of antifungal efficacy from laboratory and clinical trials. Future Microbiol 3(3):265–269

Behrens-Baumann W (ed) (1999) Mycosis of the eye and its Adnexa. Dev Ophthalmol. Basel, Karger. 32:I–X. doi:10.1159/000060776

Bernauer W, Allan BDS, Dart JKG (1998) Successful management of *Aspergillus* scleritis by medical and surgical treatment. Eye 12:311–316

Betts RF, Nucci M, Talwar D, Gareca M, Queiroz-Telles F, Bedimo RJ, Herbrecht R, Ruiz-Palacios G, Young JA, Baddley JW, Strohmaier KM, Tucker KA, Taylor AF, Kartsonis NA, Caspofungin High-Dose Study Group (2009) A multicenter, double-blind trial of a high-dose caspofungin treatment regimen versus a standard caspofungin treatment regimen for adult patients with invasive candidiasis. Clin Infect Dis 48:1676–1684

Bunya VY, Hammersmith KM, Rapuano CJ, Ayres BD, Cohen EJ (2007) Topical and oral voriconazole in the treatment of fungal keratitis. Am J Ophthalmol 143 (1):151–153

Christmas NJ, Smiddy WE (1996) Vitrectomy and systemic fluconazole for treatment of endogenous fungal endophthalmitis. Ophthalmic Surg Lasers 27 (12):1012–1018

Clinical and Laboratory Standard Institute (2009) Method for antifungal disk diffusion susceptibility testing of yeasts; approved guideline, 2nd edn. M44-A2 Clinical and Laboratory Standards Institute, Wayne

Clinical Laboratory Standard Institute (2010) Reference method for antifungal disk diffusion susceptibility testing of non-dermatophyte filamentous fungi; approved guideline. CLSI document M51-A Clinical and Laboratory Standards Institute, Villanova

Clode AB, Davis JL, Salmon J, Michau TM, Gilger BC (2006) Evaluation of concentration of voriconazole in aqueous humor after topical and oral administration in horses. Am J Vet Res 67(2):296–301

Coley KC, Crain JL (1997) Miconazole-induced fatal dysrhythmia. Pharmacotherapy 17(2):379–382

Cuenca-Estrella M, Gomez-Lopez A, Mellado E, Buitrago MJ, Monzon A, Rodriguez-Tudela JL (2006) Head-to-head comparison of the activities of currently available antifungal agents against 3,378 Spanish clinical isolates of yeasts and filamentous fungi. Antimicrob Agents Chemother 50(3):917–921

Diekema DJ, Messer SA, Hollis RJ, Jones RN, Pfaller MA (2003) Activities of caspofungin, itraconazole, posaconazole, ravuconazole, voriconazole, and amphotericin B against 448 recent clinical isolates of filamentous fungi. J Clin Microbiol 41(8):3623–3626

Dupuis A, Tournier N, Le Moal G, Venisse N (2009) Preparation and stability of voriconazole eye drop solution. Antimicrob Agents Chemother 53 (2):798–799

Dursun D, Fernandez V, Miller D, Alfonso EC (2003) Advanced *Fusarium* keratitis progressing to endophthalmitis. Cornea 22(4):300–303

Fiscella RG, Moshifar M, Messick CR, Pendland PL, Chandler JW, Viana M (1997) Polyhexamethylene biguanide (PHMB) in the treatment of experimental *Fusarium* keratomycosis. Cornea 16:447–449

Fitzsimons R, Peters AL (1986) Miconazole and ketoconazole as a satisfactory first-line treatment for keratomycosis. Am J Ophthalmol 101(5):605–608

FlorCruz NV, Evans JR (2015) Medical interventions for fungal keratitis. Cochrane Database Syst Rev 4. doi:10.1002/14651858.CD004241.pub4

Foster CS (1981) Miconazole therapy for keratomycosis. Am J Ophthalmol 91(5):622–629

Foster CS, Stefanyszyn M (1979) Intraocular penetration of miconazole in rabbits. Arch Ophthalmol 97 (9):1703–1706

Foster CS, Lass JH, Moran-Wallace K, Giovanoni R (1981) Ocular toxicity of topical antifungal agents. Arch Ophthalmol 99(6):1081–1084

Fothergill AW (2006) Miconazole: a historical perspective. Expert Rev Anti Infect Ther 4(2):171–175

Freda R (2006) Use of oral voriconazole as adjunctive treatment of severe cornea fungal infection: case report. Arq Bras Oftalmol 69(3):431–434

Friedberg ML, Pleyer U, Mondino BJ (1991) Device drug delivery to the eye. Collagen shields, iontophoresis and pumps. Ophthalmology 98:725–732

Gallis HA, Drew RH, Pickard WW (1990) Amphotericin B: 30 years of clinical experience. Rev Infect Dis 12 (2):308–329

Gao H, Pennesi M, Shah K, Qiao X, Hariprasad SM, Mieler WF, Wu SM, Holz ER (2003) Safety of intravitreal voriconazole: electroretinographic and histopathologic studies. Trans Am Ophthalmol Soc 101:183–189

Garcia- Valenzuela E, Song CD (2005) Intracorneal injection of amphothericin B for recurrent fungal keratitis and endophthalmitis. Arch Ophthalmol 123 (12):1721–1723

Gauthier GM, Nork TM, Prince R, Andes D (2005) Subtherapeutic ocular penetration of caspofungin and associated treatment failure in *Candida albicans* endophthalmitis. Clin Infect Dis 41(3):e27–e28

Giaconi JA, Marangon FB, Miller D, Alfonso EC (2006) Voriconazole and fungal keratitis: a report of two treatment failures. J Ocul Pharmacol Ther 22 (6):437–439

Goldblum D, Frueh BE, Zimmerli S, Böhnke M (2000) Treatment of postkeratitis *Fusarium* endophthalmitis with amphotericin B lipid complex. Cornea 19 (6):853–856

Goldblum D, Frueh BE, Sarra GM, Katsoulis K, Zimmerli S (2005) Topical caspofungin for treatment of keratitis caused by *Candida albicans* in a rabbit model. Antimicrob Agents Chemother 49(4):1359–1363

Green WR, Bennett JE, Goos RD (1965) Ocular penetration of amphotericin B: a report of laboratory studies and a case report of postsurgical *Cephalosporium* endophthalmitis. Arch Ophthalmol 73:769–775

Groll AH, Mickiene D, Petraitiene R, Petraitis V, Lyman CA, Bacher JS, Piscitelli SC, Walsh TJ (2001a) Pharmacokinetic and pharmacodynamic modeling of anidulafungin (LY303366): reappraisal of its efficacy in neutropenic animal models of opportunistic mycoses using optimal plasma sampling. Antimicrob Agents Chemother 45:2845–2855

Groll AH, Mickiene D, Petraitis V, Petraitiene R, Ibrahim KH, Piscitelli SC, Bekersky I, Walsh TJ (2001b) Compartmental pharmacokinetics and tissue distribution of the antifungal echinocandin lipopeptide micafungin (FK463) in rabbits. Antimicrob Agents Chemother 45:3322–3327

Hariprasad SM, Mieler WF, Lin TK, Sponsel WE, Graybill JR (2008) Voriconazole in the treatment of fungal eye infections: a review of current literature. Br J Ophthalmol 92(7):871–878

Heel RC, Brogden RN, Speight TM, Avery GS (1978) Econazole: a review of its antifungal activity and therapeutic efficacy. Drugs 16(3):177–201

Herbrecht R, Denning DW, Patterson TF, Bennett JE, Greene RE, Oestmann JW, Kern WV, Marr KA, Ribaud P, Lortholary O, Sylvester R, Rubin RH, Wingard JR, Stark P, Durand C, Caillot D, Thiel E, Chandrasekar PH, Hodges MR, Schlamm HT, Troke PF, de Pauw B, Invasive Fungal Infections Group of the European Organisation for Research and Treatment of Cancer and the Global *Aspergillus* Study Group (2002) Voriconazole versus amphotericin B for primary therapy of invasive aspergillosis. N Engl J Med 347(6):408–415

Hirose H, Terasaki H, Awaya S, Yasuma T (1997) Treatment of fungal corneal ulcers with amphotericin B ointment. Am J Ophthalmol 124(6):836–838

Ishibashi Y (1983) Oral ketoconazole therapy for keratomycosis. Am J Ophthalmol 95(3):342–345

Ishibashi Y (1986) Miconazole and ketoconazole as a satisfactory first-line treatment for keratomycosis. Am J Ophthalmol 102(4):547–548

Ishibashi Y, Kaufman HE (1985) The effects of subconjunctival miconazole in the treatment of experimental *Candida* keratitis in rabbits. Arch Ophthalmol 103(10):1570–1573

Jang GJ, Kim KS, Shin WS, Lee WK (2005) Treatment of *Candida* chorioretinitis with voriconazole. Korean J Ophthalmol 19:73–76

Johns KJ, O'Day DM (1988) Pharmacologic management of keratomycoses. Surv Ophthalmol 33:178–188

Johnson LB, Kauffman CA (2003) Voriconazole: a new triazole antifungal agent. Clin Infect Dis 36 (5):630–637

Jones D (2004) Diagnosis and management of fungal keratitis. In: Tasman W, Jaeger EA (eds) Duane's clinical ophthalmology. Lippincott Williams & Wilkins, Philadelphia

Jones DB, Sexton R, Rebell G (1970) Mycotic keratitis in South Florida: a review of thirty-nine cases. Trans Ophthalmol Soc 89:781–797

Jones DB, Forster FK, Rebell G (1972) *Fusarium solani* keratitis treated with natamycin (pimaricin): eighteen consecutive cases. Arch Ophthalmol 88(2): 147–154

Kalavathy CM, Parmar P, Kaliamurthy J, Philip VR, Ramalingam MD, Jesudasan CA, Thomas PA (2005) Comparison of topical itraconazole 1% with topical natamycin 5% for the treatment of filamentous fungal keratitis. Cornea 24(4):449–452

Katz J, Khatry SK, Thapa MD, Schein OD, Kimbrough Pradhan E, LeClerq SC, West KP (2004) A randomised trial of povidone-iodine to reduce visual impairment from corneal ulcers in rural Nepal. Br J Ophthalmol 88(12):1487–1492

Kaur IP, Rana C, Singh H (2008) Development of effective ocular preparations of antifungal agents. J Ocul Pharmacol Ther 24(5):481–493

Kaushik S, Ram J, Brar GS, Jain AK, Chakraborti A, Gupta A (2001) Intracameral amphotericin B: initial experience in severe keratomycosis. Cornea 20 (7):715–719

Kermani NK, Aggarwal SP (2000) Isolated post-operative *Aspergillus niger* endophthalmitis. Eye (Lond). 14:114–116

Khoo SH, Bond J, Denning DW (1994) Administering amphotericin B-a practical approach. J Antimicrob Chemother 33(2):203–213

Klotz SA, Zahid M, Bartholomew WR, Revera PM, Butrus S (1996) *Candida albicans* keratitis treated successfully with itraconazole. Cornea 15(1):102–104

Kobayashi D, Kondo K, Uehara N, Otokozawa S, Tsuji N, Yagihashi A, Watanabe N (2002) Endogenous reactive oxygen species is an important mediator of miconazole antifungal effect. Antimicrob Agents Chemother 46(10):3113–3117

Komadina TG, Wilkes TD, Shock JP, Ulmer WC, Jackson J, Bradsher RW (1985) Treatment of *Aspergillus fumigatus* keratitis in rabbits with oral and topical ketoconazole. Am J Ophthalmol 99(4):476–479

Kuriakose T, Kothari M, Paul P, Jacob P, Thomas R (2002) Intracameral amphotericin B injection in the management of deep keratomycosis. Cornea 21 (7):653–656

Kuse ER, Chetchotisakd P, da Cunha CA, Ruhnke M, Barrios C, Raghunadharao D, Sekhon JS, Freire A, Ramasubramanian V, Demeyer I, Nucci M, Leelarasamee A, Jacobs F, Decruyenaere J, Pittet D, Ullmann AJ, Ostrosky-Zeichner L, Lortholary O, Koblinger S, Diekmann-Berndt H, Cornely OA, Micafungin Invasive Candidiasis Working Group (2007) Micafungin versus liposomal amphotericin B for candidaemia and invasive candidosis: a phase III randomised double-blind trial. Lancet 369:1519–1527

Lalitha P, Shapiro BL, Srinivasan M, Prajna NV, Acharya NR, Fothergill AW, Ruiz J, Chidambaram JD, Maxey KJ, Hong KC, McLeod SD, Lietman TM (2007) Antimicrobial susceptibility of *Fusarium, Aspergillus*, and other filamentous fungi isolated from keratitis. Arch Ophthalmol 125(6):789–793

Lalitha P, Vijaykumar R, Prajna NV, Fothergill AW (2008) *In vitro* natamycin susceptibility of ocular isolates of *Fusarium* and *Aspergillus* species: comparison of commercially formulated natamycin eye drops to pharmaceutical-grade powder. J Clin Microbiol 46:3477–3478

Lalitha P, Prajna NV, Oldenburg CE, Srinivasan M, Krishnan T, Mascarenhas J, Vaitilingam CM, McLeod SD, Zegans ME, Porco TC, Acharya NR, Lietman TM (2012) Organism, minimum inhibitory concentration, and outcome in a fungal corneal ulcer clinical trial. Cornea 31:662–667

Lee SJ, Lee JJ, Kim SD (2009) Topical and oral voriconazole in the treatment of fungal keratitis. Korean J Ophthalmol 23(1):46–48

Levin LA, Avery R, Shore JW, Woog JJ, Baker AS (1996) The spectrum of orbital aspergillosis: clinicopathological review. Surv Ophthalmol 41:142–154

Li L, Wang Z, Li R, Luo S, Sun X (2008) In vitro evaluation of combination antifungal activity against Fusarium species isolated from ocular tissues of keratomycosis patients. Am J Ophthalmol 146:724–728

Louie A, Liu W, Miller DA, Sucke AC, Liu QF, Drusano GL, Mayers M, Miller MH (1999) Efficacies of high-dose fluconazole plus amphotericin B and high-dose fluconazole plus 5-fluorocytosine versus amphotericin B, fluconazole, and 5-fluorocytosine monotherapies in treatment of experimental endocarditis, endophthalmitis, and pyelonephritis due to *Candida albicans*. Antimicrob Agents Chemother 43:2831–2840

Mahashabde S, Nahata MC, Shrivastava U (1987) A comparative study of anti-fungal drugs in mycotic corneal ulcer. Indian J Ophthalmol 35(5–6):149–152

Mahdy RA, Nada WM, Wageh MM, Kader MA, Saleh MM, Alswad MM (2010a) Assessment safety and efficacy of a combination therapy of topical amphotericin B and subconjunctival fluconazole for the treatment of fungal keratitis. Cutan Ocul Toxicol 29(3):193–197

Mahdy RA, Nada WM, Wageh MM (2010b) Topical amphotericin B and subconjunctival injection of fluconazole (combination therapy) versus topical amphotericin B (monotherapy) in treatment of keratomycosis. J Ocul Pharmacol Ther 26(3):281–285

Manzouri B, Vafidis GC, Wyse RK (2001) Pharmacotherapy of fungal eye infections. Expert Opin Pharmacother 2:1849–1857

Marangon FB, Miller D, Giaconi JA, Alfonso EC (2004) In vitro investigation of voriconazole susceptibility for keratitis and endophthalmitis fungal pathogens. Am J Ophthalmol 137(5):820–825

Matsumoto Y, Dogru M, Goto E, Fujishima H, Tsubota K (2005) Successful topical application of a new antifungal agent, micafungin, in the treatment of refractory fungal corneal ulcers: report of three cases and literature review. Cornea 24(6):748–753

Matsumoto Y, Murat D, Kojima T, Shimazaki J, Tsubota K (2011) The comparison of solitary topical micafungin or fluconazole application in the treatment of *Candida* fungal keratitis. Br J Ophthalmol 95 (10):1406–1409

Mithal K, Pathengay A, Bawdekar A, Jindal A, Vira D, Relhan N, Choudhury H, Gupta N, Gupta V, Koday NK, Flynn HW Jr (2015) Filamentous fungal endophthalmitis: results of combination therapy with intravitreal amphotericin B and voriconazole. Clin Ophthalmol 9:649–655

Mohan M, Gupta SK, Kalra VK, Vajpayee RB, Sachdev MS (1987) Silver sulphadiazine in the treatment of mycotic keratitis. Indian J Med Res 85:572–575

Mohan M, Gupta SK, Kalra VK, Vajpayee RB, Sachdev MS (1988) Topical silver sulphadiazine: a new drug for ocular keratomycosis. Br J Ophthalmol 72:192–195

Mohan M, Panda A, Gupta SK (1989) Management of human keratomycosis with miconazole. Aust N Z J Ophthalmol 17(3):295–297

Mora-Duarte J, Betts R, Rotstein C, Colombo AL, Thompson-Moya L, Smietana J, Lupinacci R, Sable C, Kartsonis N, Perfect J, Caspofungin Invasive Candidiasis Study Group (2002) Comparison of caspofungin and amphotericin B for invasive candidiasis. N Engl J Med 347(25):2020–2029

Morris MI, Villmann M (2006a) Echinocandins in the management of invasive fungal infections, part 1. Am J Health Syst Pharm 63(18):1693–1703

Morris MI, Villmann M (2006b) Echinocandins in the management of invasive fungal infections, part 2. Am J Health Syst Pharm 63(19):1813–1820

Motukupally SR, Nanapur VR, Chathoth KN, Murthy SI, Pappuru RR, Mallick A, Sharma S (2015) Ocular infections caused by *Candida* species: type of species, *in vitro* susceptibility and treatment outcome. Indian J Med Microbiol 33(4):538–546

Müller GG, Kara-José N, Castro RS (2013) Antifungals in eye infections: drugs and routes of administration. Revista Brasileira de Oftalmologia 72(2):132–141

Nulens E, Eggink C, Rijs AJ, Wesseling P, Verweij PE (2003) Keratitis caused by Scedosporium apiospermum successfully treated with a cornea transplant and voriconazole. J Clin Microbiol 41 (5):2261–2264

O'Brien TP (1999) Therapy of ocular fungal infections. Ophthalmol Clin North Am 12:33–50

O'Day DM (1987) Selection of appropriate antifungal therapy. Cornea 6(4):238–245

O'Day DM, Ray WA, Head WS, Robinson RD (1984) Influence of the corneal epithelium on the efficacy of topical antifungal agents. Invest Ophthalmol Vis Sci 25(7):855–859

O'Day DM, Head WS, Robinson RD, Stern WH, Freeman JM (1985) Intraocular penetration of systemically administered antifungal agents. Curr Eye Res 4 (2):131–134

O'Day DM, Head WS, Robinson RD, Clanton JA (1986) Corneal penetration of topical amphotericin B and natamycin. Curr Eye Res 5(11):877–882

O'Day DM, Foulds G, Williams TE, Robinson RD, Allen RH, Head WS (1990) Ocular uptake of fluconazole following oral administration. Arch Ophthalmol 108 (7):1006–1008

O'Day DM, Ray WA, Robinson RD, Head WS, Williams TE (1991) Differences in response in vivo to amphotericin B among *Candida albicans* strains. Invest Ophthalmol Vis Sci 32(5):1569–1572

Odds FC, van Gerven F, Espinel-Ingroff A, Bartlett MS, Ghannoum MA, Lancaster MV, Pfaller MA, Rex JH, Rinaldi MG, Walsh TJ (1998) Evaluation of possible correlations between antifungal susceptibilities of filamentous fungi in vitro and antifungal treatment outcomes in animal infection models. Antimicrob Agents Chemother 42:282–288

Oliveira PR, Resende SM, Oliveira FC, Oliveira AC (2001) Ceratite fúngica. Arq Bras Oftalmol 64(1):75–79

Pappas PG, Kauffman CA, Andes D, Benjamin DK Jr, Calandra TF, Edwards JE Jr, Filler SG, Fisher JF, Kullberg BJ, Ostrosky-Zeichner L, Reboli AC, Rex JH, Walsh TJ, Sobel JD, Infectious Diseases Society of America (2009) Clinical practice guidelines for the management of candidiasis: 2009 update by the

Infectious Diseases Society of America. Clin Infect Dis 48:503–535

Parchand S, Gupta A, Ram J, Gupta N, Chakrabarty A (2012) Voriconazole for fungal corneal ulcers. Ophthalmology 119(5):1083

Pearce JW, Giuliano EA, Moore CP (2009) *In vitro* susceptibility patterns of *Aspergillus* and *Fusarium* species isolated from equine ulcerative keratomycosis cases in the midwestern and southern United States with inclusion of the new antifungal agent voriconazole. Vet Ophthalmol 12(5):318–324

Polack FM, Kaufman HE, Newmark E (1971) Keratomycosis. Medical and surgical treatment. Arch Ophthalmol 85(4):410–416

Pradhan L, Sharma S, Nalamada S, Sahu SK, Das S, Garg P (2011) Natamycin in the treatment of keratomycosis: correlation of treatment outcome and *in vitro* susceptibility of fungal isolates. Indian J Ophthalmol 59:512–514

Prajna NV, John RK, Nirmalan PK, Lalitha P, Srinivasan M (2003) A randomised clinical trial comparing 2% econazole and 5% natamycin for the treatment of fungal keratitis. Br J Ophthalmol 87(10): 1235–1237

Prajna NV, Nirmalan PK, Mahalakshmi R, Lalitha P, Srinivasan M (2004) Concurrent use of 5% natamycin and 2% econazole for the management of fungal keratitis. Cornea 23(8):793–796

Prajna NV, Mascarenhas J, Krishnan T, Reddy PR, Prajna L, Srinivasan M, Vaitilingam CM, Hong KC, Lee SM, McLeod SD, Zegans ME, Porco TC, Lietman TM, Acharya NR (2010) Comparison of natamycin and voriconazole for the treatment of fungal keratitis. Arch Ophthalmol 128(6):672–678

Prajna NV, Krishnan T, Mascarenhas J, Rajaraman R, Prajna L, Srinivasan M, Raghavan A, Oldenburg CE, Ray KJ, Zegans ME, McLeod SD, Porco TC, Acharya NR, Lietman TM, Mycotic Ulcer Treatment Trial Group. (2013) The mycotic ulcer treatment trial: a randomized trial comparing natamycin vs voriconazole. JAMA Ophthalmol 131(4):422–429

Prakash G, Sharma N, Goel M, Titiyal JS, Vajpayee RB (2008) Evaluation of intrastromal injection of voriconazole as a therapeutic adjunctive for the management of deep recalcitrant fungal keratitis. Am J Ophthalmol 146(1):56–59

Qiu WY, Yao YF, Zhu YF (2005) Fungal spectrum identified by a new slide culture and in vitro drug susceptibility using E test in fungal keratitis. Curr Eye Res 30:1113–1120

Qu L, Li L, Xie H (2010) Corneal and aqueous humor concentrations of amphotericin B using three different routes of administration in a rabbit model. Ophthalmic Res 43(3):153–158

Rahman MR, Minassian DC, Srinivasan M, Martin MJ, Johnson GJ (1997) Trial of chlorhexidine gluconate for fungal corneal ulcers. Ophthal Epidemiol 4 (3):141–149

Rahman MR, Johnson GJ, Husain R, Howlader SA, Minassian DC (1998) Randomised trial of 0.2% chlorhexidine gluconate and 2.5% natamycin for fungal keratitis in Bangladesh. Br J Ophthalmol 82(8):919–925

Rajasekaran J, Thomas PA, Kalavathy CM, Joseph PC, Abraham DJ (1987) Itraconazole therapy for fungal keratitis. Indian J Ophthalmol 35(5–6):157–160

Reboli AC, Rotstein C, Pappas PG, Chapman SW, Kett DH, Kumar D, Betts R, Wible M, Goldstein BP, Schranz J, Krause DS, Walsh TJ, Anidulafungin Study Group (2007) Anidulafungin versus fluconazole for invasive candidiasis. N Engl J Med 356:2472–2482

Reddy PR, Reddy PS, Reddy AR, Saboo NK (1982) A comparative evaluation of Nystatin, Amphotericin-B and Miconazole in keratomycosis. Indian J Ophthalmol 30(4):249–250

Rex J, Pfaller M (2002) Has antifungal susceptibility testing come of age? Clin Infect Dis 35:982–989

Riddell J IV, Comer GM, Kauffman CA (2011) Reviews of anti-infective agents: treatment of endogenous fungal endophthalmitis: focus on new antifungal agents. Clin Infect Dis 52(5):648–653

Rosa RH Jr, Miller D, Alfonso EC (1994) The changing spectrum of fungal keratitis in south Florida. Ophthalmology 101:1005–1013

Sawyer PR, Brogden RN, Pinder RM, Speight TM, Avery GS (1975) Miconazole: a review of its antifungal activity and therapeutic efficacy. Drugs 9(6):406–423

Seiff SR, Choo PH, Carter SR (1999) Role of local amphotericin B therapy for sino-orbital fungal infections. Ophthalmic Plast Reconstr Surg 15:28–31

Serracarbassa PD, Peyman GA, Liang C, Calixto N Jr, Nair MG (1998) Toxicity and efficacy of intravitreal injection of spartanamicin B in the treatment of Candida endophthalmitis. Int Ophthalmol 22(1):53–58

Sharma N, Agarwal P, Sinha R, Titiyal JS, Velpandian T, Vajpayee RB (2011) Evaluation of intrastromal voriconazole injection in recalcitrant deep fungal keratitis: case series. Br J Ophthalmol 95(12):1735–1737

Sharma N, Chacko J, Velpandian T, Titiyal JS, Sinha R, Satpathy G, Tandon R, Vajpayee RB (2013) Comparative evaluation of topical versus intrastromal voriconazole as an adjunct to natamycin in recalcitrant fungal keratitis. Ophthalmology 120(4):677–681

Sharma N, Sankaran P, Agarwal T, Arora T, Chawla B, Titiyal JS, Tandon R, Satapathy G, Vajpayee RB (2015) Evaluation of intracameral amphotericin B in the management of fungal keratitis: randomized controlled trial. Ocul Immunol Inflamm 23:1–5

Sharma S, Das S, Virdi A, Fernandes M, Sahu SK, Kumar Koday N, Ali MH, Garg P, Motukupally SR (2015) Re-appraisal of topical 1% voriconazole and 5% natamycin in the treatment of fungal keratitis in a randomised trial. Br J Ophthalmol 99(9):1190–1195

Shirley SF, Little JR (1979a) Immunopotentiating effects of amphotericin B. I. Enhanced contact sensitivity in mice. J Immunol 123(6):2878–2882

Shirley SF, Little JR (1979b) Immunopotentiating effects of amphotericin B. II. Enhanced in vitro proliferative responses of murine lymphocytes. J Immunol 123 (6):2883–2889

Siatiri H, Daneshgar F, Siatiri N, Khodabande A (2011) The effects of intrastromal voriconazole injection and topical voriconazole in the treatment of recalcitrant *Fusarium* keratitis. Cornea 30(8):872–875

Singh SM, Khan R, Sharma S, Chatterjee PK (1989) Clinical and experimental mycotic corneal ulcer caused by *Aspergillus fumigatus* and the effect of oral ketoconazole in the treatment. Mycopathologia 106(3):133–141

Souri EN, Green WR (1974) Intravitreal amphotericin B toxicity. Am J Ophthalmol 78(1):77–81

Sponsel WE, Graybill JR, Nevarez HL, Dang D (2002) Ocular and systemic posaconazole(SCH-56592) treatment of invasive *Fusarium solani* keratitis and endophthalmitis. Br J Ophthalmol 86(7):829–830

Srinivasan M, Gonzales CA, George C (1997) Epidemiology and aetiological diagnosis of corneal ulceration in Madurai, South India. Br J Ophthalmol 81:965–971

Sud IJ, Feingold DS (1981) Mechanisms of action of the antimycotic imidazoles. J Invest Dermatol 76 (6):438–441

Sullivan LJ, Snibson G, Joseph C, Taylor HR (1994) *Scedosporium prolificans* sclerokeratitis. Aust N Z J Ophthalmol 22:207–209

Sun CQ, Prajna NV, Krishnan T, Mascarenhas J, Rajaraman R, Srinivasan M et al (2010) Expert prior elicitation and Bayesian analysis of the Mycotic Ulcer Treatment Trial I. Investig Ophthalmol Vis Sci 54 (6):4167–4173

Sweetman S (2004) Martindale: the complete drug reference, 34th edn. Pharmaceutical Press, London, pp 550–554

Thakar M (1994) Oral fluconazole therapy for keratomycosis. Acta Ophthalmol 72(6):765–767

Thevissen K, Ayscough KR, Aerts AM, Du W, De Brucker K, Meert EM, Ausma J, Borgers M, Cammue BP, François IE (2007) Miconazole induces changes in actin cytoskeleton prior to reactive oxygen species induction in yeast. J Biol Chem 282 (30):21592–21597

Thomas PA, Kalavathy CM, Abraham DJ, Rajasekaran J (1987) Oral ketoconazole in Keratomycosis. Indian J Ophthalmol 35(4):197–203

Torres MA, Mohamed J, Cavazos-Adame H, Martinez LA (1985) Topical ketoconazole for fungal keratitis. Am J Ophthalmol 100(2):293–298, trial. Cornea 2012; 31: 662–67

Torres HA, Hachem RY, Chemaly RF, Kontoyiannis DP, Raad II (2005) Posaconazole: a broad-spectrum triazole antifungal. Lancet Infect Dis 5(12):775–785

Tu EY, McCartney DL, Beatty RF, Springer KL, Levy J, Edward D (2007) Successful treatment of resistant ocular fusariosis with posaconazole (SCH-56592). Am J Ophthalmol 143(2):222–227

Uchida K, Abe S, Yamaguchi H (2006) The postantifungal effect (PAFE) of itraconazole, in comparison with those of miconazole and fluconazole, on *Candida* species. Microbiol Immunol 50(9):679–685

Ullmann AJ, Cornely OA, Burchardt A, Hachem R, Kontoyiannis DP, Töpelt K, Courtney R, Wexler D, Krishna G, Martinho M, Corcoran G, Raad I (2006) Pharmacokinetics, safety, and efficacy of posaconazole in patients with persistent febrile neutropenia or refractory invasive fungal infection. Antimicrob Agents Chemother 50(2):658–666

Urbak SF, Degn T (1994) Fluconazole in the management of fungal ocular infections. Ophthalmologica 208 (3):147–156

Vermes A, Guchelaar HJ, Dankert J (2000) Flucytosine: a review of its pharmacology, clinical indications, pharmacokinetics, toxicity and drug interactions. J Antimicrob Chemother 46(2):171–179

Wagner C, Graninger W, Presterl E, Joukhadar C (2006) The echinocandins: comparison of their pharmacokinetics, pharmacodynamics and clinical applications. Pharmacology 78(4):161–177

Weinstein M, Reller L, Murphy J, Lichtenstein KA (1983) The clinical significance of positive blood cultures: a comprehensive analysis of 500 episodes of bacteremia and fungemia in adults. I. Laboratory and epidemiologic observations. Rev Infect Dis 5(1):35–53

Wellington M, Gigliotti F (2001) Update on antifungal agents. Pediatr Infect Dis J 20(10):993–995

Wood TO, Williford W (1976) Treatment of keratomycosis with amphotericin B 0.15%. Am J Ophthalmol 81(6):847–849

Xuguang S, Zhixin W, Zhiqun W, Shiyun L, Ran L (2007) Ocular fungal isolates and antifungal susceptibility in northern China. Am J Ophthalmol 143(1):131–133

Yamaguchi H, Abe S, Tokuda Y (1993) Immunomodulating activity of antifungal drugs. Ann N Y Acad Sci 685:447–457

Yee RW, Cheng CJ, Meenakshi S, Ludden TM, Wallace JE, Rinaldi MG (1997) Ocular penetration and pharmacokinetics of topical fluconazole. Cornea 16 (1):64–71

Yilmaz S, Maden A (2005) Severe fungal keratitis treated with subconjunctival fluconazole. Am J Ophthalmol 140(3):454–458

Yoon KC, Jeong IY, Im SK, Chae HJ, Yang SY (2007) Therapeutic effect of intracameral amphotericin B injection in the treatment of fungal keratitis. Cornea 26(7):814–818

Antifungal Susceptibility Testing of Dermatophytes

10

Indira Gadangi

Abstract

The cases of dermatophytoses have increased over the past few decades. In the last few years, a number of newer less toxic antifungal drugs have become available for clinical use. The increased use of antifungal, often for prolonged periods, has led to the recognition of the phenomenon of acquired antifungal resistance among previously susceptible strains or species and to the increased incidence of infections with less common species. Our study mainly focused on the in vitro susceptibility of clinical isolates of dermatophytes. The microbroth dilution method was performed according to CLSI standards. In the present study, antifungal susceptibility testing was done by microdilution method of dermatophytes against five antifungal agents namely, ketoconazole (imidazoles) fluconazole, itraconazole (triazoles), griseofulvin and terbinafine and their activity against significant number of strains, representing a wide spectrum of dermatophyte species is assessed. Dermatophytic strains: A total of 119 strains of dermatophytes belonging to 10 species were tested. They were *T. rubrum* ($n = 40$), *T. mentagrophytes* ($n = 19$), *T. violaceum* ($n = 15$), *M. gypseum* ($n = 12$), *E. flocossum* ($n = 9$), *M. audouinii* ($n = 8$), *T. schoenleinii* ($n = 5$), *M. canis* ($n = 5$), *T. tonsurans* ($n = 4$) and *T. verrucosum* ($n = 2$). The MIC ranges of all the 119 isolates of dermatophytes tested for antifungal susceptibility showed that terbinafine had the lowest MIC range 0.001–0.64 μg/ml followed by ketoconazole at a MIC range 0.01–3.84 μg/ml. The itraconazole showed a MIC range 0.082–20.45 μg/ml whereas the griseofulvin and fluconazole showed a highest MIC range 0.32–5.12 μg/ml. The MIC 50 of terbinafine was low at 0.02 μg/ml followed by ketaconazole 0.24 μg/ml. The MIC 50 of itraconazole and griseofulvin was 1.28 μg/ml .The highest MIC 50 with

I. Gadangi (✉)
Kakatiya University, Hanamkonda, India
e-mail: gadangi.indira@gmail.com

A. Basak et al. (eds.), *Recent Trends in Antifungal Agents and Antifungal Therapy*,
DOI 10.1007/978-81-322-2782-3_10

2.56 μg/ml was recorded for fluconazole. The MIC90 of terbinafine was low at 0.32 μg/ml followed by ketaconazole with 1.92 μg/ml. The MIC 90 itraconazole was 2.50 μg/ml and for griseofulvin it was 2.56 μg/ml. The highest MIC 90 of flucanozole was high at 10.24 μg/ml. In our study, we observed that terbinafine had the lowest MIC values compared to ketoconazole, itraconazole, griseofulvin and fluconazole.

10.1 Introduction

10.1.1 Historical Review

Historically, medical mycology, specifically relating to human disease, began with the discovery of the fungal etiology of favus and centered around three European physicians in the mid-nineteenth century: Robert Remak, Johann L. Schönlein, and David Gruby. According to Seeliger (1985), Remak (1842) first observed peculiar microscopic structures appearing as rods and buds in crusts from favic lesions. The real founder of dermatomycology was David Gruby on the basis of his discoveries during 1841–1844, his communications to the French Academy of Sciences, and his publications during this period. Independently, and unaware of the work of Remak and Schönlein, he described the causative agent of favus, both clinically and in microscopic details of the crusts, and established the contagious nature of the disease. He has described ectothrix invasion of the beard and scalp, naming the etiologic agent of the latter *Microsporum* (referring to the small spores around the hair shaft) *audouinii*, and described endothrix hair invasion by *Herpes* (*Trichophyton*) *tonsurans*. In addition to these observations on dermatophytes, he has also described the clinical and microscopic appearance of thrush in children.

Raymond Sabouraud (1970), one of the best known and most influential of the early medical mycologists, began his scientific studies of the dermatophytes around 1890, culminating in the publication of his classic volume, "Les Teignes", in 1910. Sabouraud's contributions included his studies on the taxonomy, morphology, and methods of culturing the dermatophytes and the therapy of the dermatophytoses. He classified the dermatophytes into four genera, *Achorion*, *Epidermophyton*, *Microsporum*, and *Trichophyton*, primarily on the basis of the clinical aspects of the disease, combined with cultural and microscopic observations. The medium that he has developed is in use till date for culturing fungi (although the ingredients are modified) and is named in his honor, Sabouraud glucose (dextrose) agar.

In 1934, Chester Emmons modernized the taxonomic scheme of Sabouraud and others and established the current classification of the dermatophytes on the basis of spore morphology and accessory organs. He eliminated the genus *Achorion* and recognized only the three genera *Microsporum*, *Trichophyton*, and *Epidermophyton* on the basis of mycological principles.

10.1.2 Medical Mycology

Medical mycology is the study of mycoses of man and their etiologic agents. Mycoses are the diseases caused by fungi. Among thousands of species of fungi that are known, less than 100 are pathogenic to man. In addition to these species which are generally recognized as pathogenic to man, it is firmly established that under unusual circumstances of abnormal susceptibility of patients or the traumatic implantation of the fungus, other fungi are capable of causing lesions and are known as opportunistic fungi. These circumstances may be:

(i) A debilitating condition of the host, as diabetes
(ii) A concurrent disease such as leukemia

(iii) Prolonged treatment with corticosteroids
(iv) Immunosuppressive drugs or an antibiotic treatment for long duration

Systemic and subcutaneous mycoses are caused by fungi which are essentially free-living dermatophytes in nature. The mycoses are not contagious and infection in man follows inhalation of spores or traumatic implantation of fungi. Certain fungi cause diseases and death in man, and these diseases vary from superficial skin infections to subcutaneous or generalized systemic deep mycoses. According to site of infection, the fungal diseases are classified into five types (Chander 2002):

10.1.2.1 Superficial Mycoses

The superficial (cutaneous) mycoses are usually confined to the outer layers of the skin, hair, and nails and do not invade living tissues. The fungi are called dermatophytes. Dermatophytes, or more properly keratinophilic fungi, produce extracellular enzymes (keratinases) which are capable of hydrolyzing keratin.

10.1.2.2 Subcutaneous Mycoses

These are chronic, localized infections of the skin and subcutaneous tissue following the traumatic implantation of the etiologic agent. The causative fungi are all soil saprophytes of regional epidemiology whose ability to adapt to the tissue environment and elicit disease is extremely variable.

10.1.2.3 Systemic Mycoses

These diseases involve the internal organs like the lungs and brain due to dissemination through blood. The causative fungi are usually dimorphic, which are endemic in some parts of America.

10.1.2.4 Opportunistic Mycoses

These diseases are caused by the nonpathogenic fungi or contaminants in persons whose immunological defense mechanisms are weakened by endogenous causes like cancer, leukemia, AIDS, or other exogenous causes like immunosuppressive therapy and aggressive use of drugs and corticosteroids.

10.1.2.5 Miscellaneous Mycoses

These are the diseases that include ones that could not be grouped under any of the above diseases.

10.2 Superficial Mycoses and Dermatophytes

The superficial mycoses are caused by dermatophytes which are included in *Deuteromycetes* depending upon the following features of *Deuteromycetes*:

1. Fungi have no demonstrable sexual reproductive cycle.
2. All fungi lacking sexual process are gathered in this group.
3. It includes many disease-producing fungi as dermatophytes.

Dermatophytes are fungi that can cause infections of the skin, hair, and nails due to their ability to utilize keratin. They colonize the keratin tissues and inflammation is caused by host response to metabolic by-products. These infections are known as ringworm or tinea, in association with the infected body part. Occasionally, the organisms do invade the subcutaneous tissues, resulting in kerion development. The organisms are transmitted by either direct contact with infected host (human or animal) or direct or indirect contact with infected exfoliated skin or hair in combs, hairbrushes, clothing, furniture, theater seats, caps, bed linens, towels, hotel rugs, and locker room floors.

Depending on the species, the organism may be viable in the environment for up to 15 months. There is an increased susceptibility to infection when there is a preexisting injury to the skin such as scars, burns, and wounds and during marching, high temperature, and humidity (Table 10.1).

The dermatophytes are included in three fungal genera, namely:

Table 10.1 Classification of dermatophytes

Anthropophilic	Zoophilic	Geophilic
Epidermophyton floccosum	*Microsporum canis* (cats, dogs, etc.)	*Microsporum gypseum*
Microsporum audouinii	*Microsporum equinum* (horses)	*Trichophyton ajelloi*
Microsporum ferrugineum	*Microsporum nanum* (pigs)	*Trichophyton terrestre*
Trichophyton concentricum	*Microsporum persicolor*(rodents)	
Trichophyton kanei	*Trichophyton equinum* (horses)	
Trichophyton megnini	*Trichophyton mentagrophytes*, granular (rodents, rabbits, hedgehogs, etc.)	
Trichophyton mentagrophytes (cottony and velvety)	*Trichophyton simii* (monkeys)	
Trichophyton raubitschekii	*Trichophyton verrucosum* (cattle)	
Trichophyton rubrum		
Trichophyton schoenleinii		
Trichophyton soudanense		
Trichophyton tonsurans		
Trichophyton violaceum		
Trichophyton yaoundei		

1. *Epidermophyton*: Produces only macroconidia, no microconidia. The macroconidia are abundant and born in clusters with smooth, thick walls and two to seven septa. This genus consists of two species, one of which is a pathogen.
2. *Microsporum*: Both microconidia and rough-walled macroconidia characterize *Microsporum* species. The macroconidia are abundant and spindle shaped or fusiform shaped with three to ten septa. There are 19 described species, but only nine are involved in human or animal infections.
3. *Trichophyton*: Macroconidia of *Trichophyton* species are smooth walled. They produce microconidia abundantly that are globose or pyriform and are born singly along the sides of hyphae or in the form of grape clusters. Macroconidia are rare and are elongated pencil-shaped structures. There are 22 species, most causing infections in humans or animals.

At the National Centre for Mycology, about 58 % of the dermatophyte species frequently isolated are *Trichophyton rubrum*, 27 % are *T. mentagrophytes*, 7 % are *T. verrucosum*, and 3 % are *T. tonsurans*. Infrequently isolated species (less than 1 %) are *Epidermophyton floccosum*, *Microsporum audouinii*, *M. canis*, *M. equinum*, *M. nanum*, *M. persicolor*, *Trichophyton equinum*, *T. kanei*, *T. raubitschekii*, and *T. violaceum*.

10.3 Dermatophytes

10.3.1 Distribution of Dermatophytes (Etiology and Ecology)

Epidermophyton floccosum is anthropophilic and worldwide in distribution. It infects the groin, body, epidemic athlete's foot, and occasionally nails but not hair. *M. audouinii* is also anthropophilic, worldwide in distribution, and rare except in Africa and Asia. It mostly infects the scalp and body, causing epidemic tinea capitis in prepubescent children, rarely spread by guinea pigs and dogs. *M. canis* is zoophilic and infects cats and dogs and less commonly monkeys, guinea pigs, horses, mice, cows, and rabbits. It is mostly worldwide in distribution, but less common in North America, the UK, and Scandinavia than the rest of the world. It infects the body in adults, scalp in children, and rarely nails. *M. equi*num is zoophilic, infects horses, and is worldwide in distribution but rare in man. *M. ferrugineum* is anthropophilic and found in Asia, Africa, and Eastern Europe, but rare in the western hemisphere except Brazil. *M. gypseum* is geophilic and worldwide, but rare in North America and Europe and common in South America. It infects the feet, hand, body, scalp, and rarely nails and is usually acquired from soil but occasionally animals, including flies. *M. nanum* is zoophilic, present in pigs, and distributed worldwide but infects the scalp and body of man and shows ectothrix infection of hair. *M. persicolor* is zoophilic, with respect to bank vole and mice, but not found in soil. It is sporadic in Europe, especially the UK; there are reports of infection of the skin, not hair. *T. equinum* is zoophilic, seen in horses worldwide, and very rare in man. *T. kanei* is anthropophilic in the body and feet, while incidence in nails is very rare. *T. megnini* is anthropophilic and present in Spain, Portugal, and rarely Africa, while in the Mediterranean, it infects mostly the body, scalp, and beard and shows ectothrix infection. *T. mentagrophytes* is both anthropophilic and zoophilic – rodents and small and large mammals – worldwide, and found in soil and can infect the feet, body, nails, beard, scalp, hand, groin. The zoophilic ones are ectothrix; anthropophilic ones do not infect hair. *T. raubitschekii* is anthropophilic and seen in southern Asia and India. Mediterranean strain of which infects mostly the body and rarely the scalp. *T. rubrum* is anthropophilic, worldwide in distribution, seldom isolated from animals, and never found in soil. It infects the feet, nails, body, groin, and rarely scalp, is both endothrix and ectothrix, and is the most frequently isolated dermatophyte. *T. soudanense* is also anthropophilic; appears in Africa and occasionally North America, the UK, and Brazil; and infects the scalp and body primarily "shower sites." *T. terrestre* is geophilic, worldwide, and a nonpathogenic species. *T. tonsurans* is anthropophilic; is worldwide; infects the scalp, body primarily "shower sites," and occasionally nails; and is endothrix, and outbreaks are not rare. *T. verrucosum* is zoophilic, is worldwide seen in cattle and other domestic and wild animals, and infects the scalp, beard, body, and occasionally nails, and ectothrix hair infection is seen. *T. violaceum* is anthropophilic reported from areas seen in Near and Middle East, Eastern Europe, North Africa, occasionally Latin America, and the Mediterranean, imported to North America. Institutional outbreaks of *T. violaceum* have been reported in Western Europe. Infection is on the scalp, body primarily "shower sites", rarely feet, and nails. Incidence of endothrix infection is seen in most cases.

10.3.2 Laboratory Identification

Specimens are collected for laboratory identification of dermatophytes by scraping the skin from the margin of the lesion onto folded black paper. Hairs are plucked, not cut, from the edge of the lesion and are cut into short segments. Hairs that fluoresce under a Wood's lamp are chosen; if they do not fluoresce, choose broken or scaly ones. Nails are scraped or minced into small pieces. Nail scrapings are obtained from the nail bed or from infected areas after discarding outer layers. Each specimen is divided between at least two types of culture media. For direct examination, a small sample of the specimen is selected for direct microscopic examination and investigated for the presence of fungal elements. The specimen is mounted in a small amount of potassium hydroxide or calcofluor-white. The KOH slides are gently heated and allowed to clear for 30–60 min before examining on a light or phase contrast microscope. Calcofluor-white slides are examined on a fluorescent microscope. When present in the direct examination, dermatophytes appear as hyaline (nonpigmented), septated elements. Hyphae rounding up into arthroconidia are diagnostic of dermatophyte involvement. If arthroconidia are found absent, the elements could also be due to a non-dermatophyte agent of onychomycosis or a small segment of a contaminating organism. When hair is involved, the arthroconidia may be found on the periphery of the hair shaft (ectothrix) or within the shaft (endothrix). *Malassezia furfur* infections (tinea versicolor) are diagnosed by the presence of spherical yeast cells with a single bud and a collar and short curved hyphal strands.

10.3.2.1 Culture Media

The following media are generally employed for culturing dermatophytes.

10.3.2.1.1 BCP

Bromocresol purple-milk solids-glucose agar is a differential media useful in the characterization of dermatophyte species. The growth pattern of restricted or profuse is determined by comparison to a tube of nutrient media such as SDA. Some species produce an alkaline reaction (change media to purple); others do not produce a pH change (leave the media a sky-blue color). Hydrolysis of the milk solids results in a zone of clearing around the colony.

10.3.2.1.2 PDA

Potato dextrose agar is used to observe pigment formation.

10.3.2.1.3 PYE

Phytone yeast extract agar is a nutritionally enriched media that supports luxurious growth of most fungi. It contains antibiotics to inhibit bacterial growth.

10.3.2.1.4 SDA

Sabouraud dextrose agar (Emmons modification) is a nonselective medium which supports the growth of most fungi. Species of the dermatophytes isolated during this study were identified on the basis of their growth characteristics.

10.3.2.2 Biochemical Tests

10.3.2.2.1 Slide Culturing

The slide culture is a method of examining the microscopic structures of a fungus. The organism is grown on a glass coverslip placed on a block of agar. When sufficient growth has occurred, the coverslip is placed on a drop of mounting media on a microscope slide and examined by phase contrast microscopy.

10.3.2.2.2 Scotch Tape Mount

The scotch tape mount is used for examining the microscopic structures of filamentous fungi. A piece of clear, transparent tape held with sterile forceps is touched on the surface of the colony. The tape is placed on a drop of mounting media on a slide, followed by another drop and a coverslip over it. The phase contrast microscope is used to examine the sample.

10.3.2.2.3 Hair Perforation

Many dermatophytes have the ability to degrade hair. The hair perforation test determines whether the organism simply erodes the hair shaft or produces an enzyme for penetration and invasion of the shaft resulting in perforating bodies or cones.

10.3.2.2.4 Rice Grain Test

Microsporum audouinii grows poorly on rice grains and produces a dark discoloration useful in differentiating it from atypical *M. canis* strains. The rice grains also enhance the production of macroconidia in some species.

10.3.2.2.5 Urease

Christensen's urea broth indicates the presence of the enzyme urease, which splits urea into ammonia, resulting in an alkaline environment. The phenol red indicator turns the media from a straw yellow to fuchsia at pH 8.4.

10.3.2.2.6 Vitamin Requirements

Certain species of dermatophytes have distinctive nutritional requirements that may be beneficial to differentiate from similar species. The agar base is vitamin free and various vitamins are added to the basal media. The growth on the vitamin-enriched media is compared to the vitamin-free media to determine enhancement, if any, of aerial mycelium.

10.3.3 Epidemiology

Dermatophytes are by far the most significant fungi because of their widespread involvement of population at large and their prevalence all over the world. They are assuming greater significance both in developed and developing countries particularly due to the advent of immunosuppressive drugs and disease. Hot and humid climate in the tropical and subtropical countries like India makes dermatophytosis a very common superficial fungal skin infection.

The prevalence of dermatophytic infections are governed by environmental conditions, personal hygiene (Oyeka 1990), and individual susceptibility from place to place. The isolation of different dermatophytes also varies markedly from one ecological niche to another niche depending on their primary habitat (Aly 1994). Many saprophytic soil fungi are closely related to dermatophytes, sharing the ability to utilize the keratin as growth substrate, so it is believed that dermatophytes might have been evolved from these keratinophilic soil fungi. During the evolutionary process, they led to the development of epidemiological groups of anthropophilic, zoophilic, and geophilic species (Emmons et al. 1997). Some species of dermatophytes are endemic in certain parts of the world and have limited geographic distribution (Ajello 1968 on Epidemiologic profile of dermatophytosis in Stockholm, Sweden). Laboratory records comprising direct microscopy and culture results of 37,503 specimens from skin, hair, and nail scrapings collected from January 2005 through December 2009 were retrospectively analyzed in the mycology laboratory at Karolinska University Hospital. Onychomycosis had, over time, the highest overall prevalence of 14.1 %, followed by tinea pedis (4.4 %). *Trichophyton rubrum* was the predominant pathogen isolated from these cases (83.2 %), followed by *T. mentagrophytes* (7.4 %). In contrast, *T. violaceum* and *T. soudanense* accounted for 81.6 % of the isolates from patients with tinea capitis. Now *Trichophyton soudanense*, *T. gourvilii*, and *T. yaoundei* are geographically restricted to Central and West Africa (Singh and Beena 2003). *Microsporum ferrugineum* predominates in Japan and surroundings. *T. concentricum* is confined to the islands of South Pacific and in Central and South America; however, the increasing mobility of world population is disrupting several of these epidemiological patterns (Badillet 1991).

In recent times, *T. tonsurans* is replaced by *M. audouinii* as the principal causative agent of tinea capitis infections in the USA. This may be due to the mass migration of population from Mexico and other Latin American countries where the *T. tonsurans* was predominant. The most common etiological agents of dermatophytoses in the Western countries are *T. rubrum* and *M. canis*. *Microsporum distortum* is a rare case of tinea capitis in New Zealand and Australia. The prevalence of dermatophytoses varies in India. In 1900, Dr. Powell reported the first case of dermatophytoses from Assam, India. The commonest clinical types of dermatophytosis of man are tinea corporis (58.84 %), followed by tinea cruris (12.3 %), which concurs with reports from other parts of India (Kanwar et al. 2001). The incidence of tinea capitis was 6.92 %. Tinea capitis is less common in India than in other countries (Kaur 1970; Vasu 1966; Malik et al. 1978). This may be attributable to the use of hair oils (particularly mustard oil) which are customarily used by Indians and have been shown to have an inhibitory effect on dermatophytes in vitro (Hajini et al. 1970, Garg and Muller 1992). The reported incidence of tinea pedis varies from 26.4 % from Pune (Anand et al. 2001) to 0.4 % from Ahmedabad (Shah et al. 1975), and it is 11.53 % in our study.

The predominance of tinea pedis in Western countries could be because of the regular use of shoes and socks, predisposing to perspiration and maceration (Bhaskaran et al. 1977). *Trichophyton* species were more commonly isolated than *Epidermophyton* and *Microsporum*. *T. rubrum* is the main dermatophyte reported from India and other countries (Kanwar et al. 2001). Many other species of dermatophytes like *T. schoenleinii*, *T. tonsurans*, *T. verrucosum*, *T. ferrugineum*, *T. concentricum*, and *M. audouinii* are also isolated besides *T. rubrum*, *T. mentagrophytes*, *T. violaceum*, and *E. floccosum* (Varenker et al., 1991). *T. rubrum* has been found to be the main causative agent of tinea corporis, whereas tinea cruris is mainly caused by *E. floccosum* and tinea capitis by *T. violaceum* (Kanwar et al. 2001). A higher incidence of dermatophytosis in males than in females has been reported both in India and abroad (Kanwar et al. 2001). Philpot suggested that males may be more vulnerable to infection due to higher exposures in the army, schools, and sporting activities and due to the type of shoes and socks they use (Philpot 1977). This is especially true for tinea cruris. Differences in the incidence of other clinical types were also observed in the present study, e.g., tinea corporis, tinea capitis, and tinea manuum are more common in males, while tinea pedis and tinea unguium are more common in females.

10.3.4 Immunology of Dermatophytes

Human infection is the result of a complex interplay of factors pertaining to the invading organism, the host, and the environment. This is best shown in human dermatophyte infections. Acute infections are usually short lived and easy to treat when the patient has good cell-mediated immunity, short-term antidermatophyte antibodies, and delayed hypersensitivity. In chronic infections, the infection is long term and resistant to therapy, and patients have poor in vitro assessed cell-mediated immunity and immediate hypersensitivity to fungal antigens. Antidermatophyte antibodies usually do not disappear quickly. The dermatophytes have the ability to invade keratinized tissue (skin, hair, and nails) but are usually restricted to the nonliving cornified layer of the epidermis because of their inability to penetrate viable tissue of an immunocompetent host. However, invasion does elicit as host response ranging from mild to severe. Acid proteinases, elastase, keratinases, and other proteinases reportedly act as virulence factors.

The development of cell-mediated immunity correlated with delayed hypersensitivity and an inflammatory response is associated with clinical cure, whereas the lack of or a defective cell-mediated immunity predisposes the host to chronic or recurrent dermatophyte infection. Chronic dermatophytosis is mostly caused by *Trichophyton rubrum*, and there is some evidence that mannan produced by this fungus suppresses or diminishes the inflammatory response. Dermatophyte colonization is characteristically limited to the dead keratinized tissue of the stratum corneum and results in either a mild or intense inflammatory reaction. Although the cornified layers of the skin lack a specific immune system to recognize this infection and rid itself of it, nevertheless, both humoral and cell-mediated reactions and specific and nonspecific host defense mechanisms respond and eventually eliminate the fungus, preventing invasion into the deeper viable tissue. This array of defense mechanisms thought to be active against dermatophytes consists of α2-macroglobulin keratinase inhibitor, unsaturated transferrin, epidermal desquamation, lymphocytes, macrophages, neutrophils, and mast cells. There are two major classes of dermatophyte antigens: glycopeptides and keratinases. The protein portion of the glycopeptides preferentially stimulates cell-mediated immunity (CMI), whereas the polysaccharide portion preferentially stimulates humoral immunity. Keratinases, produced by the dermatophytes to enable skin invasion, elicit delayed-type hypersensitivity (DTH) responses when injected intradermally into the skin of animals. Although the host develops a variety of antibodies to dermatophyte infection, i.e., immunoglobulin M (IgM), IgG, IgA, and IgE, they apparently do not help eliminate the infection since the highest level of antibodies is found in those patients with chronic infection. IgE, which mediates immediate hypersensitivity, appears to play no role in the defense process. Rather, the development of CMI which is correlated with DTH is usually associated with clinical cure and ridding the stratum corneum of the offending dermatophyte. In contrast, the lack of CMI or defective CMI prevents an effective response and predisposes the host to chronic or recurrent dermatophyte infections.

10.3.5 Treatment of Dermatophytosis

Topical antifungal preparations should be effective in treating small, uncomplicated tinea infections located in areas other than the scalp. These include topical clotrimazole and iconazole (available over the counter) and terbinafine cream. The topical antifungal drugs such as ointments, lotions, powders, and sprays are used for tinea infections on the body. The creams are applied on affected parts twice or thrice daily for at least 1 month for better results. Powder and sprays are generally used for athlete's foot. Topical medications applied once or twice daily are the primary treatment indicated for tinea corporis/cruris and tinea pedis/manuum. In topical therapy, the improvement is more with allylamines than azoles.

Treatment of dermatophyte infection involves primarily oral and/or topical formulations of azoles or allylamines, particularly itraconazole and terbinafine. The use of oral antifungals may be practical where the tinea involvement is extensive or chronic or where application of a topical is not feasible. For tinea unguium (onychomycosis) and tinea capitis, oral therapies are the primary treatments provided. Recently, topical amorolfine and ciclopirox formulations have been approved for use in milder onychomycosis cases, and their role in the treatment of the different clinical forms of onychomycosis is currently being defined. Relapse of infection remains a problem, particularly with tinea pedis/unguium. Griseofulvin is the drug of choice for the treatment of tinea capitis till today, and azoles like imidazole and fluconazoles are used for the treatment of onychomycosis. But terbinafine is the effective antifungal drug for ringworm infections, which is less expensive, and the cure rate is more.

Appropriate follow-up duration and education of patients on proper foot hygiene are also important components in providing effective therapy. Sometimes, oral antifungal medication may be required if the condition is severe. Medications may include griseofulvin, itraconazole, terbinafine, and fluconazole. Tinea capitis (scalp), regardless of severity, is usually treated with oral antifungal medication, since topical antifungals do not penetrate hair follicles well. Corticosteroids may sometimes be used for the treatment of severely inflamed or potentially scarring lesions, such as scalp infections. Fungal infections involving the nails (onychomycosis) require oral treatment as well, because the dermatophyte is found deep in the nail. Tinea versicolor may be treated with selenium sulfide lotion or ketoconazole shampoo. Occasionally, the question arises as to whether a concurrent bacterial infection is complicating a fungal infection. This situation most commonly occurs when highly inflamed scalp lesions are draining purulent (pus) material. The lesions usually resolve with systemic (oral) antifungal and systemic corticosteroid therapy.

10.4 Antifungal Susceptibility Tests: (NCCLS Document M27-A, 1997; CLSI Standards M38-A, 2002)

In the last two decades, the incidence of infections caused by dermatophytes and other fungi has increased considerably (Weitzman and Summerbell 1995). With an increasing variety of drugs available for the treatment of dermatophytoses, the need for a reference method for the testing of the antifungal susceptibilities of dermatophytes has become apparent (Ghannoum and Rice 1997). Establishment of a reference susceptibility testing method may allow the clinician to select the appropriate therapy for the treatment of infections caused by dermatophytic fungi. Recently, a standard method for antifungal susceptibility testing of yeasts has been established by the National Committee for Clinical Laboratory Standards (NCCLS M27-A document). This reference method for yeast is the first step in the establishment of a reliable, standardized, and clinically useful technique for the susceptibility testing of filamentous and dermatophytic fungi.

10.4.1 Culture Medium

Yeast nitrogen broth (YNB) supplemented with the following composition was used.

YNB base	6.7 g
Glucose	10.0 g
Distilled water pH 6.5	100 ml

This medium was filter sterilized and used as basal medium. It was diluted to 1:10 sterile (autoclaved) distilled water just before use.

10.4.2 Antifungal Agents: Antifungal Drugs

Antifungal drugs were donated as follows: ketoconazole by Janssen Pharmaceuticals, fluconazole by Hydex Chemicals Pvt. Ltd., terbinafine named "Terbicip" produced by Cipla Ltd., and griseofulvin (also known as *Grisovin*, a proprietary name of Glaxo Laboratories). Itraconazole was used in its commercial formulation (Sri Pharmacare, IndiaMART). All drugs were dissolved in 100 % dimethyl sulfoxide (Gibco) following the protocol of NCCLS and were prepared in stock solutions of 1000 µg/ml, and fluconazole was prepared in sterile distilled water and kept at -20^0 C until use. They were subsequently prepared as stock solution, and serial twofold dilutions were performed. Final concentrations ranged from 0.125 to 64 µg/mL for fluconazole; 0.03–16 µg/mL for ketoconazole, itraconazole, and terbinafine; and 0.03–8 µg/mL for griseofulvin.

10.4.3 Preparation of Inoculum

Testing was performed by a broth macrodilution method following the recommendation of the NCCLS M27-A (1997). In brief, stock inocula of dermatophytic stains were prepared from 7- to 14-day cultures grown on Sabouraud dextrose agar (SDA) with chloramphenicol. After the appearance of the sufficient growth, the fungal colonies were covered with 5 ml of sterile saline (0.9 %), and the suspensions were made by gently probing the surface with the tip of a sterile Pasteur pipette. The resulting suspended mixture was withdrawn and transformed to a sterile tube. Heavy particles of the suspension, when present, were allowed to settle for 15 min at room temperature, and the upper homogenous suspension was used for further testing. The suspensions were mixed with a vortex mixer for 15 s and adjusted with sterile normal saline to match the opacity of 0.5 McFarland standard.

10.4.4 Turbidity Standard for Inoculum Preparation

To standardize the inoculum density for a susceptibility test, a $BaSO_4$ turbidity standard, equivalent to a 0.5 McFarland standard or its optical equivalent (e.g., latex particle suspension), should be used. A $BaSO_4$ 0.5 McFarland standard may be prepared as follows: a 0.5-ml aliquot of 0.048 mol/L $BaCl_2$ (1.175 % w/v $BaCl_2 \cdot 2H_2O$) is added to 99.5 ml of 0.18 mol/L H_2SO_4 (1 % v/v) with constant stirring to maintain a suspension.

1. The correct density of the turbidity standard should be verified by using a spectrophotometer with a 1-cm light path and matched cuvette to determine the absorbance. The absorbance at 625 nm should be 0.008–0.10 for the 0.5 McFarland standards.
2. The barium sulfate suspension should be transferred in 4–6-ml aliquots into screw-cap tubes of the same size as those used in growing or diluting the bacterial inoculum.
3. These tubes should be tightly sealed and stored in the dark at room temperature.
4. The barium sulfate turbidity standard should be vigorously agitated on a mechanical vortex mixer before each use and inspected for a uniformly turbid appearance. If large particles appcar, thc standard should be replaced. Latex particle suspensions should be mixed by inverting gently, not on a vortex mixer.
5. The barium sulfate standards should be replaced or their densities verified monthly.

The inoculum size was adjusted to between 1.0×10^6 and 5.0×10^6 spores/ml by microscopic enumeration with a cell counting hemocytometer (Neubauer chamber). In some instance, where fungi do not readily produce conidia, a small portion of the mycelial growth was harvested and gently homogenized in 2 ml of sterile saline using Tenbroeck tissue grinder, and resulting suspensions were adjusted to opacity of 0.5 McFarland standard by adding sterile saline. Inoculum quantification was made by counting microconidia in a hemocytometer and by plating 0.01 ml of suspensions in SDA. The plates were incubated at 28 °C and were examined daily for the presence of fungal colonies before the test to check the viability of the fungus.

10.4.5 Method and Test Procedure

The NCCLS broth medium macrodilution method for yeasts which was modified for mold testing (NCCLS M27-A) was used for determination of the antifungal susceptibilities of dermatophytes. Twelve test tubes for each drug, i.e., fluconazole, itraconazole, ketoconazole, and griseofulvin, were arranged in a rack as per the requirement of different MIC ranges. An additional tube in the beginning was kept for griseofulvin and was later removed after antifungal dilutions were put up, such that the final range of the drug would be from 0.03 to 81.25 μg/ml. A set of 15 tubes were arranged for terbinafine MIC testing. All the tubes were arranged in ascending order with tube containing highest concentration on the left side. In addition, four control tubes were kept and labeled as C_1–C_4.

C_1	Sterility control (3 ml of YNB)
C_2	Positive control (2.7 ml of YNB + 0.3 ml of test inoculum)
C_3	Drug control (2.7 ml of YNB + 0.3 ML of drug stock solution)
C_4	Solvent control (2.7 ml of YNB + 0.3 ml of solvent used)

The stock solutions of antifungal agents were removed from the freezer and thawed to room temperature. The YNB was diluted 1 in 10 with distilled water under sterile conditions just before use. 2.7 ml of diluted YNB was dispensed in all the tubes. The stock solutions of antifungal drug tubes were vortex mixed for a few seconds to have a uniform suspension of the drug. 2.7 ml of the antifungal agent was pipetted and mixed in the first tube on the left, containing 2.7 ml of YNB. Serial twofold dilutions were made by pipetting 2.7 ml from the first tube to the second tube to the third tube and so on until the final tube (text figs. II–V). The final reservoir suspension was discarded. 0.3 ml of fungal inocula was added to the different drug dilutions and also into a positive (C_2) control tube for each test. The final dilution of the fungal inoculums was 1:10.

10.4.6 Incubation

Tubes in the rack were incubated at 35 °C in BOD incubator until growth appeared in the drug-free control tube. Incubation ranged 6–20 days. Control tubes were observed daily for the presence or absence of visible growth. When the growth was visible, each tube was vortexed for 10 s immediately prior to being scored, which allowed the detection of even a small amount of growth. The growth in each tube was compared with the growth of control tube (C2). Each tube was given a numerical score as follows:

O	Optically clear or the absence of growth
1	Slight turbidity compared to that of the fungus-free control tube (C_1)
2	Visible turbidity as compared to C_1
3	Clear-cut turbidity with or without formation of small hyphal fragments on the surface of the broth
4	Turbidity with the formation of a surface pellicle on the surface of the broth

The highest dilution of the drug, which inhibited the fungal growth, was taken as the MIC. MIC50 was calculated by taking the drug concentration, where 50 % of isolates are inhibited. Similarly, MIC90 was noted with drug concentration where 90 % of the isolates were inhibited.

10.4.7 Antifungal Susceptibility Test by CLSI Standards M38-A Method (Modified Method of NCCLS M27-A)

This is the modified method of NCCLS M27-A and was released as a document in 2002 as M38-A microdilution test. In this procedure, all the parameters are similar with macrodilution test, but instead of using tubes, the microtiter plate is used; hence, the size of inoculum also differs.

The samples from patients were collected in aseptic conditions from infected areas such as the skin, nail, and hair (Debono and Gordee 1994; Degreef 2008). Culturing of organisms from skin scraping was done on selective medium as Sabouraud dextrose agar for identification of dermatophytic species. For antifungal susceptibility testing, these species were used after identifying them on cultural, morphological, and biochemical characteristics (Favre et al. 2003). Five antifungal drugs were used for testing. The microbroth dilution method was performed according to CLSI standards M38-A (Fernández-Torres et al. 2002).

10.4.7.1 Culture Medium

Yeast nitrogen broth (YNB) supplemented with following composition was used:

YNB base 6.7 g
Glucose 10.0 g
Distilled water 100 ml and adjusting the pH at 6.5

This medium was filtered, sterilized, and used as basal medium (autoclaved). It was diluted to 1:10 with sterile distilled water just before use.

10.4.7.2 Antifungal Agents

Antifungal drugs used in this study were supplied from various firms, as follows: ketoconazole by Janssen Pharmaceuticals, fluconazole by Hydex Chemicals Pvt. Ltd., terbinafine named “Terbicip” produced by Cipla Ltd., and griseofulvin (also known as Grisovin, a proprietary name of Glaxo Laboratories). Itraconazole was used in its commercial formulation (Sri Pharmacare, IndiaMART). All drugs were dissolved in 100 % dimethyl sulfoxide (Gibco) following the protocol of CLSI and were prepared in stock solutions of 1000 μg/ml, and fluconazole was prepared in sterile distilled water and kept at −200 °C until used. They were subsequently prepared as stock solution and serial twofold dilutions were performed. Final concentrations ranged from 0.125 to 64 μg/mL for fluconazole; 0.03–16 μg/mL for ketoconazole, itraconazole, and terbinafine; and 0.03–8 μg/mL for griseofulvin.

10.4.7.3 Preparation of Inoculum

Testing was performed by a broth microdilution method following the recommendation of the CLSI M38-A. All the strains were obtained from the patient’s samples of tinea infections. The species identification was based on morphological and biochemical characteristics and was used in inoculum preparation. In brief, stock inocula of dermatophytic stains were prepared from 7- to 14-day cultures grown on Sabouraud dextrose agar (SDA) with chloramphenicol. After the appearance of the sufficient growth, the fungal colonies were covered with 5 ml of sterile saline (0.9 %), and the suspensions were made by gently probing the surface with the tip of a sterile Pasteur pipette. The resulting suspended mixture was withdrawn and transformed to a sterile tube. Heavy particles of the suspension, when present, were allowed to settle for 15 min at room temperature, and the upper homogenous suspension was used for further testing. The suspensions were mixed with a vortex mixer for 15 s and adjusted with sterile normal saline to match the opacity of 0.5 McFarland standard.

10.4.7.4 Turbidity Standard for Inoculum Preparation

To standardize the inoculum density for a susceptibility test, a BaSO4 turbidity standard, equivalent to a 0.5 McFarland standard or its optical equivalent (e.g., latex particle suspension), should be used. The inoculum size was adjusted to between 1.0 × 106 and 5.0 × 106 spores/ml by microscopic enumeration with a cell counting hemocytometer (Neubauer chamber). In some instance, where fungi do not readily produce conidia, a small portion of the mycelial growth was harvested and gently homogenized in 2 ml of sterile saline using Tenbroeck tissue grinder, and resulting suspensions were adjusted to opacity of 0.5 McFarland standards by adding sterile saline. Inoculum quantification was made by counting microconidia in a hemocytometer and by plating 0.01 ml of suspensions in SDA. The plates were incubated at 28 °C and were examined daily for the presence of fungal colonies before the test to check the viability of the fungus.

10.4.7.5 Test Procedure

The tests were performed in polystyrene microtiter plates with flat bottom wells. By using a multichannel pipette, the aliquots of 100 μl of twofold drug dilutions were inoculated into the wells. Then, the microtiter plates were stored at −50 °C in a deep freezer until used. The microplate was inoculated with 100-μl fungal inoculum to maintain the dilutions with 0.5 × 104 to 5 × 104 spores ml-1. The plates were incubated at 28 °C for 7 days (Georgopapadakou and Tkacz 1995) for growth of the fungi. Growth and sterility control wells also maintained for each assay, and all the tests were performed in duplicate. The highest dilution of the drug, which inhibited the fungal growth, was taken as the MIC. MIC50 was calculated by taking the drug concentration, where 50 % of isolates are inhibited. Similarly, MIC90 was noted with drug concentration where 90 % of the isolates were inhibited. The MIC values were noted basing on the rate of growth inhibition.

10.4.7.6 Antifungal Susceptibility Investigations

The fungal infections are not completely cured with antifungal drugs. The treatment is less successful than that of bacterial infections because the fungal cells are eukaryotic and much more similar to human than the bacteria (Ghannoum and Rice 1999). Many drugs that inhibit or kill fungi are therefore quite toxic for humans also. Moreover, the fungal cells are equipped with a detoxifying system, which is able to modify many antibiotics, probably by hydroxylation (Gupta and Kohli 2003). Hence, the antibiotics used to treat the fungal infection will remain fungistatic for a period of time, and repeated usage of antibiotics is advised. The effective antifungal drugs may extract membrane sterols (da Silva Barros et al. 2007) or prevent their synthesis (Murray et al. 1999). Most antifungal compounds target the formation or the function of ergosterol, an important component of the fungal cell membrane (Nimura et al. 2001).

In the present study, a total of 119 strains of dermatophytes belonging to ten species were tested. All the strains were obtained from patient samples and were used in the tests. They were *T. rubrum* ($n = 40$), *T. mentagrophytes* ($n = 19$), *T. violaceum* ($n = 15$), *M. gypseum* ($n = 12$), *E. floccosum* ($n = 9$), *M. audouinii* ($n = 8$), *T. schoenleinii* ($n = 5$), *M. canis* ($n = 5$), *T. tonsurans* ($n = 4$), and *T. verrucosum* ($n = 2$).

10.4.7.7 Comparison of MICs of Five Antifungal Agents

The minimum inhibitory concentrations (MIC50, MIC90) of griseofulvin, ketoconazole, fluconazole, itraconazole, and terbinafine are compared, and the comparison of MIC values is used in determining the efficacy and the dosage of drug for the treatment of dermatophytosis.

10.5 Conclusion

In conclusion, it may be useful to undertake periodical screening programs to detect the antifungal susceptibility of newer antifungal agents. Our data on the antifungal susceptibility of dermatophyte isolates may contribute to a choice of antifungal treatment to ringworm infections. Terbinafine is considered as most potent drug followed by ketoconazole. But still the efficacy of ketoconazole drug was totally dependent upon the variation of causative dermatophytic strains of particular tinea infections. We consider that our study on the antifungal susceptibility of dermatophytes can be beneficial for investigation of in vitro resistance of dermatophytic species and for management of cases clinically unresponsive to treatment.

References

Ajello L (1968) A taxonomic review of the dermatophytes and related species. Sabouraudia 6:147–159

Aly R (1994) Ecology and epidemiology of dermatophyte infections. J Am Acad Dermatol 31(3):S21–S25

Anand AR, Madhavan HN, Sudha NV, Therese KL (2001) Use of PCR in the diagnosis of fungal endophthalmitis. Ophthalmology 108:386–390

Badillet G (1991) Dermatophytieset Dermatophytes-Atlas clinique et biologique, 3èmeth edn. ÉditionsVaria, Paris, p 303

Bhaskaran CS, Shyam Sunder Rao P, Krishanamoorthy T, Tarachand P (1977) Dermatophytosis in Tripathi. Indian J Pathol Microbiol 20:251–260

Chander J (2002) Text book of medical mycology, 2nd edn. Mehta Publishers, New Delhi

Debono M, Gordee RS (1994) Antibiotics that inhibit fungal cell wall development. Annu Rev Microbiol 48:471–497

Degreef H (2008) Clinical forms of dermatophytosis (ringworm infection). Mycopathologia 166:257–265

Emmons CW, Bindford CH, Utz JP, Kwon-Chung KL (eds) (1977) Chapter 10: Dermatophytoses. In: Medical mycology, 3rd edn. Lea & Febiger, Philadelphia, pp 117–167

Favre B, Hofbauer B, Hildering KS, Ryder NS (2003) Comparison of in vitro activities of 17 antifungal drugs against a panel of 20 dermatophytes by using a microdilution assay. J Clin Microbiol 41:4817–4819

Fernández-Torres B, Cabañes FJ, Carrillo-Muñoz AJ, Esteban A, Inza I et al (2002) Collaborative evaluation of optimal antifungal susceptibility testing conditions for dermatophytes. J Clin Microbiol 40:3999–4003

Garg AP, Muller J (1992) Inhibition of growth of dermatophytes by Indian hair oils. Mycoses 35:363–369

Georgopapadakou NH, Tkacz JS (1995) The fungal cell wall as a drug target. Trends Microbiol 3:98–104

Ghannoum MA, Rice LB (1999) Antifungal agents: mode of action, mechanisms of resistance, and correlation of these mechanisms with bacterial resistance. Clin Microbiol Rev 12:501–517

Gupta AK, Kohli Y (2003) In vitro susceptibility testing of ciclopirox, terbinafine, ketoconazole and itraconazole against dermatophytes and nondermatophytes, and in vitro evaluation of combination antifungal activity. Br J Dermatol 149:296–305

Hajini GH, Kandhari KC, Mohapatra LN (1970) Effect of hair oils and fatty acids on the growth of dermatophytes and their in vitro penetration of human scalp hair. Sabouradia 8:174

Kanwar AJ, Mamta, Chander J (2001) Superficial fungal infection. In: Valia RG, Valia AR (eds) IADVL textbook and atlas of dermatology, 2nd edn. Bhalani Publishing House, Mumbai, pp 215–258

Kaur S (1970) Incidence of dermatophytosis in Chandigarh and surrounding areas. Indian J Dermatol Venereol 36:143–145

Malik AK, Chugh TD, Prakash K (1978) Dermatophytosis in North India. Indian J Pathol Microbiol 21:53–59

Maria Elisabete da Silva Barros, Daniel de Assis Santos, Junia Soares Hamdan (2007) Evaluation of susceptibility of Trichophyton mentagrophytes and Trichophyton rubrum clinical isolates to antifungal drugs using a modified CLSI microdilution method (M38-A). J Med Microbiol 56(4):514–518

Murray PR, Baron EJ, Pfaller MA, Tenouer FC, Youlken RH (1999) Trichophyton, microsporum, epidermophyton and agents of superficial mycoses. In: Manual of clinical microbiology. ASM Press, Washington

Nimura K, Niwano Y, Ishiduka S, Fukumoto R (2001) Comparison of in vitro antifungal activities of topical antimycotics launched in 1990s in Japan. Int J Antimicrob Agents 18:173–178

Oyeka CA (1990) Tineacapitis in Awka local Government Area of Anambra state. West Afr J Med 9:120–123

Philpot CM (1977) The use of nutritional tests for the differentiation of dermatophytes. Sabouraudia 15:141–150

Remak R (1842) GelungeneInpfung des Favus. Med Z 11:37

Sabouraud R (1970) (1864–1938) French dermatologist. JAMA 214(2):363–364

Seeliger HPR (1985) The discovery of Achorion schoenleinii: facts and "stories.". Mykosen 28:161–182

Shah HS, Amin AG, Kavinde MS et al (1975) Analysis of 2000 cases of dermatomycoses. Indian J Pathol Bacteriol 18:32–37

Singh S, Beena PM (2003) Profile of dermatophyte infections in Baroda. Indian J Dermatol Venereol Leprol 69:281–283

Vasu DRBH (1966) Incidence of dermatophytosis in Warangal, Andhra Pradesh. India. Indian J Med Res 54:468–474

Weitzman I, Summerbell RC (1995) The dermatophytes. Clin Microbiol Rev 8:240–259